D0643912

Geodynamics Series

Geodynamics Series

1. Dynamics of Plate Interiors
   A. W. Bally, P. L. Bender, T. R. McGetchin, and R. I. Walcott (Editors)

2. Paleoreconstruction of the Continents
   M. W. McElhinny and D. A. Valencio (Editors)

3. Zagros, Hindu Kush, Himalaya, Geodynamic Evolution
   H. K. Gupta and F. M. Delany (Editors)

4. Anelasticity in the Earth
   F. D. Stacey, M. S. Patterson, and A. Nicholas (Editors)

5. Evolution of the Earth
   R. J. O'Connell and W. S. Fyfe (Editors)

6. Dynamics of Passive Margins
   R. A. Scrutton (Editor)

7. Alpine-Mediterranean Geodynamics
   H. Berckhemer and K. Hsü (Editors)

8. Continental and Oceanic Rifts
   G. Pálmason, P. Mohr, K. Burke, R. W. Girdler, R. J. Bridwell, and G. E. Sigvaldason (Editors)

9. Geodynamics of the Eastern Pacific Region, Caribbean and Scotia Arcs
   Ramón Cabré, S. J. (Editor)

10. Profiles of Orogenic Belts
    N. Rast and F. M. Delany (Editors)

11. Geodynamics of the Western Pacific-Indonesian Region
    Thomas W. C. Hilde and Seiya Uyeda (Editors)

12. Plate Reconstruction From Paleozoic Paleomagnetism
    R. Van der Voo, C. R. Scotese, and N. Bonhommet (Editors)

13. Reflection Seismology: A Global Perspective
    Muawia Barazangi and Larry Brown (Editors)

14. Reflection Seismology: The Continental Crust
    Muawia Barazangi and Larry Brown (Editors)

15. Mesozoic and Cenozoic Oceans
    Kenneth J. Hsü (Editor)

16. Composition, Structure and Dynamics of the Lithosphere-Asthenosphere System
    K. Fuchs and C. Froidevaux (Editors)

17. Proterozoic Lithospheric Evolution
    A. Kröner (Editor)

Circum-Pacific Orogenic Belts and Evolution of the Pacific Ocean Basin

Edited by J. W. H. Monger
J. Francheteau

Geodynamics Series
Volume 18

American Geophysical Union
Washington, D.C.

Geological Society of America
Boulder, Colorado

1987

Publication No. 0132 of the International Lithosphere Program

Published under the aegis of AGU Geophysical Monograph Board.

Library of Congress Cataloging-in-Publication Data

Circum-Pacific orogenic belts and evolution of the
Pacific Ocean Basin

(Geodynamics series, ISSN 0277-6669 ; v. 18)
(Publication no. 0132 of the International Lithosphere
Program)
Papers presented at the 27th International Geological
Congress in Moscow in 1984.
1. Geology—Pacific Ocean—Congresses. 2. Island
arcs—Pacific Ocean—Congresses. 3. Orogeny—Pacific
Ocean—Congresses. I. Monger, J. W. H.
II. Francheteau, Jean. III. International Geological
Congress (27th : 1984 : Moscow, R.S.F.S.R.) IV. Series.
V. Series: Publication . . . of the International Litho-
sphere Program ; no. 0132.
QE350.4.C338 1987 551.46'08'094 86-28814
ISBN 0-87590-519-6
ISSN 0277-6669

Printed in the United States of America.

CONTENTS

FOREWORD

Raymond A. Price

Past-President, International Lithosphere Program
and
Director General, Geological Survey of Canada,
601 Booth Street, Ottawa, Ontario, K1A OE8

The International Lithosphere Program was launched in 1981 as a ten-year project of inter-disciplinary research in the solid earth sciences. It is a natural outgrowth of the Geodynamics Program of the 1970's, and of its predecessor, the Upper Mantle Project. The Program — "Dynamics and Evolution of the Lithosphere: The Framework of Earth Resources and the Reduction of Hazards" — is concerned primarily with the current state, origin and development of the lithosphere, with special attention to the continents and their margins. One special goal of the program is the strengthening of interactions between basic research and the applications of geology, geophysics, geochemistry and geodesy to mineral and energy resource exploration and development, to the mitigation of geological hazards, and to protection of the environment; another special goal is the strengthening of the earth sciences and their effective application in developing countries.

An Inter-Union Commission on the Lithosphere (ICL) established in September 1980, by the International Council of Scientific Unions (ICSU), at the request of the International Union of Geodesy and Geophysics (IUGG) and the International Union of Geological Sciences (IUGS), is responsible for the overall planning, organization and management of the program. The ICL consists of a seven-member Bureau (appointed by the two unions), the leaders of the scientific Working Groups and Coordinating Committees, which implement the international program, the Secretaries-General of ICSU, IUGG and IUGS, and liaison representatives of other interested unions or ICSU scientific committees. National and regional programs are a fundamental part of the International Lithosphere Program and the Chairman of the Coordinating Committee of National Representatives is a member of the ICL.

The Secretariat of the Commission was established in Washington with support from the U.S., the National Academy of Sciences, NASA, and the U.S. Geodynamics Committee.

The International Scientific Program initially was based on nine International Working Groups.

- WG-1 Recent Plate Movements and Deformation
- WG-2 Phanerozoic Plate Motions and Orogenesis
- WG-3 Proterozoic Lithospheric Evolution
- WG-4 The Archean Lithosphere
- WG-5 Intraplate Phenomena
- WG-6 Evolution and Nature of the Oceanic Lithosphere
- WG-7 Paleoenvironmental Evolution of the Oceans and Atmosphere
- WG-8 Subduction, Collision, and Accretion
- WG-9 Process and Properties in the Earth that Govern Lithospheric Evolution

Eight Committees shared responsibility for coordination among the Working Groups and between them and the special goals and regional groups that are of fundamental concern to the project.

- CC-1 Environmental Geology and Geophysics
- CC-2 Mineral and Energy Resources
- CC-3 Geosciences Within Developing Countries
- CC-4 Evolution of Magmatic and Metamorphic Processes
- CC-5 Structure and Composition of the Lithosphere and Asthenosphere
- CC-6 Continental Drilling
- CC-7 Data Centers and Data Exchange
- CC-8 National Representatives

Both the Bureau and the Commission meet annually, generally in association with one of the sponsoring unions or one of their constituent associations. Financial support for scientific symposia and Commission meetings has been provided by ICSU, IUGG, IUGS, and UNESCO. The constitution of the ICL requires that membership of the Bureau, Commission, Working Groups, and Coordinating Committees change progressively during the life of the project, and that the International Lithosphere Program undergo a mid-term review in 1985. As a result of this review there has been some consolidation and reorganization of the program. The reorganized program is based on six International Working Groups:

- WG-1 Recent Plate Movements and Deformation
- WG-2 The Nature and Evolution of the Continental Lithosphere
- WG-3 Intraplate Phenomena
- WG-4 Nature and Evolution of the Oceanic Lithosphere

WG-5 Paleoenvironmental Evolution of the Oceans and the Atmosphere
WG-6 Structure, Physical Properties, Composition and Dynamics of the Lithosphere-Asthenosphere System

and six Coordinating Committees:

CC-1 Environmental Geology and Geophysics
CC-2 Mineral and Energy Resources
CC-3 Geosciences Within Developing Countries
CC-4 Continental Drilling
CC-5 Data Centers and Data Exchanges
CC-6 National Representatives
Sub-Committee 1 - Himalayan Region
Sub-Committee 2 - Arctic Region

This volume is one of a series of progress reports published to mark the completion of the first five years of the International Geodynamics Project. It is based on a symposium held in Moscow on the occasion of the 26th International Geological Congress.

Further information on the International Lithosphere Program and activities of the Commission, Working Groups and Coordinating Committees is available in a series of reports through the Secretariat and available from the President — Prof. K. Fuchs, Geophysical Institute, University of Karlsruhe, Hertzstrasse 16, D-7500 Karlsruhe, Federal Republic of Germany; or the Secretary-General — Prof. Dr. H.J. Zwart, State University Utrecht, Institute of Earth Sciences, P.O. Box 80.021, 3508 TA Utrecht, The Netherlands.

R.A. Price, President
Inter-Union Commission on the Lithosphere, 1981-85

PREFACE

The International Lithosphere Program (ILP) was established in 1981 to encourage interdisciplinary, international research on the nature, dynamics and origin of the lithosphere. To do this various working groups and coordinating committees were set up to focus on specific aspects of ILP.

The mandate of Working Group 2 (WG2), which was chaired by Rob Van der Voo of the University of Michigan, was "Phanerozoic plate motions and orogenesis." Its direct contribution to the objectives of ILP was organization and/or sponsorship of the following scientific meetings:

1) In 1982, a symposium held by the American Geophysical Union in Philadelphia had proceedings which were published in 1984 by AGU as volume 12 of the Geodynamics Series entitled "Plate reconstruction from Paleozoic paleomagnetism."

2) In 1983, a symposium held at the International Union of Geophysics and Geodesy meeting in Hamburg, concerned "Appalachian and Hercynian Fold Belts" and its proceedings were published in volume 109 (1/2) of Tectonophysics;

3) In 1984, the Geodynamics Research Symposium at Texas A&M University on "Collision tectonics and deformation of the continental lithosphere," was co-sponsored by WG2 and the proceedings appeared in 1985 as volume 119 of Tectonophysics;

4) In 1984, at the 27th International Geological Congress in Moscow, a symposium was convened by J. Francheteau, Yu.A. Kosygin and R. Van der Voo, and organized by J.W.H. Monger, on "Circum-Pacific orogenic belts and the evolution of the Pacific Ocean basin." Papers presented at the symposium form this volume.

Following the Moscow meeting, the various ILP working groups and committees were reorganized into their present form, and WG2 was merged with other groups to form part of a single large working group on "The nature and evolution of the continental lithosphere."

The period spanned by the short lifetime of WG2 was one characterized largely by testing, confirmation, or modification of the many revolutionary geotectonic concepts engendered by the plate-tectonic hypothesis. So far, the hypothesis has stood up remarkably well to nearly 20 years of rigorous examination, so much so that it has been embraced by many geoscientists as the general model to explain crustal phenomena. New concepts of the nature of deeper parts of the earth, based partly on studies of extraterrestrial bodies and partly on new geophysical techniques, will doubtless modify our understanding of the causes of plate movements and their possible long-term cyclicity, but crustal tectonic processes seen today are well-explained by the hypothesis, and no serious challengers to it have arisen so far.

As initially proposed, the plate tectonic hypothesis was derived largely from marine geophysical studies. It depicted a world whose crust is composed of a small number of rigid plates, and in which "orogenic" deformation takes place at convergent, divergent, and transform plate boundaries. This simple picture soon became modified as attempts were made to reconcile the hypothesis with other, mainly, geological observations and with the historical record of past tectonic events. It is now recognized that plates can be deformed internally at distances that are removed by thousands of kilometres from convergent plate margins. Convergent margins observed today have very different characteristics, which are interpreted as resulting from different relative densities of the converging plates, their rates of motion, and absolute plate motions. In early plate tectonic interpretations of orogenic belts, Andean-type, subduction-related orogens and Alpine-type, collision-related, orogens were distinguished. Recognition of the presence in many orogenic belts of far-travelled crustal fragments (allochthonous or suspect terranes), carries the concomitant implication that collisions may occur within orogenic belts where the dominant process is one of subduction of oceanic lithosphere, and that Alpine and Andean-type orogens are merely parts of a continuum.

The mandate of WG2 concerned plate movements and orogeny. The relationship is very clear in present day and Neogene global tectonics, but for the earlier history of the earth the required knowledge of plate positions and their relationships with other plates, through time, is generally lacking, as too much of the crustal record has been destroyed. For the last 100 Ma or so, the positions of a few plates with respect to deeper parts of the mantle are recorded by hotspot traces. The magnetic patterns in the floors of present ocean basins reflect the growth and translations of plates, and when tied to hotspots, map out plate positions at various times. In a few places it is possible to link these ocean floor plate movements with tectonic phenomena in adjacent orogenic belts, but unfortunately most pre-Jurassic ocean floors are preserved only as disrupted, ophiolitic slivers within orogenic belts. Pre-Jurassic plate positions are recorded only within the continents by such latitudinal indicators as paleomagnetic inclinations with respect to the paleohorizontal, and paleoclimatic interpretations based on fossil and sedimentary data, and the preserved records are unevenly distributed in space and time. Longitudinal positions can only be inferred by fitting continental masses together on an earth of constant size, and by interpreting meridional, long-lived accretionary prisms and pelagic sediments in ophiolite belts as traces of large, closed ocean basins. Different paleomagnetic or paleontological records in juxtaposed elements within an orogenic system, when coupled with a demonstrable relationship between the times that these elements came together and the structural, magmatic and metamorphic evolution of the system, are perhaps the most readily available evidence of the relationship between ancient plate movements and orogeny.

The Pacific Ocean basin and the Circum-Pacific orogenic belts provide perhaps the finest global laboratory in which to examine the relationship between plate movements and orogeny throughout the Phanerozoic. The geological record suggests that an ocean, the Pacific or its pre-Jurassic ancestor, Panthalassa or the Paleo-Pacific Ocean, was contiguous with most of the presently peripheral continents throughout Phanerozoic time, and that no major collisions involving continents interrupted this relationship. Within and around the present Pacific are examples of many different types of convergent, divergent and transform plate margins, and in addition to these the region provides examples of long-lived passive intraplate continent-ocean boundaries, such as that of the North American miogeocline.

This symposium volume can do no more than provide a sampling of the types and results of research activities taking place within and around the Pacific Ocean that bear on the mandate of WG2. Uyeda and Ben-Avraham discuss Recent and near-Recent plate phenomena within the present ocean basins, at, respectively, convergent margins and ridges. Engebretson and Zonenshain trace the movements of oceanic plates in later Mesozoic to Recent time in eastern and western parts of the Pacific; their findings directly address the question of plate movements and orogeny. Scotese presents the paleogeography of early Paleozoic Panthalassa. The last five papers summarize the evolution of segments of the Circum-Pacific orogenic belts within a plate tectonic framework. Coney discusses the development of the North American Cordillera, with its extensive accreted terranes but emphasizes that much of its structural development takes place within the plate margin after accretion. Mégard and Hervé et al. describe, respectively, the evolution of northern and southern parts of the Andes; the northernmost part contains accreted terranes similar to those in western North America, but in the central and southern Andes any terrane accretion was restricted to Paleozoic or Precambrian time, and most later differences along the length of the belt can be interpreted within the range of variations seen today along the Andes above the subducting oceanic lithosphere. Spörli provides a concise summary of the evolution of New Zealand within the setting of the southwestern Pacific region. Finally, Scheibner describes the long and extremely complex Paleozoic development of eastern Australia largely within the framework of the kinds of variations to be expected between convergent plates. The last four papers contain valuable, extensive bibliographies of their respective regions.

The editors would like to thank those who contributed papers to the volume. Francheteau distributed the mainly marine papers to reviewers, Monger the on-land papers. They would like to thank Rich Allmendinger, Dick Armstrong, Clark Blake, Clark Burchfiel, Tom Feininger, Ted Irving, Dave Scholl, Jack Souther and John Wheeler, as well as several anonymous reviewers, for their efforts on behalf of the volume. Steve Friday assisted in editing the papers and Bev Vanlier re-typed manuscripts.

J.W.H. Monger and Jean Francheteau
Editors

CHILEAN VS. MARIANA TYPE SUBDUCTION ZONES WITH REMARKS ON ARC VOLCANISM AND COLLISION TECTONICS

Seiya Uyeda

Earthquake Research Institute, University of Tokyo, Tokyo Japan 113

Abstract. The variety of geophysical and geological phenomena associated with subduction zones, such as backarc extension or compression, and different types of arc volcanism and seismicity, is difficult to explain in terms of a single model for downgoing slabs of cold oceanic lithosphere. Comparative studies of different subduction zones are instructive in this regard and lead to the recognition of two fundamentally contrasting modes controlled by the strength of mechanical coupling between downgoing and overriding plates. High-stress Chilean-type subduction zones are characteristic of continental arcs, whereas island arcs with actively spreading backarc basins are typified by low-stress Mariana-type subduction. Possible factors controlling the degree of mechanical coupling between downgoing and overriding plates are the differences in the nature, mainly density, of the subducting plate, and the motion of the overriding plate relative to the position of the trench line. Rollback of the subducting plate and trenchward advancement of the overriding plate cause extension and closure of backarc basins, respectively. Since all these factors can change with time, the mode of subduction at any particular subduction zone can also change with time. This may account for opening and closing of backarc basins in ancient subduction zones. Regional high heat flow in backarc basins can be explained by the same model used for ordinary seafloor spreading. However, arc volcanism and more local high heat flow in arc volcanic zones, found in both types of arcs, may be caused by flow in the mantle wedge induced by the subducting slab. Collision of buoyant features puts part of both Chilean and Mariana type arcs under extremely high compressional stress, and shuts off both volcanism and seismicity for truly great earthquakes. Upon collision, it is usually the overriding plate that yields, because it has been weakened by the continued subduction that preceded collision.

Introduction

In the framework of plate tectonics, subduction zones are regions where oceanic plates are consumed, and the counterpart of spreading centers where oceanic plates are generated. It is generally accepted that the subducting slabs of oceanic plates play the most important role in driving plate motions by their negative buoyancy (e.g. Forsyth and Uyeda, 1975; Chapple and Tullis, 1977). Moreover, tectonic and geodynamic processes are at their most active in subduction zones, as exemplified by the seismicity, crustal movements, heat flow, volcanism, and in places, backarc spreading. These processes are difficult to explain by a single model of subduction of a relatively cold oceanic plate. For instance, it is against one's intuition that a cold slab plunged into the mantle produces generally high heat flow in backarc regions and arc volcanism, and that convergence of plates generates extensional tectonics in backarc spreading. I wish to point out that the various subduction related processes are very complex and vary from one arc to another and require more sophisticated models.

One way to cope with this situation would be the approach that the author calls "comparative subductology" (Molnar and Atwater, 1978; Chase, 1978; Uyeda and Kanamori, 1979; Dewey, 1980; Uyeda, 1981, 1982, 1983, 1984). It is the intent of this short article to briefly review "comparative subductology" as published elsewhere, and then to make additional remarks on the origin of backarc high heat flow and arc volcanism, and on the effects of collision of buoyant features at subduction zones.

Comparative Subductology

Two Types of Subduction Zones

Subduction zones can be classified according to whether active backarc spreading is in progress, or not, as shown in Table 1. This classification pertains to the existence or non-existence of active extensional tectonics in backarc

TABLE I. Classification of arcs

Classification				Stress regime	Typical examples
Arc	Continental arc (without back arc basins)			Compressive (or neutral)	Peru-Chile Alaska
	Island arc (with back-arc basin)	Back-arc inactive	Back-arc inactivated	Compressive (or neutral)	Kuril, Japan Shikoku, Parece Vela Basins
			Back-arc trapped	Compressive (or neutral)	Bering Sea
		Back-arc active	Back-arc spreading	Tensional	Mariana Trough Scotia Sea, Lau Basin
			Leaky transform	Tensional with shear	Andaman Sea

regions. Since the end members are the Mariana arc and the Chilean arc, I call the respective types of subduction Mariana-type and Chilean-type subduction zones. Their contrasting characteristic features are summarized in Figure 1. There are basic differences in the stress regime, as revealed by the exclusive occurrences of truly great interplate thrust earthquakes and compressive intraplate earthquakes in the overriding plate in Chilean-type subduction zones, and the reverse situation in Mariana-type subduction zones (Uyeda and Kanamori, 1979). In addition, generally associated with Mariana-type subduction zones are low island arcs with abundant piles of basaltic volcanics overlying oceanic basement crust, whereas with Chilean-type are well-developed Cordilleran-type mountain ranges topped by andesitic strato-volcanoes. Other features, such as the nature of arc-related ore genesis (porphyry copper vs. massive sulphide), vertical motion (uplift vs. subsidence of the forearc region) and forearc history (accretion vs. erosion), and possibly the heat flow distribution and development of graben structures of the subducting plate, appear to be closely correlated with the two types of subduction.

Possible Causes for the Two Types of Subduction Zones

The remarkable difference in stress regimes in the two types of subduction is caused by the difference in degree of mechanical coupling between subducting and overriding plates. As lucidly illustrated by Dewey (1980), this is due to the relative motion between the overriding plate and the position of the trench or hingeline of the subducting plate (Fig. 2). If the absolute velocity, that is velocity relative to the deep mantle or hotspots, of the overriding plate and the absolute velocity of the hingeline (Dewey's rollback velocity), have a convergent component, Chilean-type, high-stress subduction will occur. Conversely, if their velocities have a divergent component, Mariana-type, low-stress subduction will result. Kanamori (1971) suggested that rapid rollback begins to take place after subduction has been going on for a long time by the development of a long tongue of heavy slab and "lubricating effects" at the plate interface. Alternatively, Molnar and Atwater (1978) proposed that more rapid rollback will occur when the subducting oceanic plate is older and hence denser, and vice versa. Another and probably more important factor controlling the degree of coupling between plates may be the absolute velocity of the over-

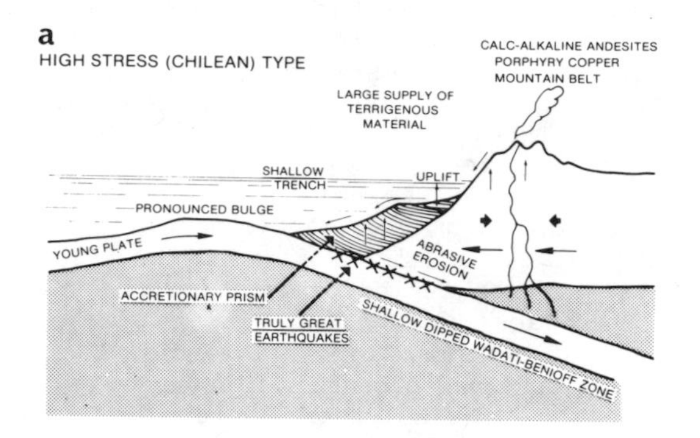

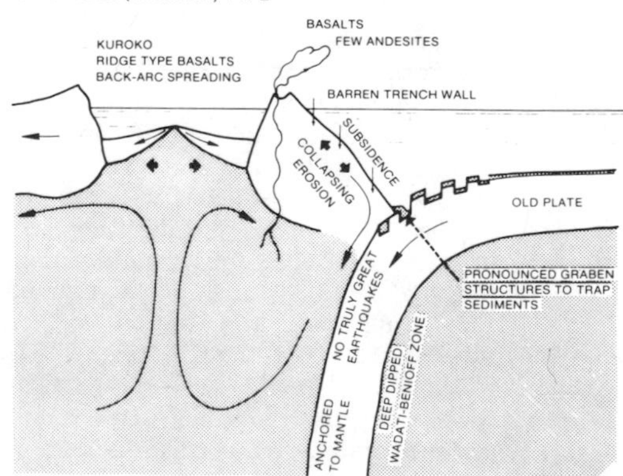

Fig. 1. Schematic diagrams showing the two typical models of subduction with possible tectonic implications and causes (not to scale). a) High stress (Chilean) type; b) Low stress (Mariana) type (Uyeda, 1984).

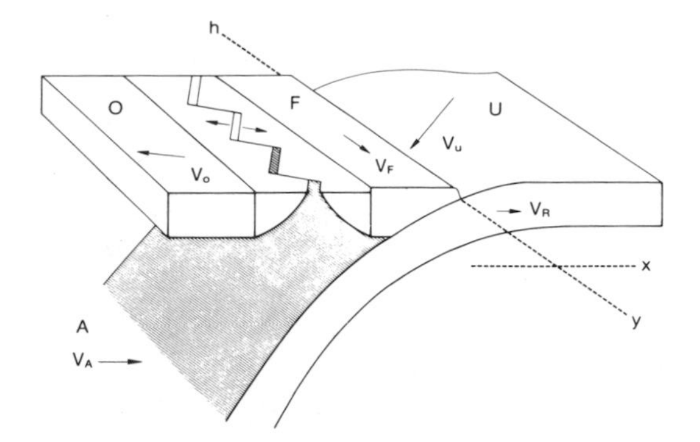

Fig. 2. Schematic diagram of subduction geometry. O - overriding plate with velocity Vo; U - underthrusting plate with velocity Vu; F - arc platelet with velocity V_F; A - asthenosphere with velocity V_A; V_R - rollback velocity; h - trench hingeline (Uyeda, 1983).

riding plate (Chase, 1978; Uyeda and Kanamori, 1979) because the subducted slab is likely to be anchored to the mantle (e.g. Hyndman, 1972), which at first approximation is essentially stationary, so that the hinge line also tends to stay stationary. In 1980, Ruff and Kanamori showed that, in addition to the youth of the subducting plate, a higher convergence rate also enhances stronger coupling. Peterson and Seno (1984) found from the moment release rates of thrust interplate earthquakes that the age of the subducting plate and the absolute velocity of the overriding plate are the two important factors.

Time Variations in the Type of Subduction

Dependence of rollback on the age of the subducting slab leads to a rational explanation of the almost exclusive occurrences of active backarc spreading in the western Pacific margin, where the subducting Pacific plate is old (Molnar and Atwater, 1978). However it is important to note that backarc spreading is generally ephemeral. For example, backarc spreading of the Japan and Kuril basins took place during geologically short periods of time, or else was episodic, as is the case for the Philippine Sea (Karig, 1971; Uyeda and Ben-Avraham, 1972). Moreover, it has recently been suggested that the Japan Sea is now beginning to close by subduction along its eastern margin (Nakamura, 1983), and the central Okinawa Trough is probably starting to spread (Kimura, 1985). These variations are difficult to explain if the age of the subducting slab is the sole governing factor, although Seno (1983) proposed otherwise. However, the changes are also difficult to explain by changes of absolute motion of the huge, overriding Eurasian plate. An alternative suggestion is that overriding plate motions are of a more local nature. For the South China Sea, Okinawa Trough, Japan Sea and even the Kuril Basin, intraplate deformations and ruptures may relate to collision of India with Eurasia (Tapponnier et al., 1982; Eguchi et al., 1979; Kimura and Tamaki, 1985), and for the Okinawa Trough, collision of the Philippine arc with Taiwan (Letouzey and Kimura, 1985). In the case of episodic spreading of the Philippine Sea, collisions in the Philippines are suggested to be the cause (Uyeda and McCabe, 1983; Yamano and Uyeda, 1985a). If we take the local variations combined with the age control, time variations in the mode of subduction may well be explained. Yet another alternative to explain time variations is the evolutionary model originated by Kanamori (1971) and developed by Niitsuma (1978) and Kobayashi (1983), in which the subducting slab detaches from time to time. The relative importance of these models may well be finally evaluated in the near future.

Thermal Aspects of Comparative Subductology

Mechanical aspects of subduction zones have been, to some extent, clarified by the approach of comparative subductology reviewed briefly in the preceding section. In this section, thermal aspects of subduction zones, which are equally enigmatic, are considered. Why is heat flow in the backarc region high, and why is arc volcanism generated at all when a cold lithospheric slab subducts?

Heat Flow

Since backarc spreading is the result of the parting of plates (Fig. 2), at least as observed at the surface, the basic process should be the same as ordinary seafloor spreading and plate creation. In fact, the same heat flow vs. age relationship seen in ocean basins, also appears to hold true for backarc basins (e.g. Anderson, 1980). The data available, especially on the ages of backarc basins, are definitely insufficient to detect if there is any systematic difference in the heat flow vs. age relationship between backarc and ocean basins. Extremely high heat flow values ($\gtrsim$1000 mW^2) have been found locally within axial zones of backarc basins, and are comparable with those of axial zones of mid-oceanic ridges, (e.g. Hobart et al., 1979, for the Mariana Trough; Yamano and Uyeda, 1985b, for the Okinawa Trough. Although active hydrothermal venting and mineralization have not been directly observed by submersible studies in backarc basins, the probability of finding them seems to be quite high if dives are made in appropriate areas. Thus, the regionally high heat flow in backarc basins of present and past Mariana-type subduction zones may be explained, to the first approximation, by the same model as that of seafloor spreading.

It is also true, however, that the geodynamic setting of backarc basins is quite different from that of new ocean basins, in that subducting slabs may exist at the margin of the former. The

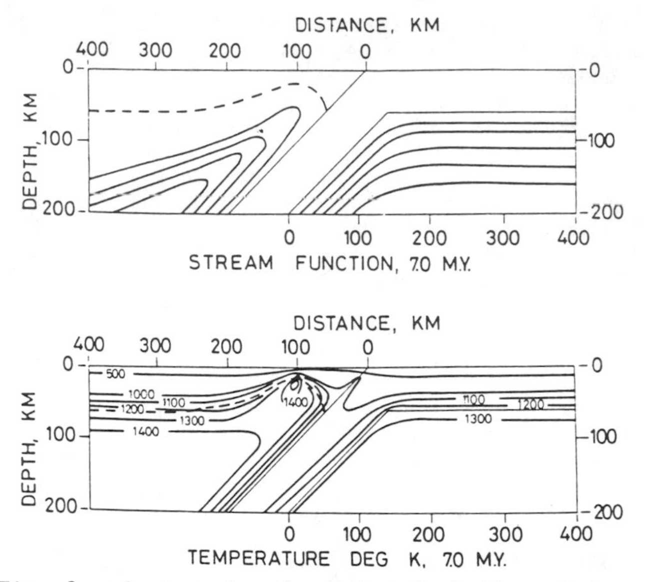

Fig. 3. An example of model calculation of induced wedge flow and isotherms under subduction zones. At 7 Ma after initiation of subduction with velocity of 3 cm/yr. (Bodri and Bodri, 1978).

cooling effect of a cold subducting slab can be recognized by low heat flow in the trench and forearc regions of many subduction zones. In this respect, it is instructive that the Nankai Trough bottom shows high heat flow (~150 mW/m^2) and the subducting plate there is the young (24-18 Ma) and hot Shikoku Basin plate (Yamano et al. 1984).

Insofar as material composing the mantle wedge above the slab is regarded as a viscous fluid, a flow must be induced by the downgoing motion of the slab (e.g. McKenzie, 1969). Such a flow should exist even under Chilean-type arcs, except in some segments of the South American system where the absence of a significant mantle wedge is suspected (e.g. Isacks and Barazangi, 1977), and accordingly should exert mechanical and thermal effects (e.g. Sleep and Toksoz, 1973; Andrews and Sleep, 1974; Bodri and Bodri, 1978; Honda and Uyeda, 1983). In this example, the flow may help extensional tectonics within the overriding plate but probably is not strong enough to cause backarc spreading by itself. Here, the thermal effect of the wedge is not clear, neither observationally or theoretically, and heat flow distributions in backarc regions of Chilean-type subduction zones are not well known, except for the Aleutian and Indonesian arcs (Langseth et al., 1980; Thamrin, 1985). Average heat flow in the Aleutian Basin is 55 mW/m^2 (60 mW/m^2 when corrected for sedimentation). These values appear to be a little too high for a trapped Mesozoic ocean basin material, and excess heat flow may be attributable to induced wedge flow. However, Langseth et al. (1980) suggested that the excess heat flow could be due to thermal rejuvenation of the basin by subduction of the Kula spreading center. In the Indonesian arc, high heat flow values are concentrated in the Sumatran backarc, which is less Chilean-like, and the backarc of the Java arc has only moderately high heat flow. In the Chilean-arc, the measurements are too sparse to draw any conclusion, but the area directly east of the Chilean Andes seems to have heat flow that is only moderately high (Uyeda and Watanabe, 1982). In this case, the heat contribution from thick continental crust may be significant. Tentatively, we infer that the backarc areas related to Chilean-type subduction zones do not have regionally high mantle heat flow as do those related to the Mariana-type subduction zones. Thus it appears that the regional thermal effect of presumed mantle wedge flow is much smaller than that of the backarc spreading, which agrees with various model calculations in which thick lithosphere is assumed for the overriding plate (e.g. Honda and Uyeda, 1983).

On the Origin of Arc Volcanism

Although backarc regional high heat flow is apparently missing in the case of arcs without backarc basins, arc volcanism and associated localized high heat flow exist above almost all subduction zones. Frictional heating generated above the subducting slab does not seem to cause this, for, as pointed out by Andrews and Sleep (1974) the high stress required to generate frictional heat cannot be sustained at the interface when the temperature is raised even moderately. It may be possible, however, that the temperature of the upper portion of the subducted slab becomes high enough to dehydrate the hydrous minerals.

A major role appears to be played here by induced wedge flow (Andrews and Sleep, 1974; Bodri and Bodri, 1978) as shown, for example, in Figure 3. Downward flow along the slab would thermally tap the hot, lower mantle and then rise more or less adiabatically. During ascent along stream lines, melting would take place because the melting temperature curve is steeper than the adiabat. The degree of melting would increase upwards and reach a maximum at the highest point of the stream line. Into this corner region, stress is concentrated so as to generate heat locally. The heated lithosphere thins, as observed under the Japanese Islands (e.g. Yoshii, 1979) and a volcanic front probably results (line in plan view). As soon as the flow begins to decrease, solidification starts and eruption ceases.

This simple system must involve many more factors, such as the effects of materials released from the slab including water from dehydration and possibly some melt of subducted sediments or crust. Although it is generally considered difficult to melt the slab, Hsui et al. (1983)

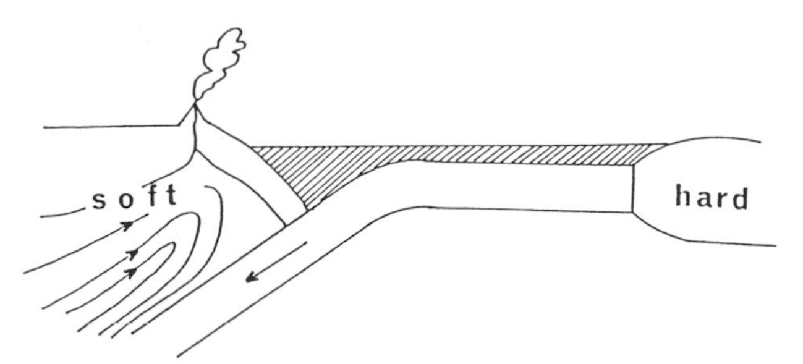

Fig. 4. Schematic diagram showing why collided side is generally softer and yields upon collision.

maintain that it is possible, and isotopic data, in particular ^{10}Be data (Brown et al., 1982) seem to support this proposal, which remains one of the major outstanding problems in plate tectonics. The tectonic stress state in the arc lithosphere may be another important factor in controlling the abundance, composition and mode of volcanic activity. This aspect, however, seems to be relatively straightforward in terms of "comparative subductology", when compared with the problem related to the melting of the slab.

Effects of Collision

In the framework of "comparative subductology" collision of buoyant features against the overriding plate plays a unique role in raising the compressional stress in the overriding plate. In some cases, collision makes a segment of a Mariana-type subduction zone look like Chilean-type, as for example, the uplift of forearcs (Yonekura, 1983), and the occurrence of porphyry copper in some portions of arcs that otherwise are of Mariana-type (Uyeda and Nishiwaki, 1980). In addition, as noted by Nur and Ben-Avraham (1981), collision may shut off active volcanism. The author concurs with the view that this can happen because plumbing systems for magma ascent are closed as a result of very high compressive stress.

Another very important observation is that seismicity is reduced at collision boundaries (Vogt et al., 1976). The effect is most pronounced in the reduction of numbers of great earthquakes (McCann et al., 1979) rather than in changes in numbers of small and moderate sized earthquakes. This is particularly interesting because one might expect that greater earthquakes occur under higher stress, as is the case in the Chilean-type subduction zones. At collision boundaries, it appears that stress is built up higher than in the Chilean-type boundaries, because slip cannot take place between the overriding plate and the unsubductable element. When the colliding feature is large in comparison with the length of the plate boundary, the collision influences plate motions (Yamano and Uyeda, 1985a), but when the feature is small, the plate motion is not affected, and with time a new subduction zone starts seaward of the colliding feature, and the collision is finished. Until that happens, namely during collision, plate motions on the spot have to be taken up largely by anelastic deformation of the overriding plate as evidenced by cusps of arcs (Vogt et al., 1976). Notable examples are the Himalayas (Molnar and Tapponnier, 1975) and Izu in Japan (Matsuda, 1978). The deformation can of course be seismic, but does not seem to generate truly great earthquakes that would require large displacement on a large slip plane, which is only possible in subduction.

Finally, the author would like to conclude with the following important point. In collision, it is the overriding plate that suffers more deformation, probably because the collided side tends to be softer than the colliding feature for the following reason. For a collision to take place, it must be preceded by subduction to bring the colliding element to the plate boundary. The subduction process weakens the upper plate by backarc spreading or mantle wedge flow and arc volcanism (Fig. 4).

References

Anderson, R. N., Update of heat flow in the East and Southeast Asian Seas, in: Amer. Geophys. Un., Mongr. "Tectonic/Geologic Evolution of Southeast Asia", Hayes, D., ed., 319-326, 1980.

Andrews, D. J., and N. H. Sleep, Numerical modeling of tectonic flow behind island arcs, Geophys. J. Roy. Astr. Soc., 38, 237-251, 1974.

Bodri, L., and B. Bodri, Numerical investigation of tectonic flow in island-arc areas, Tectonophys., 50, 163-175, 1978.

Brown, L., J. Klein, R. Middleton, I. S. Sacks, and F. Tero, ^{10}Be in island-arc volcanoes and implications for subduction, Nature, 299, 718-720, 1982.

Chapple, W. M., and T. E. Tullis, Evaluation of the forces that drive the plates, J. Geophys. Res., 82, 1967-1984, 1977.

Chase, C., Extension behind island arcs and motions relative to hot-spots, J. Geophys. Res., 83, 5385-5387, 1978.

Dewey, J. L., Episodicity, sequence and style at convergent plate boundaries, in: "The Continental Crust and its Mineral Resources", Strangway, D., ed., Geol. Assoc. Canada, Spec. Pap., 20, 553-574, 1980.

Eguchi, T., S. Uyeda and T. Maki, Seismotectonics and Tectonic history of Andaman Sea, Tectonophys., 57, 35-51, 1979.

Forsyth, D. W., and S. Uyeda, On the relative importance of driving forces of plate motion, Geophys. J. Roy. Astron. Soc., 43, 163-200, 1975.

Hobart, M., R. N. Anderson, and S. Uyeda, Heat transfer in the Mariana Trough, EOS, 60, 383, 1979.

Honda, S., and S. Uyeda, Thermal process in sub-

duction zones, in: "Arc Volcanism: Physics and Tectonics", Shimozuru, D., and I. Yokoyama, eds., 117-140, Terra Sci. Pub. Com., Tokyo, 1983.
Hsui, A., B. Marsh, and N. Toksöz, On melting of the subducted ocean crust: effects of subduction induced flow, Tectonophys., 99, 207-220, 1983.
Hyndman, R. D., Plate motions relative to the deep mantle and the development of subduction zones, Nature, 238, 263-264, 1972.
Isacks, B. L., and M. Barazangi, Geometry of Benioff zones: Lateral segmentation and downward bending of the subducted lithosphere, in: "Island Arcs, Deep Sea Trenches and Back-arc Basins", Talwani, M., and W. C. Pitman III, eds., Maurice Ewing Series, 1, 99-114, Amer. Geophys. Un., 1977.
Kanamori, H., Seismological evidence for a lithospheric normal faulting-The Sanriku Earthquake of 1933, Phys. Earth Planet. Inter., 4, 289-300, 1971.
Karig, D. E., Origin and development of the marginal basins in the western Pacific, J. Geophys Res., 76, 2542-2561, 1971.
Kimura, G. and K. Tamaki, Collision, rotation and back-arc spreading: the case of the Okhotsk and Japan Seas, Tectonophys., (in press), 1985.
Kimura, M., Backarc rifting in the Okinawa Trough Marine and Petroleum Geology, (in press), 1985.
Kobayashi, K., Cycles of subduction and Cenozoic arc activity in the northwestern Pacific margin, in: "Geodynamics of the Western Pacific - Indonesian Region", Hilde, T., and S. Uyeda, eds., Geodynamic Series, 11, Amer. Geophys. Un./Geol. Soc. Amer., 287-301, 1983.
Langseth, M., M. A. Hobart, and K. Horai, Heat flow in Bering Sea, J. Geophys. Res., 85, 3740-3750, 1980.
Letouzey, J., and M. Kimura, The Okinawa Trough: Genesis structure and evolution of a backarc basin developed in a continent (preprint), 1985.
Matsuda, T., Collision of the Izu-Bonin Arc with Central Honshu: Cenozoic tectonics of the Fossa Magna, Japan, J. Phys. Earth, 26, Suppl., S409-421, 1978.
McCann, W. R., S. P. Nishenko, L. R. Sykes, and J. Krause, Seismic gaps and plate tectonics: seismic potential for major boundaries, Pure Appl. Geophys., 117, 1082-1147, Birkhäuser Verlag, Basel, 1979.
McKenzie, D. P., Speculations on the consequences and causes of plate motions, Geophys. J. Roy. Astron. Soc., 18, 1-32, 1969.
Molnar, P., and T. Atwater, Interarc spreading and Cordilleran tectonics as alternates related to the age of subducted oceanic lithosphere, Earth Planet. Sci. Lett., 41, 330-340, 1978.
Molnar, T., and T. Tapponier, Cenozoic tectonics of Asia: effects of a continental collision, Science, 189, 419-426, 1975.
Nakamura, K., Possible nascent trench along the Eastern Japan Sea as the convergent boundary between Eurasian and North American plates, Bull. Earthq. Res. Inst., 58, 711-722, (in Japanese with English Abstract), 1983.
Niitsuma, N., Magnetic stratigraphy of the Japanese Neogene and the development of the island arcs of Japan, J. Phys. Earth 26, Suppl., S637-S378, 1978.
Nur, A., and Z. Ben-Avraham, Volcanic gaps and the consumption of aseismic ridges in South America, Mem. Geol. Soc. Amer., 154, 729-740, 1981.
Peterson, E., and T. Seno, Factors affecting seismic moment release rates in subduction zones, J. Geophys. Res., 89, 10.233-10.248, 1984.
Ruff, L., and H. Kanamori, Seismicity and the subduction process, Phys. Earth Planet. Inter., 23, 240-252, 1980.
Seno, T., Age of subducting plate and tectonics, Marine Science Monthly, 15, 761-766, (in Japanese), 1983.
Sleep, N., and M. N. Toksöz, Evolution of marginal basins, Nature, 233, 548-550, 1973.
Tapponnier, P., G. Peltzer, A. Y. Le Dain, and R. Armijo, Propogating extrusion tectonics in Asia: New insights from simple experiments with plasticine, Geology, 10, 609-688, 1982.
Thamrin, M., An investigation on the relationship between geologic phenomena and heat flow value as observed in oil fields in Indonesian sedimentary basins, Tectonophy. (in press), 1985.
Uyeda, S., Subduction zones and back arc basins - A review, Geol. Rundsch., 70, 552-569, 1981.
Uyeda, S., Subduction zones: an introduction to comparative subductology, Tectonophys., 81, 133-159, 1982.
Uyeda, S., Comparative subductology, Episodes, vol. 1983, no. 2, 19-24, 1983.
Uyeda, S., Subduction Zones: Their diversity, mechanism and human impacts, Geojournal, 8.4, 381-406, 1984.
Uyeda, S. and Z. Ben-Avraham, Origin and development of the Philippine Sea, Nature, Phys. Sci., 240, 176-178, 1972.
Uyeda, S., and H. Kanamori, Back-arc opening and the mode of subduction, J. Geophys. Res., 84, 1049-1061, 1979.
Uyeda, S., and R. McCabe, A possible mechanism of episodic spreading of the Philippine Sea, in: "Accretion Tectonics in the Circum-Pacific Regions", Hashimoto, M., and S. Uyeda, eds., Terra Sci. Pub. Co., Tokyo, 291-306, 1983.
Uyeda, S., and C. Nishiwaki, Stress field, metallogenesis and mode of subduction, in: "The Continental Crust and its Mineral Deposits", Strangway, D. W., eds., Geol. Soc. Canada, Sp. Paper, 20, 323-339, 1980.
Uyeda, S., and T. Watanabe, Terrestrial heat flow in western South America, Tectonophys., 83, 63-70, 1982.

Vogt, P. R., A. Lowice, D. R. Bracey, and R. N. Hey, Subduction of aseismic ridges: Effects on shape, seismicity, and other characteristics of consuming plate boundaries, Geol. Soc. Amer., Sp. Paper, 172, p. 59, 1976.

Yamano, M., S. Honda, and S. Uyeda, Nankai Trough a hot trench? J. Marine Geophys. Res., 6, 187-203, 1984.

Yamano, M., and S. Uyeda, Possible effects of collisions on plate motions, Tectonophys., (in press), 1985a.

Yamano, M., and S. Uyeda, Heat flow in the Okinawa Trough (preprint), 1985b).

Yonekura, N., Late Quaternary vertical crustal movements in and around the Pacific as deduced from former shoreline data, in: "Geodynamics of the Western Pacific-Indonesian Region", Hilde, T. and S. Uyeda, eds., 41-50, Geodynamics Ser., 11, Amer. Geophys. Un./Geol. Soc. Amer., 1983.

Yoshii, T., A detailed cross-section of the deep seismic zone beneath northeastern Honshu, Japan Tectonophys., 55, 349-360, 1979.

EFFECTS OF COLLISIONS AT TRENCHES ON OCEANIC RIDGES AND PASSIVE MARGINS

Zvi Ben-Avraham[1] and Amos Nur

Department of Geophysics, Stanford University, Stanford, California 94305

Abstract. The process of subduction along consuming plate boundaries is occasionally interrupted by the arrival at the trenches of elevated bathymetric features. Such features are generally called oceanic plateaus and include extinct arcs, extinct spreading ridges, detached and submerged continental fragments, oceanic islands and seamounts. The oceanic plateaus are generally buoyant, and upon arrival at subduction zones may be accreted to the upper plate, rather than subducted with the oceanic lithosphere within which they are embedded. The collisions between oceanic plateaus of various sizes and the upper plates above subduction zones may be responsible for the following fundamental tectonic processes which take place along convergent margins of continents: gaps in active volcanic chains, disorder in the normal seismic pattern associated with subduction of oceanic crust, the emplacement of ophiolites, and possibly the creation of marginal basins. Furthermore, accretion of oceanic plateaus to continents is probably one of the causes of orogenic deformation during the processes of mountain building. These accreted, or allochthonous, terranes have by now been recognized in the circum-Pacific margins as well as in other margins of continents where ancient orogenies exist, such as eastern North America and western Europe. It is generally recognized that trench-pull is a major force moving plates. If so, temporary cessation of underthrusting due to collision of an oceanic plateau with a trench could result in major changes in kinematics of associated plates. A readjustment period is required, during which either breakup of the plateau and dispersion of its fragments along the continental margin takes place, or the subduction zone shifts into a new configuration. During this period the subduction process is retarded or stopped; this leads to changes in the direction and speed of plate motions. Because the direction of motion of an oceanic plate is determined by forces acting at the trench, plate motions will change when part of a trench segment is eliminated. After the dispersion/consumption process is completed the plate may resume its original direction and speed. Thus, collision of oceanic plateaus with trenches probably caused changes in plate motions in the past, and are recorded by magnetic lineations. Another event which may be responsible for cessation of subduction along trench segments is the collision of a spreading ridge with a continental margin. This process can lead to the development of a transform plate boundary, such as the San Andreas Fault system along the California continental margin, that in turn allows growth of a slab-free region beneath the part of the continental block adjacent to the transform. The stress caused by collisions at the plate boundary probably can be transmitted far into the mid-plate region. For example, discrete ridge axis jumps have been observed along the Nazca-Pacific plate boundary and the Galapagos Ridge. We suggest that these jumps probably resulted from consumption of the thick-rooted, buoyant, aseismic Nazca, Juan Fernandez, Carnegie ridges, and perhaps also the Cocos Ridge, at the subduction zones bordering Central and South America. In the eastern Mediterranean the collision of the Erastosthenes and the Anaximander seamounts with the Cyprus arc results in a unique tectonic setting, because of the small size of the eastern Mediterranean basin. As a result of the collision, some of the motion between the African plate and the Eurasian plate is taking place by thrust faulting along the North African passive margin. It seems that the stress at the plate boundary here may be transmitted southward to cause reactivation of a pre-existing fault zone along the passive margin. The collision of an oceanic plateau with a subduction zone is thus an important process controlling events taking place on the margins of continents and in ocean basins. Therefore a rough correlation might exist between tectonic events on land and in the ocean. The possibility of correlating the time of events such as the creation of volcanic gaps on land with events like ridge jumps at sea can provide us with a quantitative tool to study some of the processes which link plate tectonics to continental tectonics.

[1]On leave from the Department of Geophysics and Planetary Sciences, Tel-Aviv, Israel

Introduction

It is generally thought that of the various plate driving forces, slab-pull is the greatest (e.g. Forsyth and Uyeda, 1975). Consequently the absolute motion of any individual plate may be controlled by the positions of trenches and ridges along the plate boundary (Gordon et al., 1978). Accordingly, in order to maintain the motion of a plate, the subduction process at the trenches around it should be continuous.

Several processes or events occasionally interrupt the subduction of the oceanic plates at the trenches. These include (e.g. Dickinson and Snyder, 1979; Nur and Ben-Avraham, 1982a,b) the arrival of elevated bathymetric features at the trench and the development of a transform plate boundary along the continental margin. As a result, the direction of plate motions, the speed of plates and even plate geometries may undergo changes due to interruptions of the subduction processes in trenches surrounding the plate.

In this paper we speculate on the tectonic effects of consumption of elevated features at convergent margins of plates on the other plate boundaries, namely, at oceanic ridges and at passive continental margins within these plates. In particular we consider evidence from the Nazca plate and the eastern Mediterranean, where interrupted subduction and the consequent effects are presently taking place.

Interruption of Subduction at Trenches

The orderly subduction process at trenches is occasionally interrupted by the arrival at the trenches of various elevated bathymetric features. These bathymetric features, referred to as oceanic plateaus, are of several origins, and include extinct arcs, extinct spreading ridges, detached and submerged continental fragments, oceanic islands and seamounts, and the traces of hotspots (Fig. 1) (Ben-Avraham et al., 1981; Nur and Ben-Avraham, 1982a,b). Oceanic plateaus are usually buoyant and therefore may tend to accrete by collision to the continental margin upon arrival at subduction zones, rather than subduct with the oceanic lithosphere in which they are embedded. Such collisions between oceanic plateaus and plates overriding the subduction zones may be responsible for important tectonic effects which take place along the margins of continents. They cause gaps in active volcanic chains (Nur and Ben-Avraham, 1981), disorder the normal seismic pattern associated with subduction of oceanic crust (Barazangi and Isacks, 1976), emplace ophiolites (Ben-Avraham et al., 1982), and possibly create marginal basins (Ben-Avraham and Cooper, 1981). Furthermore the accretion of oceanic plateaus to the continents may be one of the causes of orogenic deformation during the process of mountain building (Nur and Ben-Avraham 1982a,b). These accreted plateaus, or allochthonous terranes, have by now been recognized in the Circum-Pacific margins (Coney et al., 1980) and in the Alpine mountain system as well as in other margins of continents where ancient orogenies exist, such as eastern North America (Cook et al., 1979; Williams and Hatcher, 1970) and western Europe (Leggett et al., 1983).

The outcome of the transformation of oceanic plateaus into allochthonous terranes depends on the tectonic stress regime (Uyeda and Kanamori, 1979). In the eastern Pacific, for example, where tectonic stress is high, colliding plateaus may undergo extensive deformation during accretion with only small changes in plate configuration. In contrast, plate boundary geometry may change significantly upon plateau collision in low-stress regimes such as in the west and north Pacific and in the eastern Mediterranean. Here, where the subduction zones are not adjacent to a continent but rather to island arcs and marginal seas, plateaus or rises may simply dock, causing the subduction zone to jump to the oceanic side of the plateau, or to switch polarity, or an elongate, continuous arc may be cut into small segments. Several marginal basins around the Pacific may have resulted from this docking process. For example, the creation of the Aleutian arc may be the result of the collision of the Bowers and Umnak plateaus with the Mesozoic subduction zone in the Bering Sea (Ben-Avraham and Cooper, 1981). Segmentation of arc systems can be recognized in several places around the Pacific, such as the breakup of the arc bordering the Philippine Sea on the east into the Izu, Bonin, Mariana and Yap trenches. The separation of the once continuous arc system in the eastern Mediterranean into the Hellenic and Cyprus arcs could have also resulted from the arrival of an oceanic plateau (Rotstein and Ben-Avraham, 1985), now accreted into southern Turkey.

Cessation of subduction along trench segments can occur also via the development of a transform plate boundary along the continental margin. An example is the San Andreas fault system along the California continental margin, which was formed by the collision of an active spreading center with the trench that once existed along that margin. The absence of subduction at the transform boundary along the California continental margin led to the growth of a slab-free region adjacent to the San Andreas transform (Dickinson and Snyder, 1979).

Effects of Irregular Subduction on Tectonic Events Within the Ocean Basins

The consumption of oceanic plateaus that interrupt the subduction process requires a period of readjustment. During this period, breakup of the plateau and dispersion of its fragments along the continental margin may take place, or the subduction zone may be shifted into a new configuration. As a result, the subduction process is retarded or stopped during the time of readjustment, leading to changes in the direction and

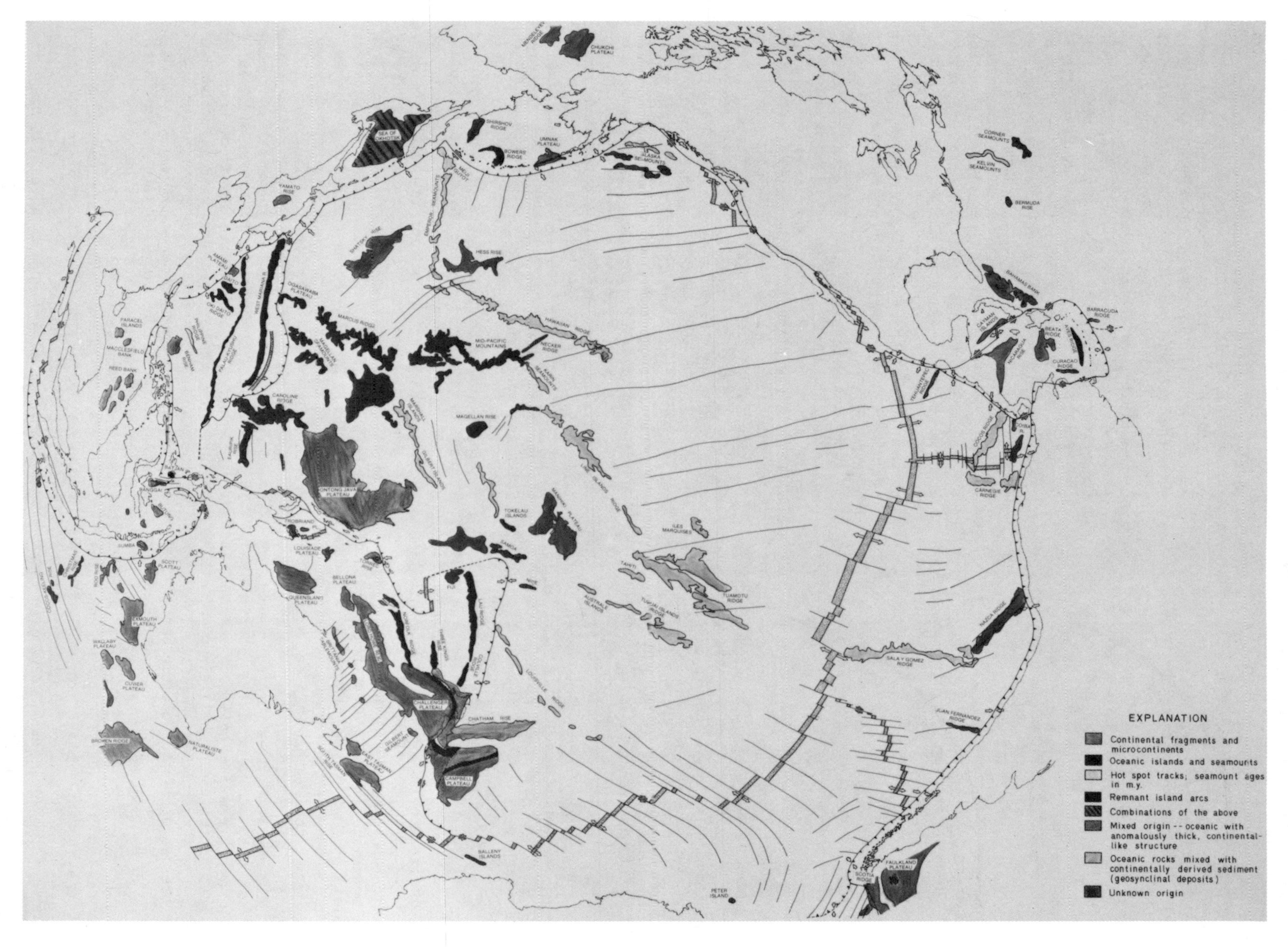

Fig. 1. Oceanic plateaus in the Pacific Ocean and adjacent seas.

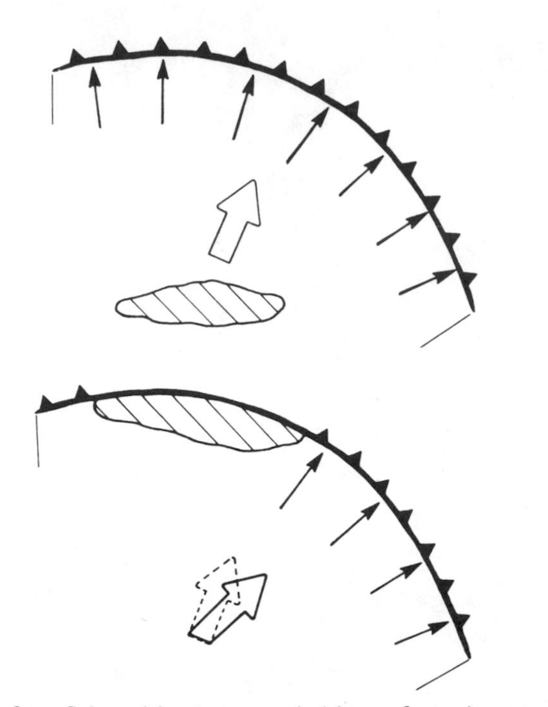

Fig. 2. Schematic representation of a change in the direction of motion of an oceanic plate due to the collision of a large oceanic plateau with a segment of the subduction zones bordering the plate. The assumption here is the direction of motion of an oceanic plate is determined by the forces acting at the trenches.

speed of plate motion. Because the direction of motion of an oceanic plate is determined by the forces acting at the trenches, plate motion will change when part of a trench segment is eliminated (Fig. 2). Thus, collisional events of oceanic plateaus with trench segments have probably caused changes in plate motions in the past, such as is evident in the pattern of magnetic anomalies of Pacific Ocean basin plates (Engebretson and Ben-Avraham, 1981).

It has often been suggested that the stress at the plate boundary can be transmitted far into the mid-plate region (e.g., Shimazaki, 1982). This raises the possibility that the collision of oceanic plateaus at trenches may be related to such phenomena as spreading ridge axis jumps which are clearly recognized in the oceans, or tectonic activity along passive continental margins far away from the collision zones. In several places in the world, collisions presently occurring at trenches cause marked effects on tectonic processes within the oceanic basin.

The Nazca Plate

We have discussed in earlier papers how plateaus that are presently being consumed at subduction zones cause profound geological effects in the continental margins, such as reduced seismicity and shifts in volcanic activity (Nur and Ben-Avraham, 1981). The best examples are the collision of the Carnegie, Nazca, and Juan Fernandez aseismic ridges and the active Chile Rise with the western margin of South America which results in a remarkable variety of combinations of seismicity, volcanism and morphology (Vogt et al., 1976; Barazangi and Isacks, 1976). We have suggested that volcanic gaps and associated shallow dips of seismic planes on the western margin of South America are closely related to the oblique consumption of the ridges, and that similar effects may exist in general where anomalously buoyant crust is being consumed. We now propose that old ridges at the eastern boundary of the Nazca plate interrupted its smooth subduction beneath the North American plate, and this caused westward jumps in the east Pacific Rise, at the western edge of the Nazca plate. These jumps are recorded by the magnetic anomalies and the bathymetry.

Along the Nazca-Pacific plate boundary several discrete and large ridge axis jumps, as well as small axial offsets, have occurred within the last 10 Ma (Rea, 1981). In particular, the Galapagos Rise (Fig. 3) on the Nazca plate marks the location of an old ridge system that has been shifted to the present location of the East Pacific Rise by three large jumps. Altogether, the spreading centre shifted westwards for 600 to 850 km, over a period of 2.5 million years, in the interval between 8.2 and 5.7 Ma (Rea, 1981). On the ridge crest five relatively small offsets of 10-15 km each have been recognized. All of these offsets have happened within roughly 0.5 million years, and have affected a part of the ridge axis only 100-200 km long (Rea, 1981). Similar axis shifts have taken place along the Galapagos spreading center in the last few million years.

Ridge axis jumps and offsets probably occurred at earlier times along the Nazca-Pacific plate boundary as well as along other segments of spreading centers in the Pacific. North of 4°S, the present East Pacific Rise was presumably created in the Pliocene by eastward jumps from the Mathematicians-Clipperton Ridge (Van Andel et al., 1975). In the northeast Pacific, ridge axis jumps have been also postulated near the Surveyor Fracture Zone (Shih and Molnar, 1975).

Another feature at the Nazca-Pacific plate boundary is the existence of the small and growing Easter plate (Herron, 1972; Forsyth, 1972) west of Easter Island. Active sea-floor spreading centers form the eastern and western boundaries of the plate. This is a young plate, only 3.2 Ma old (Handschumacher et al., 1981), which might eventually become incorporated into either Pacific or Nazca plates as one of its bounding ridges ceases to be active.

In addition to the changes in the above configuration of the Nazca-Pacific plate boundary, major reorientations of the Nazca (Farallon) and South America plate motions have also been in-

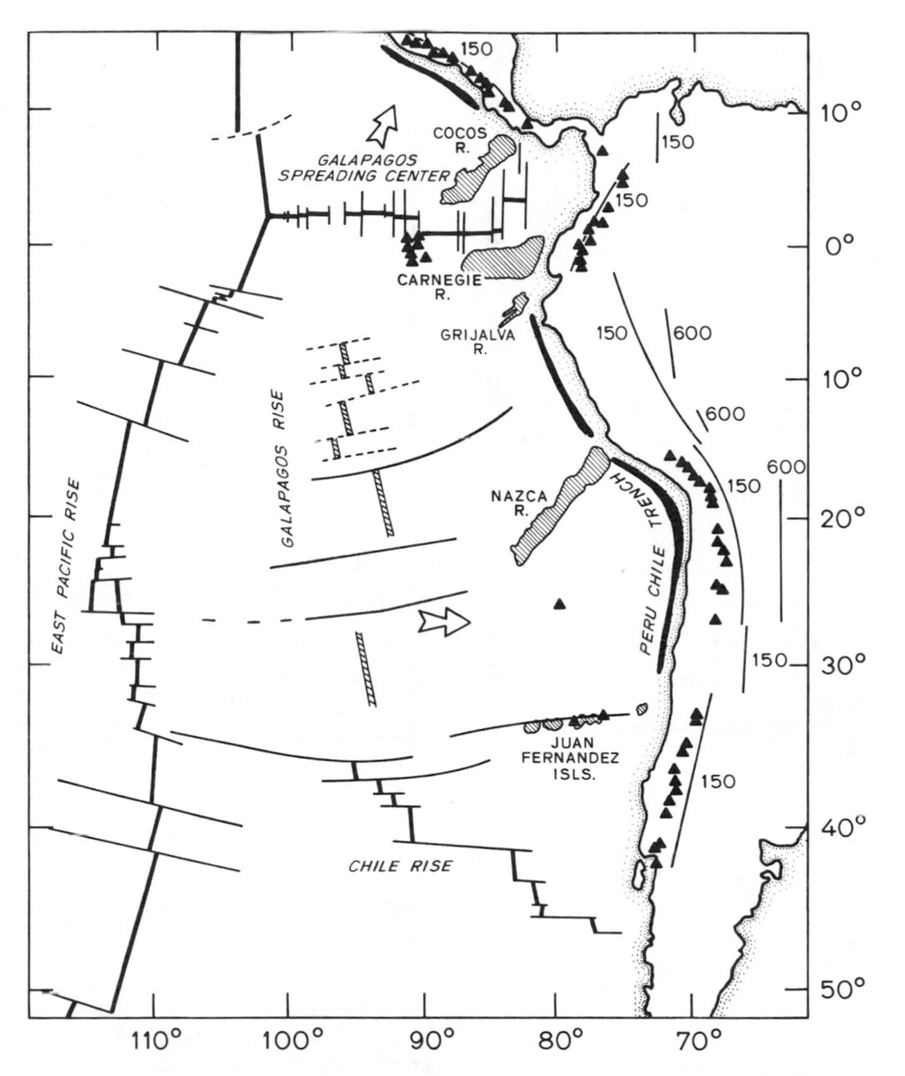

Fig. 3. Major tectonic elements in the Nazca and Cocos plates, including trench segments, spreading ridges (heavy lines), old ridge system (double lines with lined pattern), aseismic ridges (lined pattern), active volcanoes (triangles), and depth to Benioff zone (numbers, in kilometres) as shown by seismic activity.

ferred from the magnetic anomalies. Two such reorientations have been suggested at 48 and 25 Ma (Pilger, 1983). Specifically, whereas since 25 Ma the convergence between Nazca and South America plates has been essentially east-west, prior to 25 Ma convergence was approximately northeast-southwest. Before 48 Ma the relative motion between the two plates is uncertain. Furthermore, small scale perturbations in relative motions have been suggested throughout the history of the plates (Pilger, 1983), and involve changes both in the direction and rate of the relative motions.

In attempting to understand these changes, we take into account the following major events in Nazca plate history: a change in the relative motion of the Nazca (Farallon) and South America plates from a northeast-southwest direction to an east-west direction at about 25 Ma; the breakup of the Farallon plate into Nazca and Cocos plates by the formation of the Galapagos spreading center at about 25 Ma; and jumps of the spreading center at the Galapagos Rise to the East Pacific Rise south of 5°S between 8.2 and 5.7 Ma. Throughout this time small offsets of the ridge crest have also taken place.

We suggest that these events result in part, but not entirely from collisions along the Middle and South American trenches. The major change of the motion of the Nazca plate, at about 25 Ma from its northeast-southwest convergence to east-west convergence, can be correlated with events at the eastern boundary of the present Caribbean plate. Before 25 Ma, the Caribbean was part of a

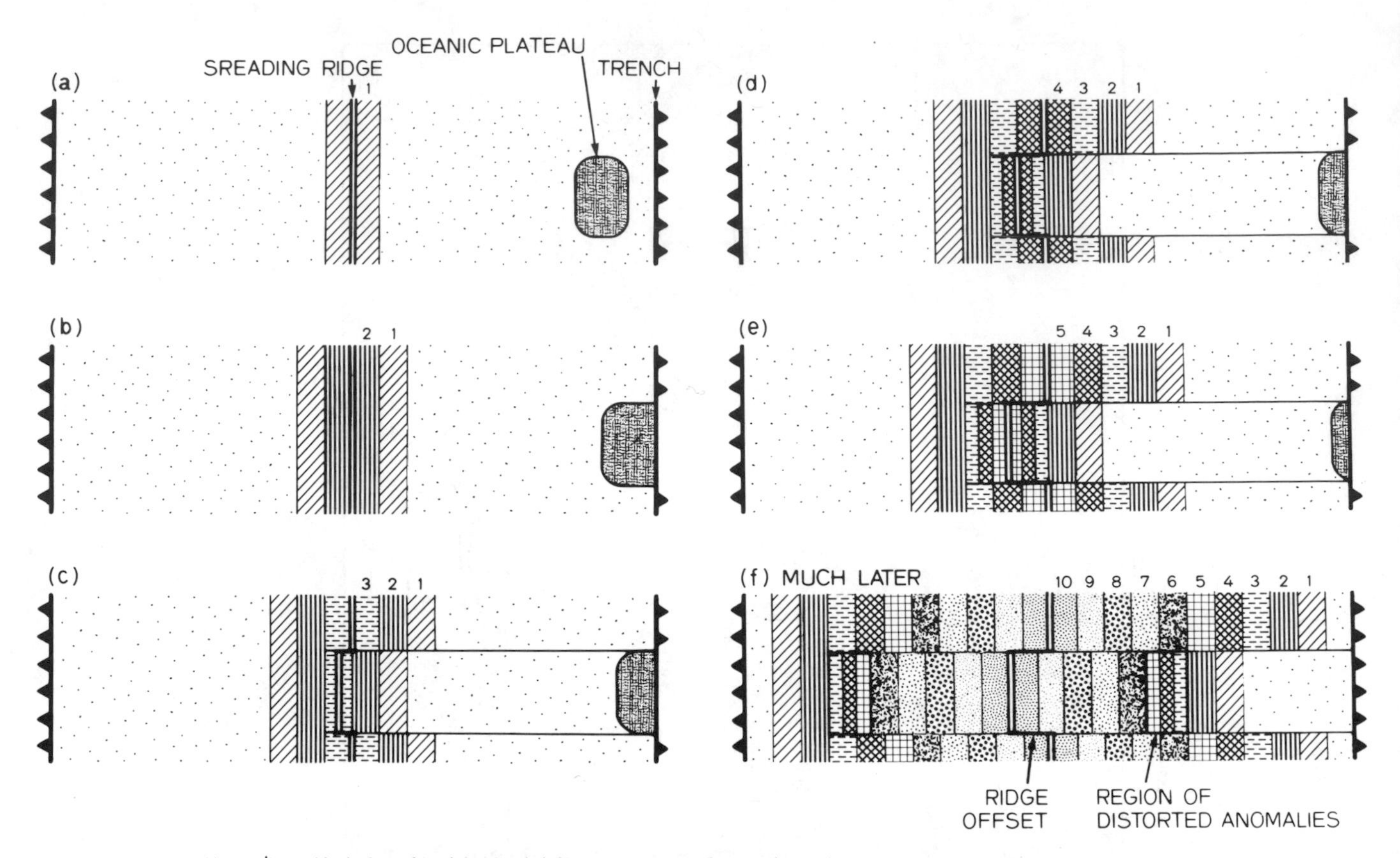

Fig. 4. Model of ridge shifts, see explanation in the text. The shift occurs due to a slowdown of the subduction process caused by a collision of an oceanic plateau with the trench. One of the consequences of the ridge shift is the existence of zones of closely spaced magnetic anomalies.

larger plate which included the east Pacific (Farallon) (Sykes et al., 1982). The Middle America trench south of Mexico was not in existence, and the motion between the east Pacific (Farallon)-Caribbean plate and the North American plate was accommodated along the Cuban arc. Upon initiation of subduction in central America, the Farallon plate could not continue to move to the northeast during an adjustment period and had to change its direction towards the Peru-Chile trench which was active throughout this period of time. Eventually subduction became fully developed at the middle America trench. The two slab pull force sets, along the Middle America trench and the Peru-Chile trench, subsequently led to the rifting of the Farallon plate along the Galapagos spreading center. As a result, the Nazca plate kept moving in its new northeast-east direction and did not move back to its former southwest-northeast direction.

The other major Nazca plate event we considered is the triple ridge axis jump from the old Galapagos Rise to the East Pacific Rise. There were three separate jumps over an interval of about 2.5 million years starting at 8.2 Ma in the region between 4.5° and 13°S (Rea, 1981). We propose that these jumps are associated with the collision of the Nazca Ridge with the Peru-Chile trench. The collision of this ridge with the subduction zone has been suggested to be the cause of the volcanic gap north of the point of collision of Nazca Ridge and the continent (Nur and Ben-Avraham, 1981; Pilger, 1981). The volcanic gap, about 1500 km long, extends from 2° to 15°S and in this region the inclination of the seismic zone is shallow in comparison with segments containing active volcanoes. We suggested (Nur and Ben-Avraham, 1981) that the point of collision of the ridge with the continent moves south along the plate boundary because the strike of the ridge is oblique to the direction of plate motion. This migration leaves behind a zone in which subduction is temporarily retarded, causing the volcanic gap. After the ridge segment becomes well embedded within the continental margin or consumed, subduction of oceanic crust is re-established. A new trench forms and a new seismic slab is developed. Possibly during the subduction hiatus, spreading at the Galapagos Rise had to stop as there was no active trench.

The small offsets on the Galapagos Rise, mentioned earlier, may have occurred in response to ongoing collision of the Nazca and Juan Fernandez aseismic ridges, as well as that of the active Chile Rise, with the South America trench. As with the large ridge jumps, the stresses at the trench on the eastern boundary of the plate are being transferred to the spreading ridge on the western boundary of the plate. In the next section is described a model for ridge jump, or shift, in response to processes at the trench.

Ridge Shifts Due to Consumption of Oceanic Rises

If an oceanic rise is capable of slowing down the rate of subduction, some interesting consequences may follow. Consider, as in Figure 4, a spreading ridge, and two associated subduction zones. An oceanic plateau is embedded in one oceanic plate which is moving towards the subduction zone (Fig. 4a). As the plateau collides with the subduction zone (Fig. 4b), we assume that it will tend to slow the subduction process down in the region of collision.

How is this localized slowdown accommodated? Three obvious possibilities can be envisioned: (1) the plateau, or rise, becomes totally detached from the underlying plate, and accretes to the overriding plate, without any further disturbance in the oceanic plate; (2) the plateau or rise is totally subducted, without any disruption of either plate; and (3) the plateau resists either subduction or detachment.

Let us assume that this resistance is sufficient to cause a slowdown in relative plate motion. The most likely way by which such slowdown may be accommodated is by differential motion along faults in the oceanic plate. The faults which could possibly accommodate this motion are the fracture zones which extend beyond the ridges. Although the basic view is that no differential slip should occur on fracture zones outside the ridge offsets, some evidence exists that transverse motion on these segments does or did take place, such as in the Azores fracture zone or the Eltanin fracture zone (Jurdy and Gordon, 1984). Often such motions are small, so that they cannot be clearly resolved.

It is true that these fractures are oriented in the most unfavourable orientation for shear motion, as they are parallel to the direction of plate motion, and hence parallel to the direction of principal stress. However, this same dilemma exists at the spreading ridges, where transform faults exhibit the same geometry (Lachenbruch and Thompson, 1972).

We will assume therefore that the fracture zones in the oceanic crust are the sites at which differential horizontal motion is accommodated. As shown in Figure 4 two such fracture zones are ready to accommodate the differential motion caused by the slowdown due to the consumption of the rise.

We assume next that the rate of relative spreading at the ridge, and hence the rate of new plate growth, tends to be always symmetrical. Consequently, as shown in Figure 4c, if plate motion slows on one side of the ridge, and yet spreading is to remain symmetrical, it follows that the ridge segment must shift towards the fast moving plate, and away from the slow moving plate. The rate of ridge segment offset should be simply proportional to the difference in velocity between the undisturbed plate and the slowed segment. For the sake of illustration, we show in Figure 4c-e the case when subduction of the segment ceased altogether. In this case the ridge segment moves at half the speed of normal spreading.

A direct consequence of the ridge shift and slowdown is the reduction in plate growth rate and the proportional narrowing of the spacing between magnetic anomalies, relative to the undisturbed situation (Fig. 4, c-e) during the period of slowdown.

Eventually, the rise is consumed - either by subduction, or by detachment and accretion to the overriding plate - and normal subduction is restored. This restoration is associated with return to normal spreading rates at the ridge, which is now permanently offset (Fig. 4f).

The processes described above for ridge shifts are highly speculative, as no specific case with data has been presented here. However the model suggests a very robust test, because it requires that four tectonic processes take place more or less conterminously: an oceanic rise is consumed, which might be reflected in a fossil volcanic gap or the presence of an accreted terrane; transverse relative motion across fracture zones takes place; a ridge segment is shifted; and the spacing of ocean floor anomalies is reduced. Such a rigorous test remains to be carried out.

One important shortcoming of the model is that it fails to suggest how ridge jumps - as opposed to continuous offsets - might be induced by the consumption of plateaus at subduction zones.

The Eastern Mediterranean

The eastern Mediterranean is a small remanent ocean basin (Fig. 5) which is being closed by the convergence of Eurasia and Africa. In this basin, many tectonic features are affected by collisional events at its trenches. Seismic refraction studies have shown that the crust under the eastern Mediterranean is oceanic (Makris et al., 1983) whereas the Sinai Peninsula is continental (Ginzburg et al., 1979). The crust between Sinai and Cyprus has a seismic velocity of 6.8 km/sec and is only 7 km thick. It is overlain by more than 12 km of sediments. Under Cyprus the crust has a P velocity of 6.0 km/sec, typical of continental crust. Besides Cyprus several oceanic plateaus are also embedded within the crust of the eastern Mediterranean. The most notable are the Erastosthenes and the Anaximander seamounts.

Several tectonic features in this small basin may be attributed to past collisional events. For example, the separation of the once continuous

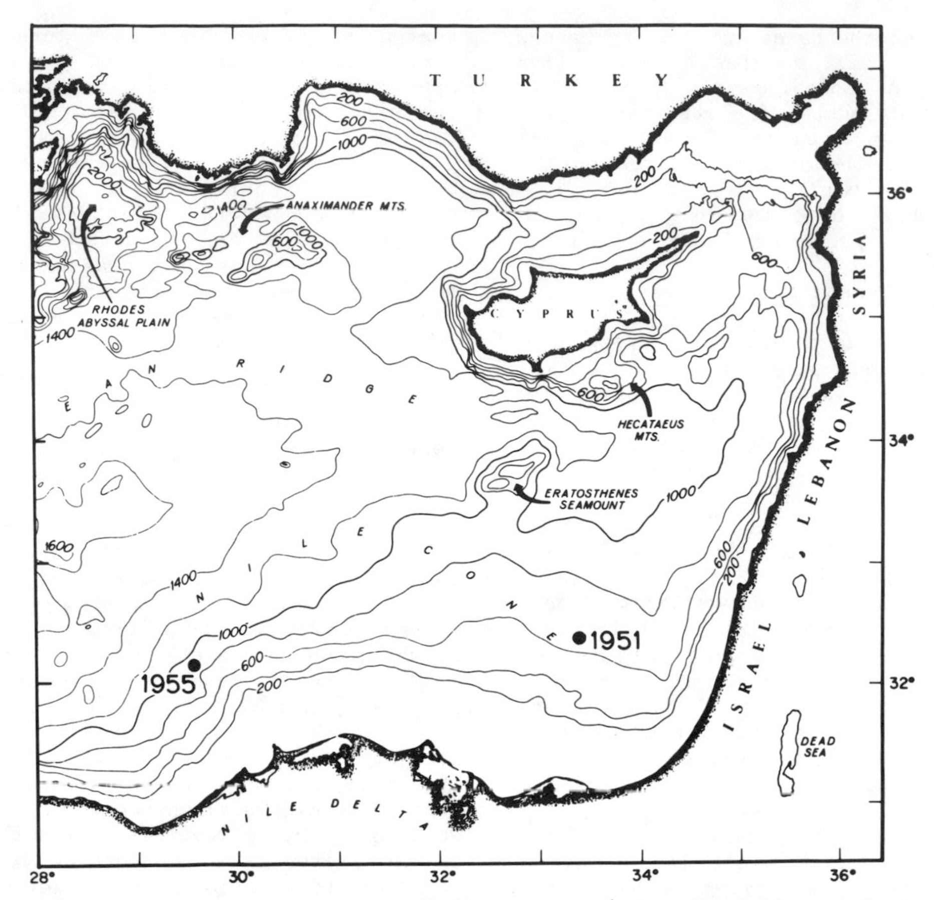

Fig. 5. The bathymetry of the eastern Mediterranean (after Woodside and Bowin, 1970). Contour interval is 200 fathoms. Dots represent large earthquakes in the North African passive margin.

arc system in the northern part of the eastern Mediterranean has been attributed to the arrival of a microcontinent (now embedded within the continental margin of southern Turkey) by Rotstein and Ben-Avraham (1985). A second example is provided by the seismic activity off Cyprus, which is much less than along Crete. In addition, intermediate depth earthquakes occur along the Cretan arc but not along the Cyprus arc. The different modes of activity along the two arcs is most probably related to the head-on collision of the Eratosthenes Seamount and the oblique collision of the Anaximander Seamount with the Cyprus arc. In contrast, there are presently no oceanic plateaus south of the Hellenic trench along the Crete arc. It is of course possible and even likely that some plateaus collided with the arc in the past. These plateaus now are incorporated with the arc itself as a result of subsequent subduction jumps to the south. Similar events may have happened also along the Cyprus arc, such as attested by the Hecataeus Seamount which now is part of the continental margin of Cyprus (Fig. 5).

The eastern Mediterranean basin represents the final stages of closure of an older and larger ocean basin. The continental margin of eastern North Africa can thus be classified as a passive margin, one not associated with a plate boundary. However, this margin is surprisingly active seismically, although much less so than the Cyprus and Crete arcs to the north. In contrast with typical passive margins, this activity consists of several seismic events which are quite substantial. The focal mechanism of the 1955 event (Fig. 5), which was of magnitude 6.75 with an epicenter located at the northwest margin of the Nile cone, was determined to be due to thrust faulting (Wang, 1981). The compressional (P) axis strikes almost exactly north-south (N3°E) and is almost horizontal (14° plunge). The sparse first motion data from the 1951 (m=5.8) event (Shirokova, 1967; Ben-Menaham et al., 1976) are also compatible with the 1955 focal mechanism although a definite reliable determination is not possible because of the lack of astrong surface wave (Wang, 1981). Thrust faulting along the eastern North Africa margin has also been proposed by Neev et al. (1976), on the basis of shallow penetrating seismic reflection profiles.

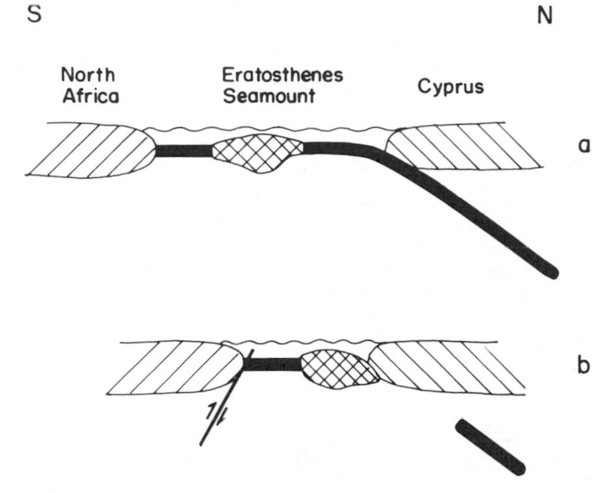

Fig. 6. Cartoon depicting the reactivation of the basement faults in the zone of transition from oceanic to continental crust along the North African margin due to the collision of the Eratosthenes and Anaximander seamounts with the Cyprus arc.

We suggest that here, again, the effects of disrupted subduction may be transmitted a considerable distance, this time to the passive margin. It has been suggested that the evolution of a passive margin is controlled by faulting and subsidence of the oceanic-continental lithosphere transition (Turcotte et al., 1977) and that these faults are sometimes reactivated (Stein et al., 1979). In the eastern North Africa margin this reactivation of the basement faults, as expressed by the seismicity, is caused by the collisions of the Eratosthenes and Anaximander seamounts with the Cyprus arc (Fig. 6). These collisions disrupt normal underthrusting, so that some of the motion between the African and Eurasian plates may be accommodated by thrust faulting along the North African passive margin, through stress transmission from the Cyprus arc in the north across the basin into the continental margin in the south. By this process, this passive margin is being transformed into an active one, possibly with the initiation of a downgoing slab beneath it, as indicated by the thrust fault plane solutions. The activity along the passive margin of eastern North Africa may thus continue until a new normal subduction zone will be established in the north along the Cyprus arc, after full incorporation of the Anaximander and Erastosthenes seamounts into Cyprus itself.

Conclusions

The collision or consumption of oceanic plateaus with subduction zones at active margins may be important in controlling events which take place some distance away at the passive margins of continents and at the ridges in ocean basins. Consequently correlation may be expected to exist between some tectonic events on land and in the ocean. The possibility of correlating the timing of such events as volcanic gaps on land with events like ridge jumps at sea, time of accretion of terranes, or reactivation of passive margins, may provide a quantitative tool to study the processes which link plate tectonics as seen in ocean basins to continental tectonics.

Changes in direction and speed of planes motion as well as spreading center jumps are now well documented for the Pacific Ocean basin plates as well as for other plates throughout the Mesozoic and Cenozoic. The history of the Nazca plate is of particular importance because some of the processes that led to the various changes in plate kinematics are still occurring today.

References

Barazangi, M., and B. Isacks, Spatial distribution of earthquakes and subduction of the Nazca plate beneath South America, Geology, 4, 686-692, 1976.

Ben-Avraham, Z. and A. K. Cooper, Early evolution of the Bering Sea by collision of oceanic rises and north Pacific subduction zones, Geol. Soc. Am. Bull., 92, 474-484, 1981.

Ben-Avraham, Z., A. Nur, D. Jones, and A. Cox, Continental Accretion: From Oceanic Plateaus to Allochthonous Terranes, Science, 213, 47-54, 1981.

Ben-Avraham, Z., A. Nur and D. Jones, Emplacement of Ophiolites by Collision, J. Geophys. Res., 87, B5, 3861-3867, 1982.

Ben-Menaham, A., A. Nur, and M. Vered, Tectonics, seismicity and structure of the Afro-Eurasian junction - the breaking of an incoherent plate, Physics of the Earth & Planetary Interiors, 12, 1-50, 1976.

Coney, P. J., D. Jones and J. W. H. Monger, Cordilleran suspect terranes, Nature, 288, 329-333, 1980.

Cook, F. A., D. S. Albaugh, L. D. Brown, S. Kaufman, and J. E. Oliver, Thin-skinned tectonics in the crystalline southern Appalachians; COCORP seismic-reflection profiling of the Blue Ridge and Piedmont, Geology, 7, 563-567, 1979.

Dickinson, W. R., and W. S. Snyder, Geometry of subducted slabs related to San Andreas transform, J. Geology, 87, 609-627, 1979.

Engebretson, D., and Z. Ben-Avraham, Collisional events and the direction of relative motion between North America and the Pacific basin plates since the Jurassic (abstract), Abs. with Prog., Geol. Soc. Am. Cordilleran Sec., Ann., Mtg., 13, 55, 1981.

Forsyth, D. and S. Uyeda, On the relative importance of the driving forces of plate motion, Geophys. J. Roy. Astron. Soc., 43, 162-200, 1975.

Forsyth, D. W., Mechanisms of earthquakes and plate motions in the east Pacific, Earth Plan. Sci. Lett., 17, 189-193, 1972.

Ginzburg, A., J. Makris, K. Fuchs and C. Prodehl, A seismic study of the crust and upper mantle of the Jordan-Dead Sea rift and their transition toward the Mediterranean Sea, J. Geophys Res., 84, 1569-1582, 1979.
Gordon, R. G., A. Cox, and C. E. Harter, Absolute motion of an individual plate estimated from its ridge and trench boundaries, Nature, 274, 752-755, 1978.
Handschumacher, D. W., R. H. Pilger, Jr., J. A. Foreman, and J. F. Campbell, Structure and evolution of the Easter plate, Geol. Soc. Am., 154, 63-76, 1981.
Herron, E. M., Two small crustal plates in the South Pacific near Easter Island, Nature, Phys. Sci., 240, 35-37, 1972.
Jurdy, D. M., and R. G. Gordon, Global plate motions relative to the hot spots 64 to 56 Ma, J. Geophys. Res., 89, 9927-9936, 1984.
Lachenbruch, A. H., and G. A. Thompson, Oceanic ridges and transform faults: their intersection and resistance to plate motion, Earth Plan. Sci. Lett., 15, 116-122, 1972.
Leggett, J. K., W. S. McKerrow, and N. J. Soper, A model for the crustal evolution of southern Scotland, Tectonics, 2, 187-210, 1983.
Makris, J., Z. Ben-Avraham, A. Behle, A. Ginzburg P. Gieze, L. Steinmetz, R. B. Whitmarsh, and S. Eleftheriou, Seismic refraction profiles between Cyprus and Israel and their interpretation Geophys, J. Roy. Astron. Soc., 75, 575-591, 1983.
Neev, D., G. Almagor, A. Arad, A. Ginzburg, and J. K. Hall, The geology of the southeastern Mediterranean, Israel Geol. Surv., Bull., 62, 51, 1976.
Nur, A., and . Ben-Avraham, Volcanic gaps and the consumption of aseismic ridges in South America, Geol. Soc. Am., Mem., 154, 729-740, 1981.
Nur, A., and Z. Ben-Avraham, Oceanic Plateaus, the Fragmentation of Continents, and Mountain Building, J. Geophys. Res., 87, 3644-3661, 1982a.
Nur, A., and Z. Ben-Avraham, Displaced terranes and mountain building in: Mountain building Processes, Hsu, K. J., ed., 73-84, Academic Press, London, 1982b.
Pilger, R. H., Jr., Plate reconstructions, aseismic ridges, and low-angle subduction beneath the Andes, Geol. Soc. Am., Bull., Pt. 1, 92, 448-456, 1981.
Pilger, R. H., Jr., Kinematics of the South American subduction zone from global plate reconstructions, in: Geodynamics of the Eastern Pacific Region, Caribbean and Scotia Arcs, Ramon Cabre, S. J., ed., 113-125, Geodynamics Series, 9 Amer. Geophys. Union, 1983.
Rea, D. K., Tectonics of the Nazca-Pacific divergent plate boundary, Geol. Soc. Am., 154, 27-62, 1981.
Rotstein, Y., and Z. Ben-Avraham, Accretionary processes at subduction zones in the eastern Mediterranean, Tectonophysics, 112, 551-561, 1985.
Shih, J. and P. Molnar, Analysis and implications of the sequence of ridge jumps that eliminated the Surveyor transform fault, J. Geophys. Res., 80, 4815-4822, 1975.
Shimazaki, K., Midplate, plate-margin, and plate-boundary earthquakes and stress transmission in Far East (abstract), Int. Sym. Continental Seismicity and Earthquake Prediction (ISESEP), Beijing, Abs., 1,16, 1982.
Shirokova, F. I., General features in the orientation of principal stresses in earthquake foci in the Mediterranean-Asian seismic belt, USSR Earth Phys. Bull, (Phys. Bull. Acad. Sci., 1, 22-36, 1967.
Stein, S., N. H. Sleep, R. J. Geller, S. C. Wang, and G. C. Kroeger, Earthquakes along the passive margin of eastern Canada, Geophys. Res. Lett., 6, 537-540, 1979.
Sykes, L., W. R. McCann, and A. L. Kafka, Motion of Caribbean Plate during last 7 million years and implications for earlier Cenozoic movements, J. Geophys. Res., 87, 656-676, 1982.
Turcotte, D. L., J. L. Ahern, and J. M. Bird, The state of stress at continental margins Tectonophysics, 42, 1-28, 1977.
Uyeda, S., and H. Kanamori, Back-arc opening and the mode of subduction, J. Geophys. Res., 84, 1049-1061, 1979.
Van Andel, Tj. H., G. R. Heath, and T. C. Moore, Jr., Cenozoic history and paleoceanography of the central Equatorial Pacific Ocean, Geol. Soc. Am., Mem., 143, 134, 1975.
Vogt, P. R., A. Lowrie, D. R. Bracey, and R. N. Hey, Subduction of aseismic oceanic ridges: Effects on shape, seismicity, and other characteristics of consuming plate boundaries, Geol. Soc. Am., Spec. Paper, 172, 1976.
Wang, S. C., Tectonic implications of global seismicity studies, Unpub. Ph.D Thesis, 70 p., Stanford Univ., Ph.D., 1981.
Williams, H., and R. D. Hatcher, Jr., Suspect terranes and accretionary history of the Appalachia Orogeny, Geology, 10, 530-536, 1982.
Woodside, J. M., and C. O. Bowin, Gravity anomalies and inferred crustal structure in the eastern Mediterranean Sea, Geol. Soc. Am. Bull. 81, 1107-1122, 1970.

RECONSTRUCTIONS, PLATE INTERACTIONS, AND TRAJECTORIES OF OCEANIC AND CONTINENTAL PLATES IN THE PACIFIC BASIN

David C. Engebretson

Department of Geology, Western Washington University, Bellingham, Washington 98225

Allan Cox and Michel Debiche

Department of Geophysics, Stanford University, Stanford, California 94305

Abstract. Displacements between oceanic and continental plates in and around the northern Pacific Basin were determined from fracture zones, magnetic isochrons, and hotspot tracks. The histories are presented as map reconstructions based on a series of Euler poles that describe motions of oceanic plates (Pacific, Farallon, Kula, and Izanagi) relative to the North American and Eurasian plates. The direction and speed of convergence between the oceanic and continental plates were estimated using the Euler poles and reconstructions of former plate geometries. An important conclusion to emerge from this analysis is that the convergence rate of oceanic plates with North America and Eurasia was fast (greater than 100 km/million years) in the Late Cretaceous and Paleogene and slow (less than 60 km/million years) in the Neogene. The amount of displacement available for transporting allochthonous terranes is presented in the form of models of trajectories of terranes carried by oceanic plates that arrived at North America at 30, 60, and 90 Ma. Trajectories for a 30 Ma arrival are short because of the close proximity of active spreading ridges to North America. For arrival times of 60 and 90 Ma, trajectories are longer than 6000 km. In general, the 60 Ma trajectories displayed large south-to-north displacements relative to North America whereas the 90 Ma trajectories show substantial west-to-east motion.

Introduction

The possibility that tectonic processes along active continental margins owe their origin to interactions between the continent and adjacent oceanic plates has been proposed by many authors. These tectonic processes include plutonism, episodes of strike-slip and normal faulting, basin formation, terrane transport, and orogeny.

The information needed to determine whether continental and oceanic tectonic events are in fact related, is a set of Euler poles describing the locations of oceanic plates relative to adjacent continents in the geologic past. Such a set of Euler poles can be used to generate maps showing the former locations of oceanic plates adjacent to the continental margin and also to calculate relative plate velocities along the margin. Thus they provide the basic framework needed to relate plate tectonics to continental geology. We report here the results of an analysis of the motions of oceanic and continental plates in the north Pacific basin. The results of the study include a set of Euler poles describing the plate motion, reconstructions of the plates at different epochs, convergence velocities of plates along the margin, and trajectories capable of transporting allochthonous terranes from their regions of origin to their present resting places. Earlier studies include those of Atwater (1970), Atwater and Molnar (1973), Coney (1972, 1978), Francheteau et al. (1970), Carlson (1982), Larson and Chase (1972), Hilde et al. (1976) and Alvarez et al. (1980).

The first step in our analysis was to determine the history of displacement between oceanic plates of the north Pacific basin. The result of this research was summarized in the form of a set of Euler poles describing displacements between the Pacific, Farallon, Kula, Izanagi I, Izanagi II, and Phoenix plates (Engebretson et al., 1984). The main results of this first phase of the study were as follows. (1) Spreading rates along the Pacific-Kula ridge decreased markedly between chrons 32b and 25 (71-56 Ma), probably in response to the arrival of buoyant Kula lithosphere at a subduction zone northwest of the Bering Sea. (2) Soon after chron 25 (56 Ma), a major reorganization is recorded along the Pacific-Farallon boundary. (3) East-west-trending Kula-Pacific anomalies north of chron 25 (56 Ma)

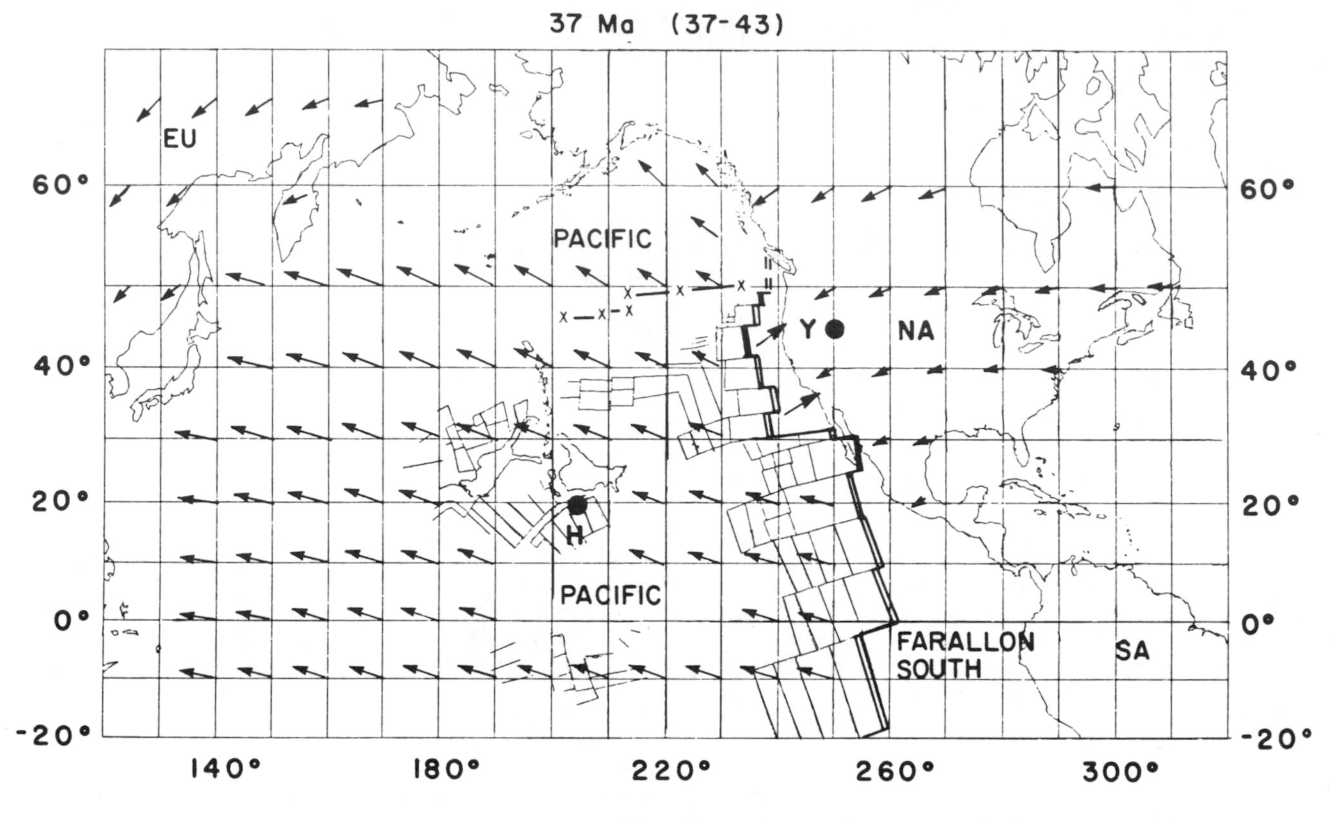

Fig. 1a.

Fig. 1. Reconstructions in fixed hotspot reference at 37 (1a) and 65 Ma (1b). Double heavy lines: ridge boundaries between oceanic plates (dashed where inferred). Arrows: motion of plates as determined from stage poles for the time intervals shown in parentheses; arrow length indicates 10 million years of motion. Pattern (x-x-x-x): abandoned Pacific-Kula ridge assuming the ridge ceased spreading at 43 Ma. Solid dots: Yellowstone (Y) and Hawaiian (H) hotspots. Diagonal shading: lithosphere where it is uncertain if it was Farallon or Kula plate. Square: location of Pacific-Kula stage pole at 65 Ma in fixed hotspot coordinates.

suggest that the Kula-Pacific spreading continued at a rapid rate after 56 Ma. (4) Probably by chron 18 (43 Ma), Pacific-Kula spreading had ceased and the Kula-Farallon ridge system had evolved into alignment with the Pacific-Farallon system.

The second step in our analysis was to determine the motion of the oceanic plates relative to North America and Eurasia (Engebretson et al., 1985). Our model for this calculation was based on the assumption that the hotspots in the Pacific and Atlantic regions have remained fixed with respect to one another. This approach was first applied by Morgan (1972) to analysis of plate motions and was first applied by Coney (1972, 1978) to an analysis of the convergence history and tectonic evolution of western North America during the Mesozoic and Cenozoic. An advantage of the hotspot approach over that of constructing a global North America-Africa-India-Antarctica-Pacific circuit (Atwater and Molnar, 1973) is that one or more of these plates may, in the past, have been divided by an undetected boundary into two sub-plates (Gordon and Cox, 1980; Suarez and Molnar, 1980; Morgan, 1981; Duncan, 1981) with an unknown amount of displacement and unknown contributions to the global circuit. Disadvantages of using a hotspot reference frame are that motions between hotspots in the Atlantic and Pacific basins, and errors in plate-hotspot motions, introduce errors into reconstructions. A comparison between reconstructions made using hotspots and the global circuit is given in Engebretson et al. (1985).

The reconstructions used as a basis for our plate model came from a variety of sources. The motion of the African plate relative to the Atlantic hotspots (i.e. Africa-hotspot motion) was taken from Morgan (1983). The ages of these reconstructions were adjusted to be consistent with the magnetic reversal time scale of Harland et al. (1982). To obtain the motion of North

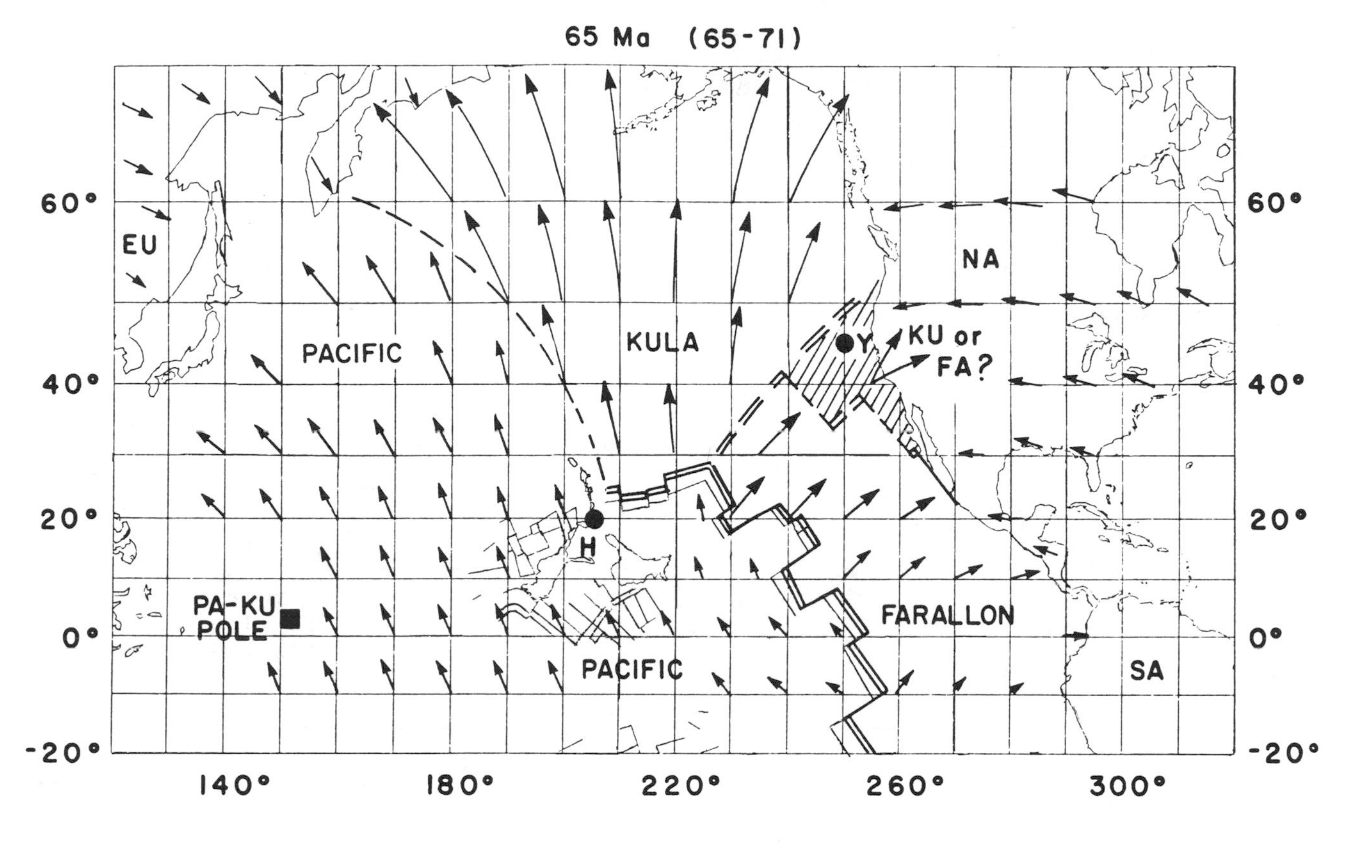

Fig. 1b.

America relative to hotspots, the Africa-North America relative motion determined by Klitgord et al. (1984 and pers. comm.) was combined with Morgan's Africa-hotspot motion. The motion of Eurasia with respect to hotspots was then determined from North America-hotspot motion and the Eurasia-North America reconstructions of Srivastava (1978), as modified by the interpretation of Cande and Kristoffersen (1977) that the anomalies Srivastava identified as anomalies 31 and 32 are anomalies 33 and 34. The timing of initial rifting of Eurasia-Greenland from North America (95 Ma) was estimated by extrapolating the spreading rate between anomalies 33 and 34 out to the ocean-continent boundary as picked by Srivastava.

Our model for the motion of the Pacific plate relative to hotspots (Engebretson et al., 1985) may be summarized as follows. From 145 to 135 Ma, the plate moved to the southwest parallel to the western Mid-Pacific Mountains. From 115 to 100 Ma, the Pacific plate moved west-southwest, parallel to the eastern portion of the Mid-Pacific Mountains. From 100-74 Ma, the Pacific plate moved rapidly to the northwest, parallel to the northern Line Islands and the three northwest-trending seamount chains just north of the Hawaiian Chain. From 74-43 Ma, the plate moved rapidly northwestward, parallel to the Hawaiian Chain. From 43 Ma to the present, the Pacific plate moved rapidly in a generally west-northwestward direction, parallel to the Hawaiian Chain, this motion being divided into three small substages: a change in the rate but not the direction of the Pacific plate motion occurred at 28 Ma and a change in both the rate and direction of motion occurred at 5 Ma (Cox and Engebretson, 1985).

Reconstructions

Reconstructions of oceanic plates within the Pacific basin were generated by rotating magnetic isochrons and fracture zones from their present locations on the Pacific plate back to their former locations using Pacific-hotspot Euler poles. For continental plates, their present coastlines were reconstructed using appropriate hotspot-fixed Euler poles. Tables containing these Euler poles can be found in Engebretson et al. (1985). Figures 1a and 1b show examples of these reconstructions.

At 37 Ma (Fig. 1a) the Pacific plate is shown to be in contact with North America near the Mendocino Fracture Zone. Since there are younger magnetic anomalies recorded on the Pacific plate to the east of the ridge shown in this reconstruction, we realize this is an impossible

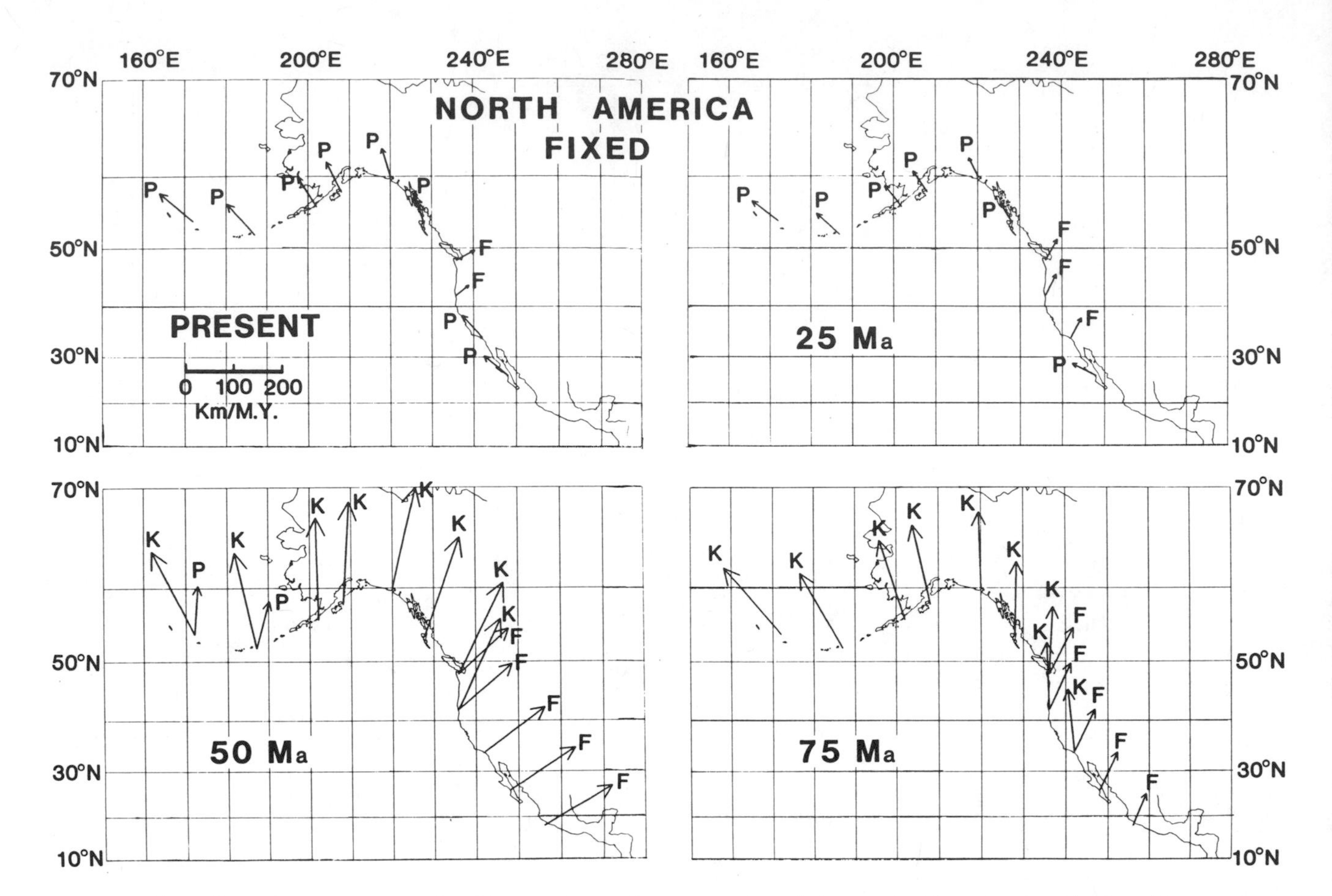

Fig. 2. Convergence velocity vectors for oceanic plates relative to North America. Each arrow at a point along the North American margin shows the convergence velocity of the oceanic plate at the time indicated. Two arrows at certain points show convergence velocities of two plates when either may have been adjacent to North America at that point. P: Pacific plate. F: Farallon plate. K: Kula plate.

geometry. However, no allowance in these reconstructions was made for Basin and Range extension or for the opening of the Gulf of California. It is more likely that the Pacific plate contacted North America at a later time, say 28 Ma (Atwater, 1970). Also note from this reconstruction that the Farallon plate was in contact with North America from northern Mexico to southern Canada. To the north and west the Pacific plate was interacting with North America, implying dominant right-lateral strike-slip motion along the British Columbia segment of the continental margin and subduction in southern Alaska. Nearly straight-on convergence of the Pacific plate is implied for northeastern Eurasia.

Figure 1b shows a hotspot-fixed reconstruction at 65 Ma. Several important features deserve discussion. All of the oceanic plates to the east and north of the Pacific plate have been consumed at subduction zones along Eurasia and North America. No direct evidence is available for the location of oceanic plate boundaries such as those that once existed between the Pacific and Kula plates or Kula and Farallon plates. The shaded pattern near western North America displays a range of possible locations for the Kula-Farallon ridge. In this model, the boundary (Pacific-Kula-Eurasia triple junction) was located near the present site of Kamchatka. The rate of convergence was high between the Kula plate and both northeastern Eurasia and northwestern North America from 71 to 65 Ma. Convergence of the Pacific plate with Eurasia near Japan was considerably slower.

Reconstructions such as those presented in Figures 1a and 1b form a basis for comparison of the geological record with displacement models. Specific predictions result directly from each reconstruction and these predictions can be tested through studying geologic indicators of nearby plate interactions. Given the great uncertainty in the geometry of the oceanic plates, the geologic record along eastern Eurasia and western North America can be used to test the plausibility of plate reconstructions.

Linear Velocities

The reconstructions given in Figures 1a and 1b include an estimate of the velocity (shown as arrows) of the plates over the hotspots. Also of great geologic importance are the velocities of convergence of oceanic plates relative to Eurasia and North America. Tables of linear velocities for oceanic plates relative to Eurasia and North America can be found in Engebretson et al. (1985). The convergence velocities were calculated using a quasi-instantaneous angular velocity derived from the appropriate stage poles. The location of the angular velocity vector is the same as that for the stage pole and the angular rate was obtained by dividing the finite angle by the time spanned by the stage. An estimate of the average instantaneous velocity of an oceanic plate relative to a fixed point on the continent was then found as the cross product of the quasi-instantaneous angular velocity vector and the position vector of the fixed point.

Vectors of convergence between oceanic plates and North America are shown in Figure 2 for four times since the Late Cretaceous. Each arrow along the continental margin shows the direction and speed with which the oceanic plates have converged with North America. Where there is some ambiguity as to which plate was offshore, velocities for both plates are shown. Numerical values of the relative velocities for Eurasia and North America can be found in Tables 5 and 6 of Engebretson et al. (1985).

Along the central portion of the western North American Cordillera, convergence was fast and in a right-lateral oblique sense in the Late Cretaceous (75 Ma in Fig. 2). Further north near Alaska the Kula plate was converging in a direction almost perpendicular to the coastline. Fast velocities were also observed for the Eocene (50 Ma) but slowed considerably by latest Oligocene (25 Ma). Note that velocities of oceanic plates after 25 Ma are considerably slower than for previous times.

Terrane Travel Histories

It has become well-established from geologic and paleomagnetic studies that large portions of land on the west coast of North America and east coast of Eurasia are made up of tectonostratigraphic terranes (Coney et al., 1980; Jones et al., 1982; Beck, 1976, 1980). Many of these terranes have been displaced from their sites of origin and thus are considered to be allochthonous with respect to cratonic North America. Terrane trajectories describe the motion of these allochthonous blocks relative to the continents where they docked. For any plate model, a set of terrane trajectories can be found which show the position of terranes as a function of time, as the terranes moved with oceanic plates or are driven by them tangentially along the continental margin.

A very important element in determining the trajectory of a terrane is its time of arrival or docking at the margin of the continent. If the time and location of arrival are known, then for any model of plate motion a terrane trajectory can be determined. This is accomplished by reconstructing the terrane to former positions using the stage poles giving the motion of oceanic plates relative to the continent. The trajectories can then be compared to maps of plate reconstructions to check the internal consistency of the trajectory calculation: the carrier plate must be older than the ages along the trajectory, and as the terrane progresses along the trajectory, the oceanic plates used to calculate the trajectory must, in the reconstruction, be located beneath the plate.

Figures 3a, 3b, and 3c show examples of terrane trajectories assuming 30, 60, and 90 Ma arrival times, respectively (see Debiche et al., in press). Segments of trajectories on different plates are indicated by different symbols, each spaced at 5 million year intervals along the trajectories. The oldest points on the trajectories for the 90 Ma arrival times are all 180 Ma, the oldest time for which oceanic plate motions have been modelled (Engebretson et al., 1985). The oldest points on the 60 and 30 Ma arrivals are marked with a star and number and give the maximum age at which a terrane could have begun its travel on that trajectory (Fig. 3). This effectively represents the time when the trajectory was initiated at an active spreading ridge.

The most striking feature in comparing Figures 3a, 3b, and 3c is their great variation in length and direction. The 30 Ma arrivals (Fig. 3a) are short because of the close proximity of spreading ridges to North America (see also Fig. 1a). This implies that the plate being subducted along western North America was, in general, quite young at 30 Ma. At 60 and 90 Ma the trajectories are much longer (Figs. 3b,3c). The main difference between the 60 and 90 Ma arrival times is that the trajectories show a greater south to north motion for 60 Ma arrivals and a greater west to east displacement for 90 Ma arrivals. Thus, greater poleward displacement would be observed in the geologic record of terranes that arrived at 60 Ma. Terranes with earlier arrival times may have travelled eastward across the paleo-Pacific, carrying fauna of Tethyan affinity.

Until now, we have presented terrane trajectories in a coordinate system that was held fixed relative to North America. In order to compare

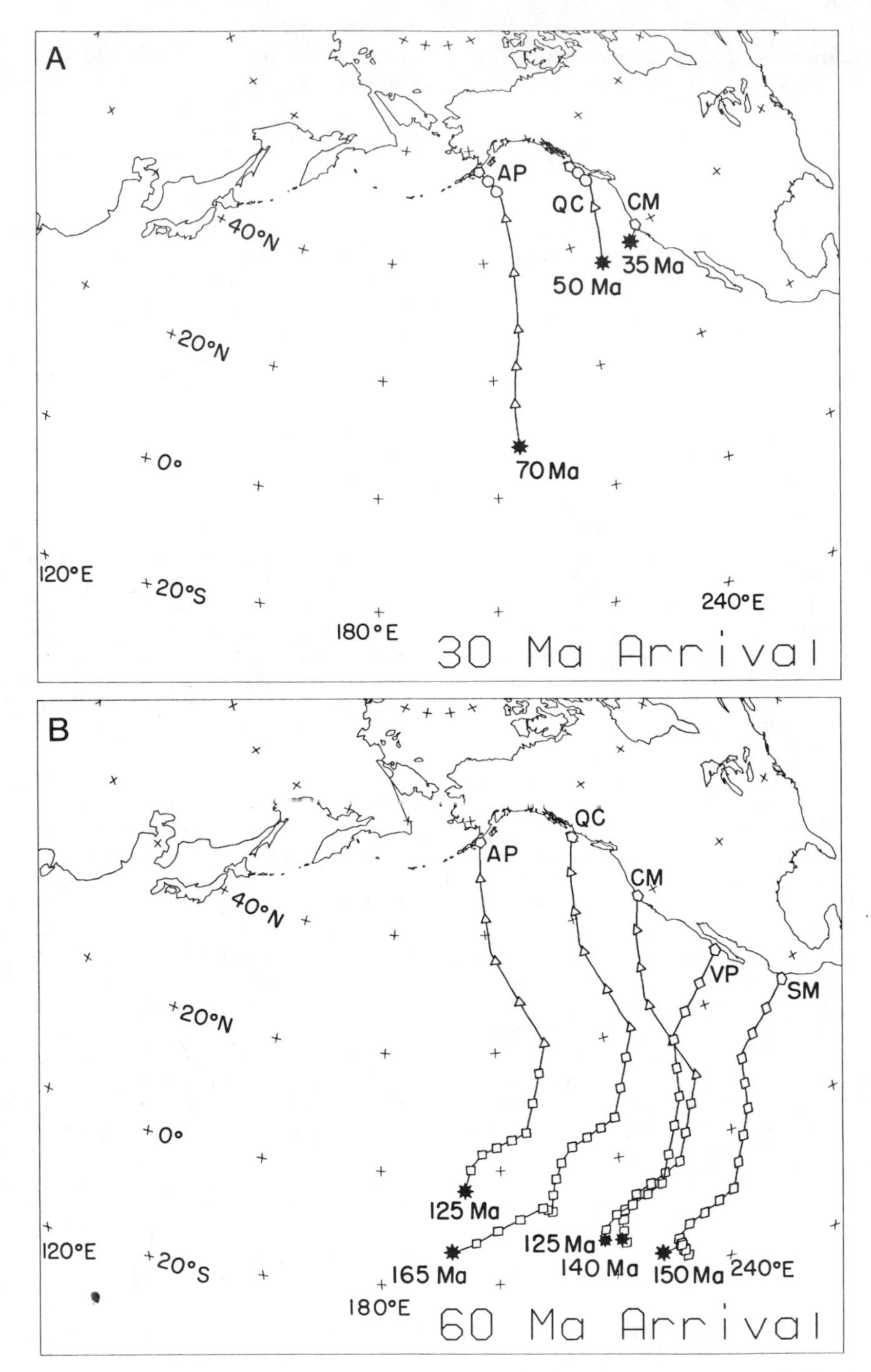

Fig. 3. Trajectories of oceanic plates in fixed North American coordinates determined by backward modelling from the arrival times shown. Symbols plotted at 5 million year intervals along trajectories are as follows. Circle: Pacific plate. Squares: Farallon plate. Triangles: Kula plate. Pentagon: docking point on North America. Star: oldest point on a trajectory that began at a spreading center at the time indicated by the adjacent number. The oldest point on the trajectories for the 90 Ma arrival time is 180 Ma and represents the earliest time for which plate motion poles are available. Eurasia is reconstructed to its position in fixed North American Coordinates at the arrival times shown for each plot. Locations: AP, Alaska Peninsula; QC, Queen Charlotte; CM, Cape Mendocino; VP, Vizcaino Peninsula; SM, Southern Mexico.

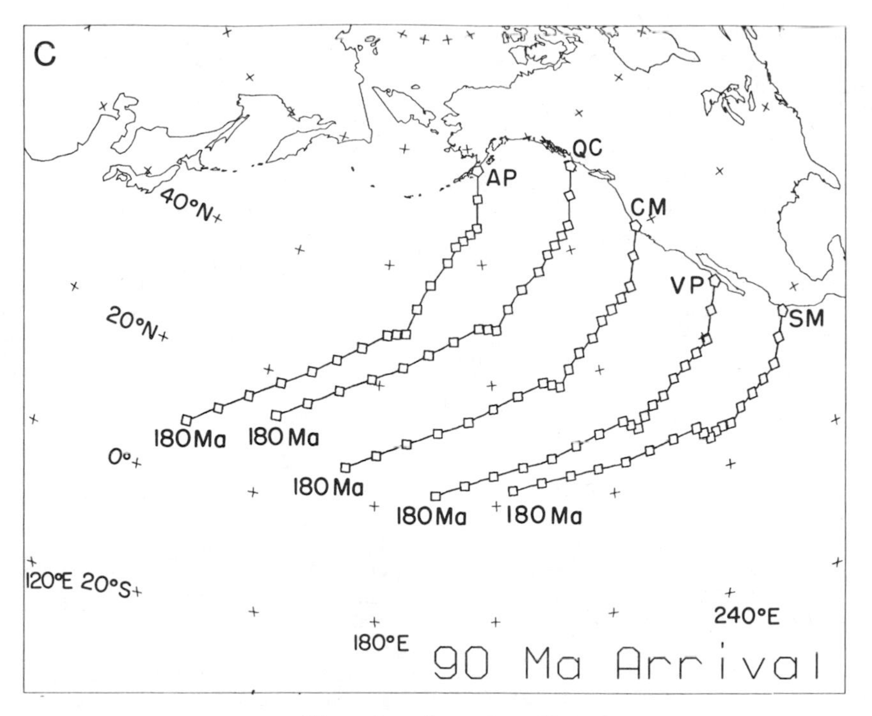

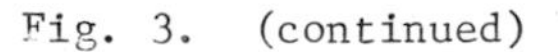
Fig. 3. (continued)

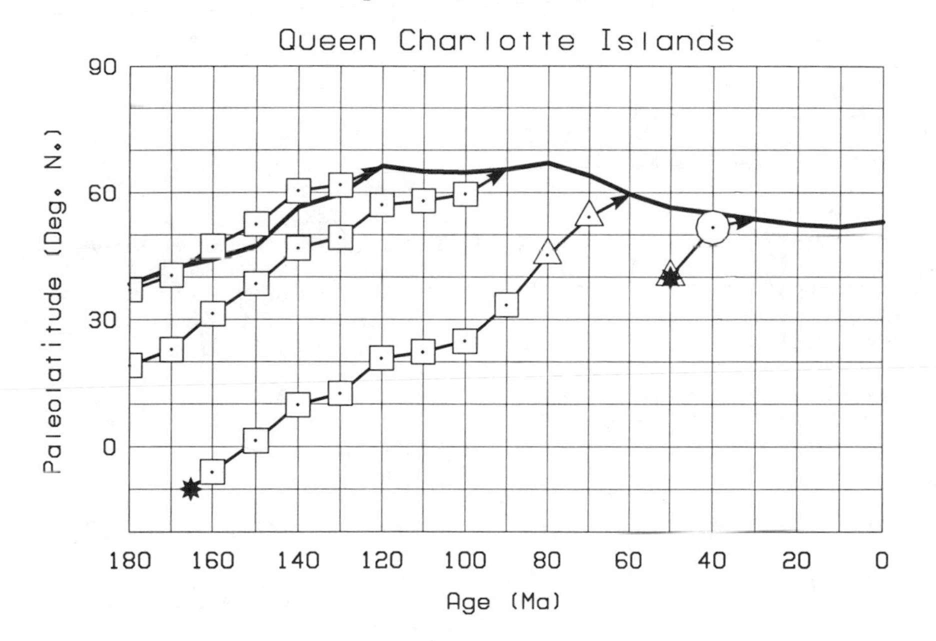

Fig. 4. Paleolatitude versus time curves for trajectories describing terranes docking at 30 Ma, 60 Ma, 90 Ma, and 120 Ma as shown in Figure 3. Solid pentagons and thick curve shows changes in paleolatitude at each docking site assuming that the site remained part of North America. Other curves show changes in paleolatitude if a terrane arrived at the docking site at the time shown by the arrowhead after it had travelled on the oceanic plate shown by the appropriate symbols. Stars mark curves based on trajectories that began at a boundary between two oceanic plates, which indicates that these curves cannot be extrapolated to older times. Curve for cratonal paleolatitudes is based on paleomagnetic poles of Irving and Irving (1982) for the Mesozoic and Diehl et al. (1983) for the Cenozoic. Squares: Farallon plate; triangles: Kula plate; circle: Pacific plate.

the trajectories with paleomagnetic data from the terranes, it is necessary to calculate for each point along a trajectory the expected paleolatitude. Debiche et al. (in press) did this in the following way. For any given age, the trajectory gives the position of the terrane relative to North America, and Irving and Irving (1982) give the position of the magnetic pole relative to North America. The expected paleolatitude of the terrane is found by measuring the angular distance from the point on the trajectory to the corresponding point on the apparent wander path for the same age. Figure 4 offers one example of such calculations. The dark single line near the top of the figure represents the paleolatitude that would be observed near Queen Charlotte Islands if that area had always been part of North America. The symbols connected by lines that intersect the North American path at the points of arrows represent paleolatitudes that would be observed as a terrane moved with an oceanic plate. The symbols are the same as those in Figure 3 and the arrows represent arrival at North America at times corresponding to the horizontal axis.

As an example, trajectory QC from Figure 3b shows a 60 Ma arrival in North America. Figure 4 predicts that if a terrane was riding on the Farallon plate (squares), transferred to the Kula plate at 85 Ma (triangles), and arrived near Queen Charlotte Island at 60 Ma, the rocks of various ages on the terrane should presently show the following paleolatitudes: 150 Ma, 0°; 120 Ma, 30°N; 90 Ma, 33°N; 60 Ma, 60°N. Similar curves for other arrival times can be found in Debiche et al. (in press).

Summary

Displacement histories of oceanic plates relative to continental plates are useful in understanding the tectonic evolution of active plate margins. Reconstructions, linear velocities, and terrane trajectory analysis can be used together for a more complete analysis of the processes responsible for the development of orogenic areas. These models of plate interactions permit specific predictions to be made that can be tested against geologic observations.

References

Alvarez, W., D. V. Kent, I. Premoli Silva, R. A. Schweikert, and R. L. Larson, Franciscan complex limestones deposited at 17° South paleolatitude, Geol. Soc. Amer. Bull., 91, 476-484, 1980.

Atwater, T., Implications of plate tectonics for the Cenozoic tectonic evolution of western North America, Geol. Soc. Amer. Bull. 81, 3513-3536, 1970.

Atwater, T., and P. Molnar, Relative motion of the Pacific and North American plates deduced from seafloor spreading in the Atlantic, Indian, and South Pacific Oceans, in: Kovach, R. L., and A. Nur, eds., Proc. of Conf. on Tectonic Problems of the San Andreas Fault System 13, Stanford, Stanford Univ. Publ., 136-148, 1973.

Beck, M. E., Jr., Discordant paleomagnetic pole positions as evidence of regional shear in the western Cordillera of North America, Amer. J. Sci., 276, 694-712, 1976.

Beck, M. E., Jr., Paleomagnetic record of plate-margin tectonic processes along the western edge of North America, J. Geophys. Res., 85, 7115-7131, 1980.

Cande, S. C., and Y. Kristofferson, Late Cretaceous magnetic anomalies in the North Atlantic, E. Plan. Sci. Lett., 35, 215-224, 1977.

Carlson, R. L., Cenozoic convergence along the California coast: A qualitative test of the hot-spot approximation, Geology, 10, 191-196, 1982.

Coney, P. J., Cordilleran tectonics and North America plate motion, Amer. J. Sci., 272, 603-628, 1972.

Coney, P. J., Mesozoic-Cenozoic Cordilleran plate tectonics, Cenozoic tectonics and regional geophysics of western Cordillera, Geol. Soc. Amer. Mem., 152, 33-50, 1978.

Coney, P. J., D. L. Jones, and J. W. H. Monger, Cordilleran suspect terranes, Nature, 286, 329-333, 1980.

Cox, A., and D. C. Engebretson, Change in motion of Pacific plate at 5 m.y.B.P., Nature, 313, 472-474, 1985.

Debiche, M. G., A. Cox, and D. C. Engebretson, Terrane trajectory analysis, Geol. Soc. Amer. Spec. Pap., in press.

Diehl, J. F., M. E. Beck, Jr., S. Beske-Diehl, D. Jacobson and B. C. Hearn, Paleomagnetism of the Late Cretaceous-early Tertiary north-central Montana alkalic province, J. Geophys. Res., 88, 10593-10609, 1983.

Duncan, R. A., Hotspots in the southern oceans - an absolute frame of reference for motion of the Gondwana continents, Tectonophysics, 75, 29-42, 1981.

Engebretson, D. C., A. Cox, and R. G. Gordon, Relative motions between oceanic plates of the Pacific basin, J. Geophys. Res., 89, 10291-10310, 1984.

Engebretson, D. C., A. Cox, and R. G. Gordon, relative motions between oceanic and continental plates in the Pacific basin, Geol. Soc. Amer. Spec. Pap., 206, 59 p., 1985.

Francheteau, J., J. G. Sclater, and H. W. Menard, Pattern of relative motion from fracture zone and spreading rate data in the northeastern Pacific, Nature, 226, 746-748, 1970.

Gordon, R. G., and A. Cox, Calculating paleomagnetic poles for oceanic plates, Geophys. J. Roy. Astron. Soc., 63, 619-640, 1980.

Harland, W. B., Cox, A., P. G. Llewellyn, C. A. G. Pickton, A. G. Smith, and R. Walters, A

Geologic Time Scale, Cambridge Univ. Press, Cambridge, 131 p., 1982.
Hilde, T. W., S. Uyeda, and L. Kroenke, Evolution of the western Pacific and its margin, Tectonophysics, 38, 145-165, 1976.
Irving, E., and G. A. Irving, Apparent polar wander paths, Carboniferous through Cenozoic, and the assembly of Gondwana, Geophys. Surv., 5, 141-188, 1982.
Jones, D. L., A. Cox, P. J. Coney and M. E. Beck, Jr., The growth of western North America, Scien. Amer., 247, 70-84, 1982.
Klitgord, K. D., P. Popenoe, and H. Schouten, A Jurassic transform plate boundary, J. Geophys. Res., 89, 7753-7772, 1984.
Larson, R. L., and C. G. Chase, Late Mesozoic evolution of the western Pacific ocean, Geol. Soc. Amer. Bull., 83, 3627-3644, 1972.
Morgan, W. J., Plate motions and deep mantle convection, Geol. Soc. Amer. Mem, 132, 7-22, 1972.
Morgan, W. J., Hotspot tracks and the opening of the Atlantic and Indian Oceans, in: The Sea, Emiliani, C., ed., 7, Wiley Interscience, New York, 443-487, 1981.
Morgan, W. J., Hotspot tracks and the early rifting of the Atlantic, Tectonophysics, 94, 123-139, 1983.
Srivastava, S. P., Evolution of the Labrador Sea and its bearing on the early evolution of the North Atlantic, Geophys. J. Roy. Astron. Soc., 52, 313-357, 1978.
Suarez, G., and P. Molnar, Paleomagnetic data and pelagic sediment facies and the motion of the Pacific plate relative to the spin axis since the Late Cretaceous, J. Geophys. Res., 85, 5257-5280, 1980.

PACIFIC AND KULA/EURASIA RELATIVE MOTIONS DURING THE LAST 130 MA AND THEIR BEARING ON OROGENESIS IN NORTHEAST ASIA

L.P. Zonenshain, M.V. Kononov, and L.A. Savostin

Institute of Oceanology, The U.S.S.R. Academy of Sciences, Krasikova 23, Moscow 117218, U.S.S.R.

Abstract. Independent data sources, derived from paleoclimatic, paleomagnetic, plate kinematic, and hotspot trajectory studies, were used to restore plate motions within the Pacific Ocean basin, first within the absolute frame and second relative to the Eurasian margin. Between 130 to 50 Ma the Kula plate moved, with respect to Eurasia, almost due northwards for a distance of nearly 13,000 km. Pacific/Eurasian plate motions can be divided into three episodes: (1) from 130 to 110 Ma, when the Pacific plate moved northwestwards at the rate of 6 cm/year, (2) from 110 to 43 Ma, with northward motion at the rate of 7 cm/year and (3) from 43 to 0 Ma with west northwest motion at the rate about 9 cm/year. The Kula and Pacific plate motion trajectories relative to Eurasia are used to show the paths of exotic terranes that were attached to the Eurasian margin during the Cretaceous and early Tertiary. Three large terranes, Kolymia, Okhotia and Koryakya, existed in the Pacific Ocean at 130 Ma. Kolymia at that time lay near the Eurasian margin, whereas Okhotia and Koryakya were located in the center of the ocean. Some of these landmasses travelled across the Pacific Ocean basin for distances of nearly 10,000 km after 130 Ma, before they collided with and accreted to the Eurasian margin.

Introduction

Two major types of orogenic belts can be distinguished: (a) those within continental interiors, exemplified by the Urals, that originate by continental collision (Zonenshain, 1984), and (b) those formed at continental margins, when blocks of buoyant crust enter a subduction zone and become accreted to the margin (e.g. Ben-Avraham et al., 1981; Churkin, 1974; Jones et al., 1977, 1982; Monger, 1977; Coney et al., 1980).

Both types exist in northeast Asia (Fig. 1). The Verkhoyansk-Kolymian orogenic belt was formed in the Early Cretaceous as the result of collision of the Kolymia microcontinent with Siberia (Parfenov, 1984; Zonenshain, 1984), and the Mongolian-Okhotsk orogenic belt originated during Middle and Late Jurassic time when the Amur-China continental block collided with Siberia (Parfenov, 1984). In the mid-Cretaceous a westward-dipping subduction zone developed (by overflipping?) and the very long Okhotsk-Chukotka volcanic belt of Andean type appeared above the subduction zone along a newly shaped continental margin. The remaining part of northeast Asia, east of the Okhotsk-Chukotka volcanic belt, belongs to an accretional margin that contains numerous exotic terranes, which arrived from the Pacific Ocean after the Early Cretaceous. We try in this paper to determine the movements of these accreted terranes and their former positions within the Pacific Ocean basin.

To solve this problem it is necessary to know (a) the nature of the accreted terranes, (b) the time of their welding to, or collision with, the continental margin, (c) their paleomagnetic characteristics, and (d) the relative movements between Pacific and Asian plates. At present, information on such features in northeastern Asia is sparse, and a comprehensive field study of the accreted terranes is only just beginning. Paleomagnetic data are scarce and relevant only to the Kolymo-Omolonsky massif. We are able to determine the relative motions of the Pacific, Kula, Farallon, and Phoenix plates, within the Pacific Ocean itself back to Early Cretaceous time using the magnetic lineation pattern. However, the Pacific/Eurasian relative motion can be calculated with certainty only for the last 50 Ma using plate motion closure through Antarctica and Africa. Evidence exists that the Antarctic was not a single continent prior to 50 Ma ago (e.g. Molnar et al., 1975; Cox and Gordon, 1978). For this reason, absolute plate motions can be used as the only basis with which to compute Pacific/Eurasian and Kula/Eurasian relative motions during the last 130 Ma. This paper should be considered as a preliminary approach to understanding the plate tectonic evolution of northeast Asia.

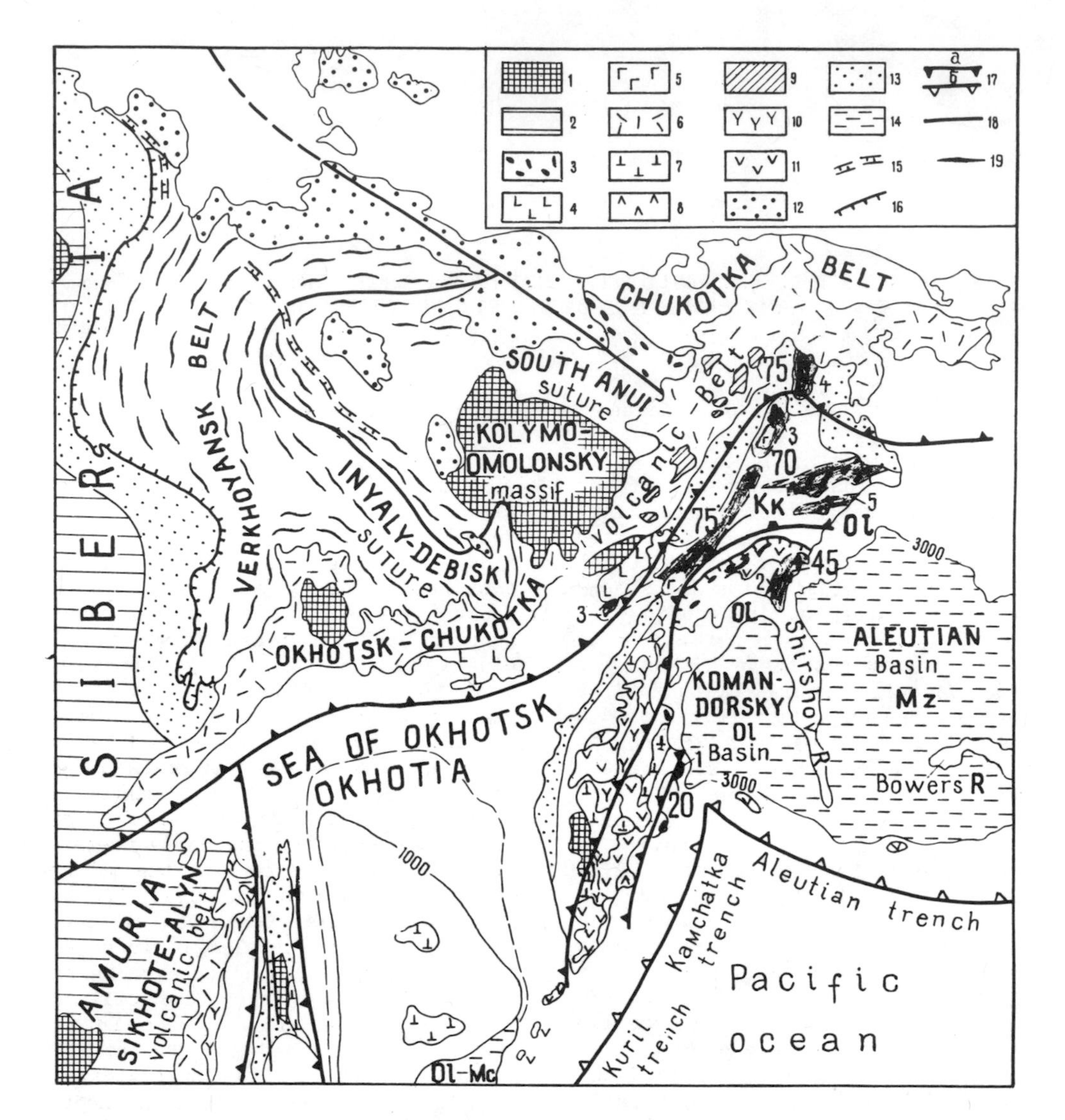

Fig. 1. Lithostratigraphic assemblages of northeast Asia. 1-Precambrian massif; 2-Siberian platform; 3-Ophiolites; 4-Triassic-Jurassic island arc volcanics; 5-Early Cretaceous island arc volcanics; 6-Mid-Cretaceous volcanic arc of Andean type; 7-Late Cretaceous island arc volcanics; 8-Latest Cretaceous and Early Tertiary island arc volcanics; 9-Cretaceous granite pluton; 10-Miocene island arc volcanics; 11-Recent island arc volcanics; 12-Molasse filled sedimentary basins; 13-Flysch filled sedimentary basins; 14-Marginal basin oceanic floor; 15-Recent continental rift (Momsky Graben); 16-Thrust; 17-Subduction zone: a-fossil (age of activity is given as a symbol (J, K, Ol, Mc) on the map); b-active; 18-Suture; 19-Fold axis. Exotic terranes are shaded; small figures are: 1-East Kamchatka, 2-Vatyn, 3-Talovo-Mainsky, 4-Pekulnei, 5-Hatyrka; KK-Koryakyan upland; Ol-Olyutorovka zone.

Exotic Terranes

Northeast Asian terranes include (1) "Kolymia", named from the Kolymo-Omolonsky massif, (2) "Okhotia", named from the Sea of Okhotsk, (3) and "Koryakia", which consists of a number of smaller terranes mainly in the Koryakian upland, within flysch-like Upper Cretaceous and Paleocene-Eocene sequences, and are separated by serpentinitic mélange zones (Fig. 1).

Within Koryakia, the following small terranes can be recognized (Fig. 1):

1. East Kamchatka, composed of Upper Cretaceous and lower Tertiary volcanics attached to the continent in late Tertiary time (Zinkevich et al., 1983; Raznitsyn et al., 1981).

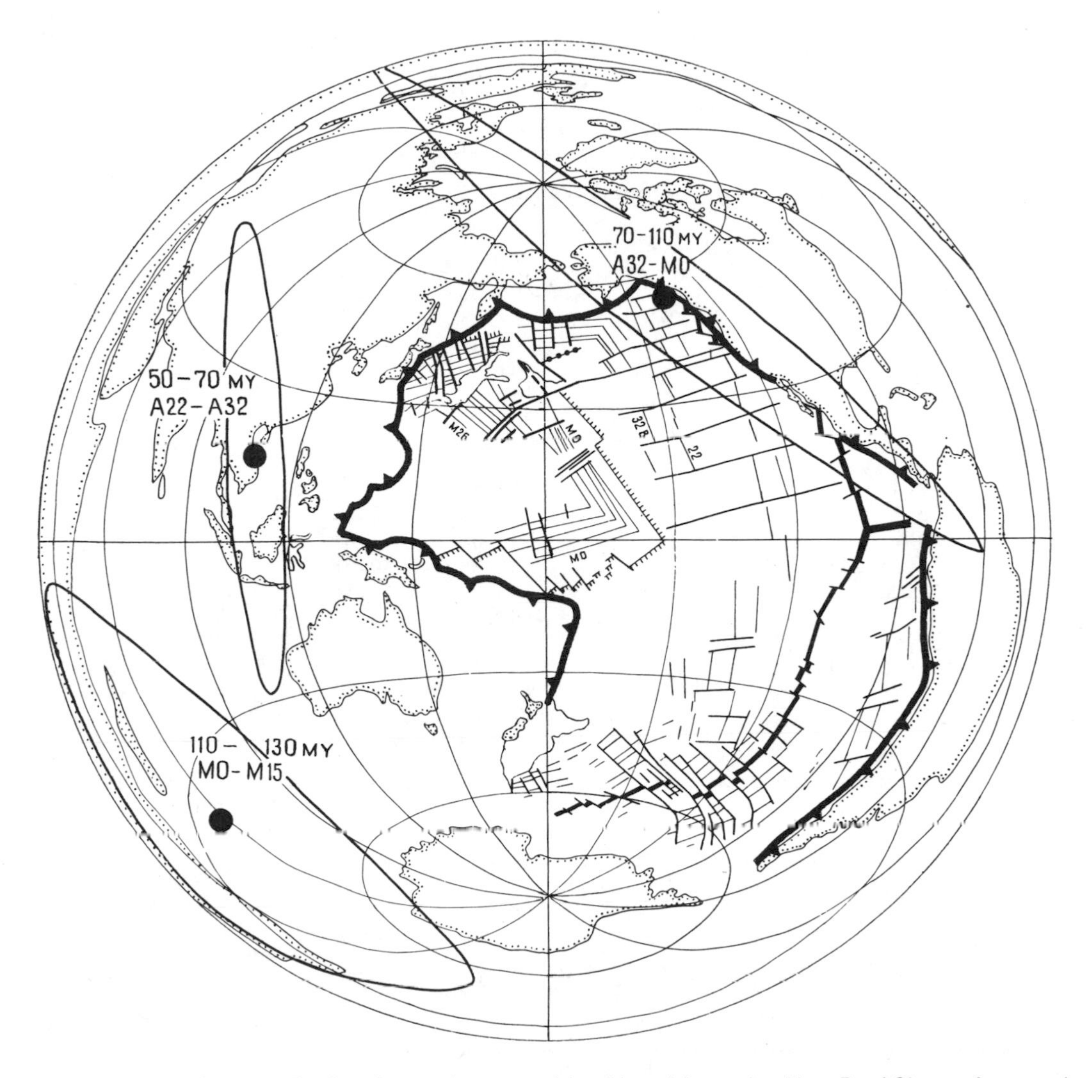

Fig. 2. Mesozoic and Early Cenozoic magnetic lineations in the Pacific and computed pole positions for the Kula/Pacific motion. Magnetic anomalies are shown as thin lines. Thin toothed line indicates the quiet magnetic zone boundary. Figures are magnetic anomaly numbers. Thick line indicates an active spreading zone, and thick toothed lines corresponds to subduction zones. The dotted line shows the Chinook zone. Dashed line indicates the Hess Rise long axis orientation. Large dots are poles of the Kula/ Pacific rotation. They are given separately for 50-70 Ma, 70-110 Ma and 110-130 Ma. 95% confidence intervals are drawn around the poles of rotation.

2. Olutorovsk (Vatyn), along the Bering Sea coast, consists of mid- and Upper Cretaceous oceanic and possibly island arc volcanics which collided with the continental margin during the Late Eocene (Aleksandrov et al., 1980; Alekseev, 1979).

3. Talovo-Mainsky, north of Penzhina Guba Bay, comprises Paleozoic (Ordovician to Permian) shelf limestones, Jurassic oceanic and island arc assemblages (Nekrasov, 1971; Aleksandrov, 1978). It collided with the Eurasian margin between earliest and mid-Cretaceous times.

4. Pekulnei, south of Chukotka, possibly has an ancient metamorphic nucleus (Markov et al., 1982) overlain by Lower and mid-Cretaceous island arc volcanics; collision possibly took place in pre-Aptian time.

5. Hatyrka, east of the Koryakya Mountains, comprises Devonian and Carboniferous shelf limestones (Zinkevich, 1981), deep water cherts with radiolaria of tropical provenance (Bragin pers. comm., 1985), and Permian fusulinid limestones of southern origin (Solovieva, 1985, pers. comm.); the exotic block is surrounded by Senonian flysch and overlain by autochthonous Maastrichtian sediments, which indicates collision time as late Senonian.

Finally, remnants of a Late Cretaceous vol-

TABLE 1. Available Parameters for the Pacific/Antarctic and Kula/Pacific Motions

Time (Anomaly)	North Latitude	East Longitude	Angle	References
		Pacific/Antarctic		
0-10 (A 5)	72.	- 70.	9.75	1
0-20 (A 6)	71.25	- 73.19	15.41	1
0-37 (A13)	74.83	- 56.86	28.01	1
0-43 (A18)	75.08	- 51.25	32.56	1
0-60 (A25)	71.61	- 57.47	40.11	1
0-70 (A31)	71.65	- 41.	53.75	1
0-84 (A34)	68.	- 50.	66.	2
		Kula/Pacific		
60(A26)- 67(A30)	19.19	117.8	- 5.26	3
59(A25)- 71(A32)	15.73	106.02	- 8.14	4
50 - 71*	15.73	106.02	-11.4**	4
71(A32)-111(MO)	53.6	-132.3	30.**	4
111(MO) -122(M12)	40.87	-128.6	15.56	4
122(M12)-132(M16)	11.89	- 88.9	5.66	4
132(M16)-140(M20)	40.2	-127	6.02	4
140(M20)-155(M26)	10.83	- 88.9	11.7	4
155(M26)-157(M29)	20.9	122.2	- 7.2	4
110(MO) -130(M15)	36.	-120.8	18.15**	4

Kula/Pacific poles were computed from magnetic lineation patterns and transform fault directions. The pole for the period of the quiet magnetic zone was computed using directions of the Emperor Fracture Zone and the long axis of the Hess Rise. References: 1-Stock and Molnar, 1982; 2-Molnar et al., 1975; 3-Cooper et al., 1976; 4-This paper.

* This pole is extrapolated from the previous parameter for 59-71 Ma

** The poles plotted on the map of Figure 2.

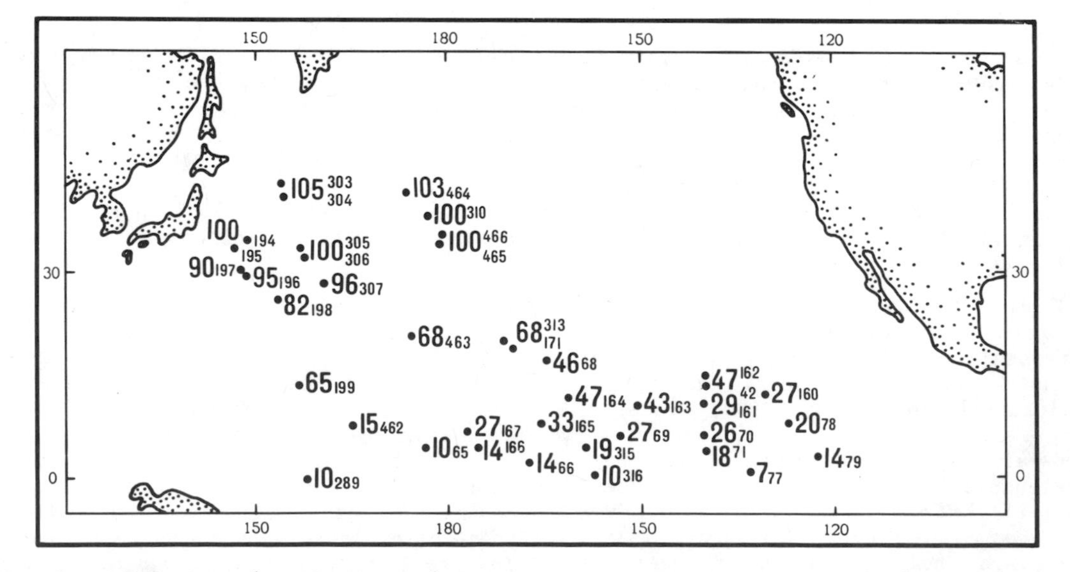

Fig. 3. DSDP sites (small figures) in the Pacific and the time in Ma (large figures) of their crossing the high productivity equatorial zone.

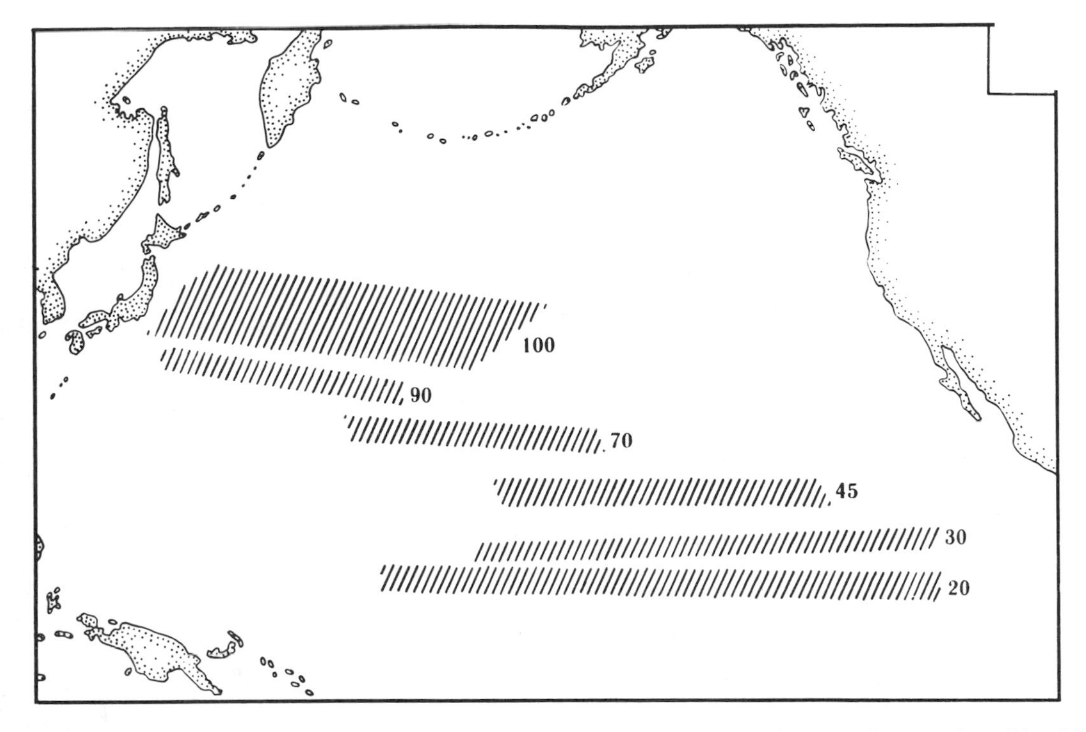

Fig. 4. Paleoequatorial positions for the Pacific plate inferred from paleoclimatic data. Width of each strip reflects scattering of the data.

canic arc were found in the Sea of Okhotsk (Geodekjan et al., 1976).

The time of collisional events in millions of years before present is shown in Figure 1. The events took place mainly between mid-Cretaceous to Miocene times although the dating of collisional events is uncertain for the Talovo-Mainsky and Pekulnei terranes and may be either Early or Late Cretaceous. In the Bering Sea realm, all collisional events had taken place before the Aleutian Basin Mesozoic floor became trapped behind the Aleutian Arc and before the Komandorsky Basin was opened in the Neogene.

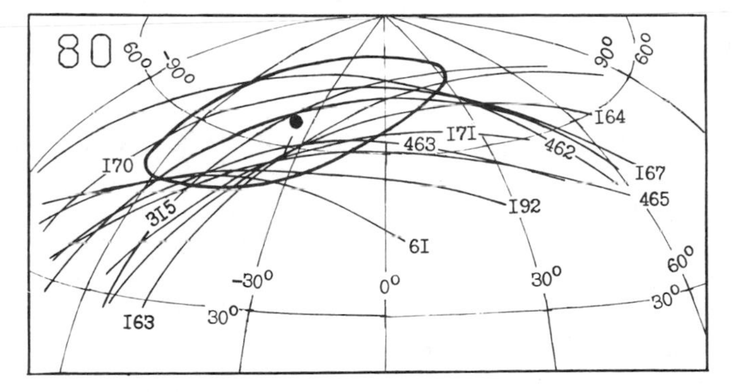

Fig. 5. Determination of the paleomagnetic pole position of the Pacific plate for 80 Ma according to paleoinclination data obtained from DSDP cores. Thin lines are small circles drawn around the DSDP sites on the distance equal to paleolatitudes. Small figures correspond to site members. The thick line shows 95% confidence interval. The dot indicates the pole position.

Relative Plate Motions Within the Pacific

Available parameters of relative plate motions (poles and angles of rotation) within the Pacific are given in Table 1. Good data are available for relative motions of the Pacific Ocean plates

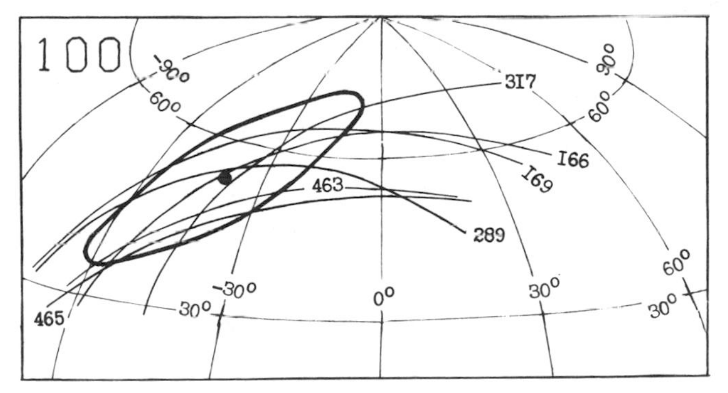

Fig. 6. Determination of the paleomagnetic pole position for 100 Ma. For explanation see Fig. 5.

TABLE 2. Paleomagnetic Pole Positions for the Pacific Plate

N	Age	Ma	Pole: North Latitude	Pole: East Longitude	95% Confidence Ellipse: Long Axis (degree)	95% Confidence Ellipse: Short Axis (degree)	95% Confidence Ellipse: Azimuth of long Axis	95 (degree)	Data	References
1	2		3	4	5	6	7	8	9	10
1		120	42.4	-31.4	19.	7.	62.		Paleoinclinations of DSDP cores	10
2		120	50.	-30.	18.	3.	77.		Skewness of magnetic lineation	1
3		100	45.	-30.	18.5*	10.*	53.		Paleoinclinations of DSDP cores	2
4		100	52.3	-40.8	24.	6.5	44.		" " " "	10
5		100	56.	-34.				8.	seamounts	3
6	Late Cretaceous	80	58.	-10.				5.	seamounts	6
7		83	59.	-25.	9.	7.	64.		Different data	4
8	Campanian	80	56.5	- 6.5	8.	3.	91.		Different data	5
9		80	64.4	-31.3	21.	6.	5.		Paleoinclinations of DSDP cores	10
10		80	61.	-45.	22.5**	11.**	65.		" " " "	2
11		85	61.	16.			13.		seamounts	3
12	Maastrichtian	69	71.	9.	6.	2.	91.		Different data	7
13	Eocene	44	78.2	16.3	5.6	4.6			seamounts	9
14		40	71.	- 6.			13.		"	3
15		30	75.	12.					"	3
16		10	87.	90.				14.	"	3
17	Hauterivian	120	41.	-31.	33.	11.	64.		Paleoinclinationa of DSDP cores	11
18	Early Cretaceous	115	53.	-25.3	5.4	23.6	70.		" " "	11
19	Early Cretaceous	115	46.1	-35.6	2.6	1.0	60.		" " "	11
20	Early Cretaceous	115	46.1	-35.6	137.	4.2	60.		" " "	11
21		90	54.	-26.	7.	6.	57.		Different data	8
22		81	57.	- 4.	5.	4.	88.		Different data	8
23		80	53.2	-34.	17.5	6.7	5.2		Paleoinclinations of DSDP cores	11
24		80	52.	-28.4	5.5	2.	56.		" " "	11
25	Late Eocene	41	77.5	21.2	2.9	1.5	74.		?Different data	12

References: 1-Larson and Chase, 1972; 2-Peirce, 1976; 3-Francheteau et al., 1970; 4-Cox, 1974; 5-Gordon and Cox, 1980; 6-Harrison et al., 1975; 7-Gordon, 1982; 8-Gordon, 1983; 9-Sager, 1982; 10-Kononov, 1984a; 11-Cox and Gordon, 1984; 12-Sager, 1983

* 73% confidence interval

** 86% confidence interval

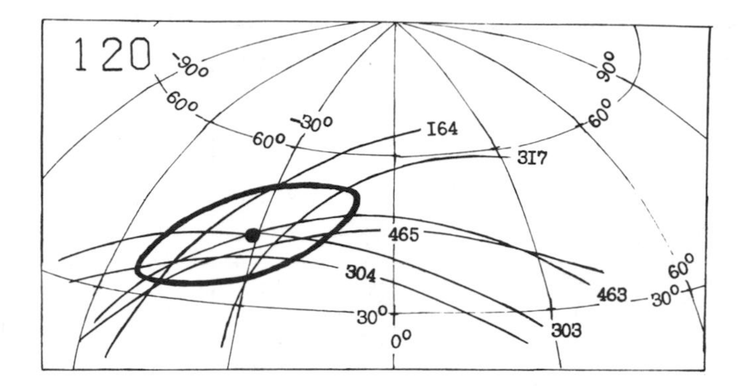

Fig. 7. Determination of the paleomagnetic pole position for 120 Ma. For explanation see Fig. 5.

after 34 anomaly time that were used to compute Pacific/Eurasian motion from Recent to 56 Ma (25 Anomaly) through the Pacific/Antarctic/Africa/ Eurasia closure (Molnar et al., 1975; Norton and Sclater, 1979; Olivet et al., 1984) (see Table 1).

The Mesozoic magnetic lineation pattern in the northwestern Pacific, as is well known, is inferred to reflect the existence of a plate adjacent to the Pacific plate in the period from 160 Ma to the Tertiary/Cretaceous boundary. This plate is the Kula plate (Larson and Chase, 1972; Hilde et al., 1976). Recently Woods and Davies (1982) and Engebretson et al. (1984) proposed that the Kula plate existed only during the second stage of the late Mesozoic plate tectonic evolution of the Pacific Ocean, after the no-reversal period (after 80 Ma). The independent Izanagi plate was created during the first stage, in Late Jurassic-Early Cretaceous time, which corresponds to the pattern of Japanese magnetic lineations. We prefer to use here for convenience, a single Kula plate for the whole late Mesozoic-early Cenozoic interval, although the Izanagi plate concept can be accepted without

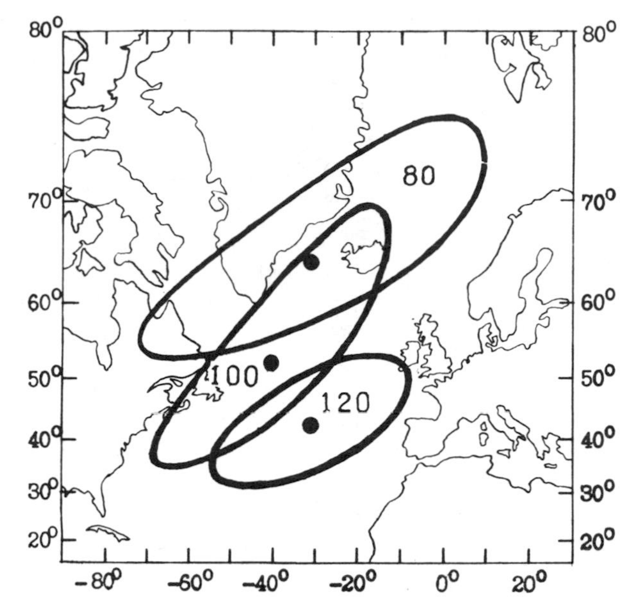

Fig. 8. Paleomagnetic pole positions of the Pacific plate for 80, 100 and 120 Ma with 95% confidence intervals.

creating significant differences for Eurasian and Pacific Ocean plate interactions.

The scenario for plate motion in the northwestern Pacific consists of three spreading events from three spreading centers (Fig. 2).

The first one was active from 160 to 110 Ma, when the Japanese lineations originated. Transform fault directions together with the one-side matching of corresponding magnetic anomalies were used to compute Kula(=Izanagi)/Pacific poles of rotation during this stage (Kononov, in prep.). Results are given in Table 1, and the pole position for the interval 130-110 Ma is shown on Figure 2. The pole obtained (the mean position 36°N, 120.8°W) is different from the Engebretson

TABLE 3. Parameters of the Absolute Pacific Plate Motion in the Hotspot Frame

Time span	Hotspot/Pacific			Ref.	Pacific/Hotspot		
	North Latitude	East Longitude	Angle		North Latitude	East Longitude	Angle
0-43	69.0	- 68.0	-34.4	1	69.0	-68.0	34.4
43-70	17.0	-107.0	-12.0	1	12.3	-78.8	12.0*
70-125	33.0	-109.5	-38.	2	28.3	-74.3	38.
70-130	33.0	-109.5	-41.4		28.3	-74.30	41.4

References: 1-Clague and Jarrard, 1973; 2-Lancelot, 1978
*This value of the rotational angle obtained by subtraction 5.1° from initial 17.1° given by (2) to take into account the hotspot displacement relative to the spin axis (Kononov, 1984b)

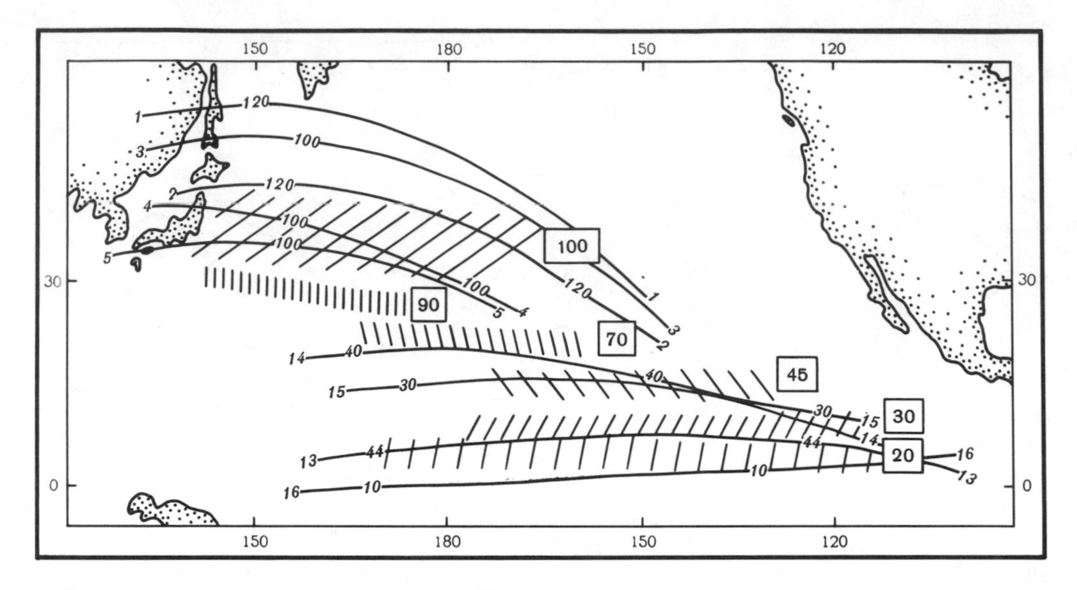

Fig. 9a.

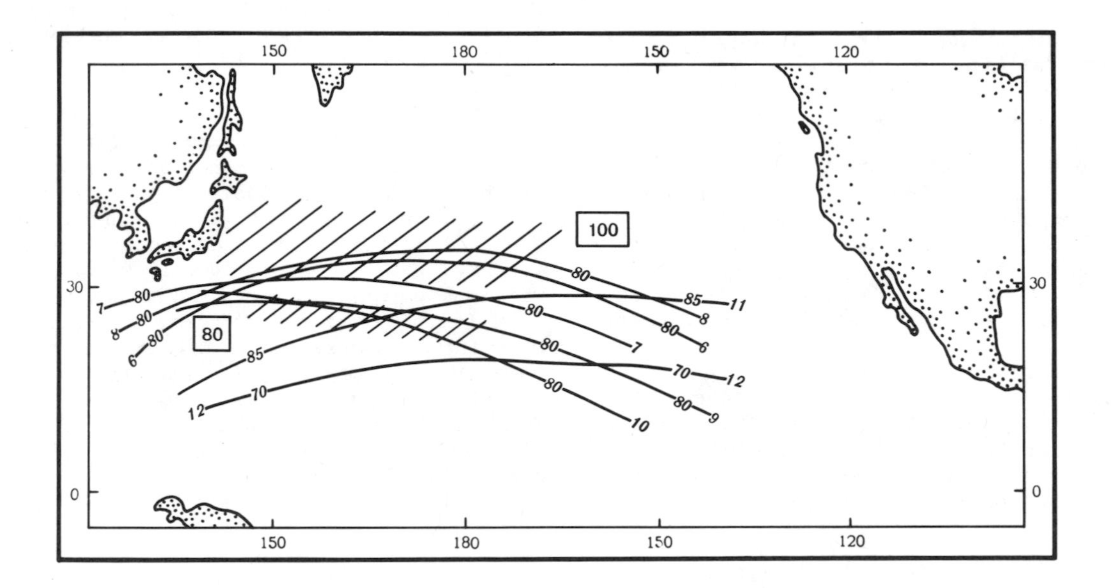

Fig. 9b.

Fig. 9. Comparison between paleoclimatic (hachured patterns) and paleomagnetic (lines) paleoequatorial positions for the 20-100 Ma interval. a) shows paleoequators for the time from 20 to 100 Ma. b) shows in detail 80 Ma time.

et al. (1984) pole (48°N,177°W), but both poles are on the same great circle which continues the direction of the Japanese lineation. The reason for the difference seems to be poor control along the Euler meridian. The half spreading rate was, during the first spreading event, nearly 5 cm/year. According to Engebretson et al. (1984) the half rate was 4 cm/year.

The second spreading system is obscure. It formed during the no-reversal interval, from 110 to 34 Ma. In our opinion, preserved remnants of this system include the Chinook Trough as a

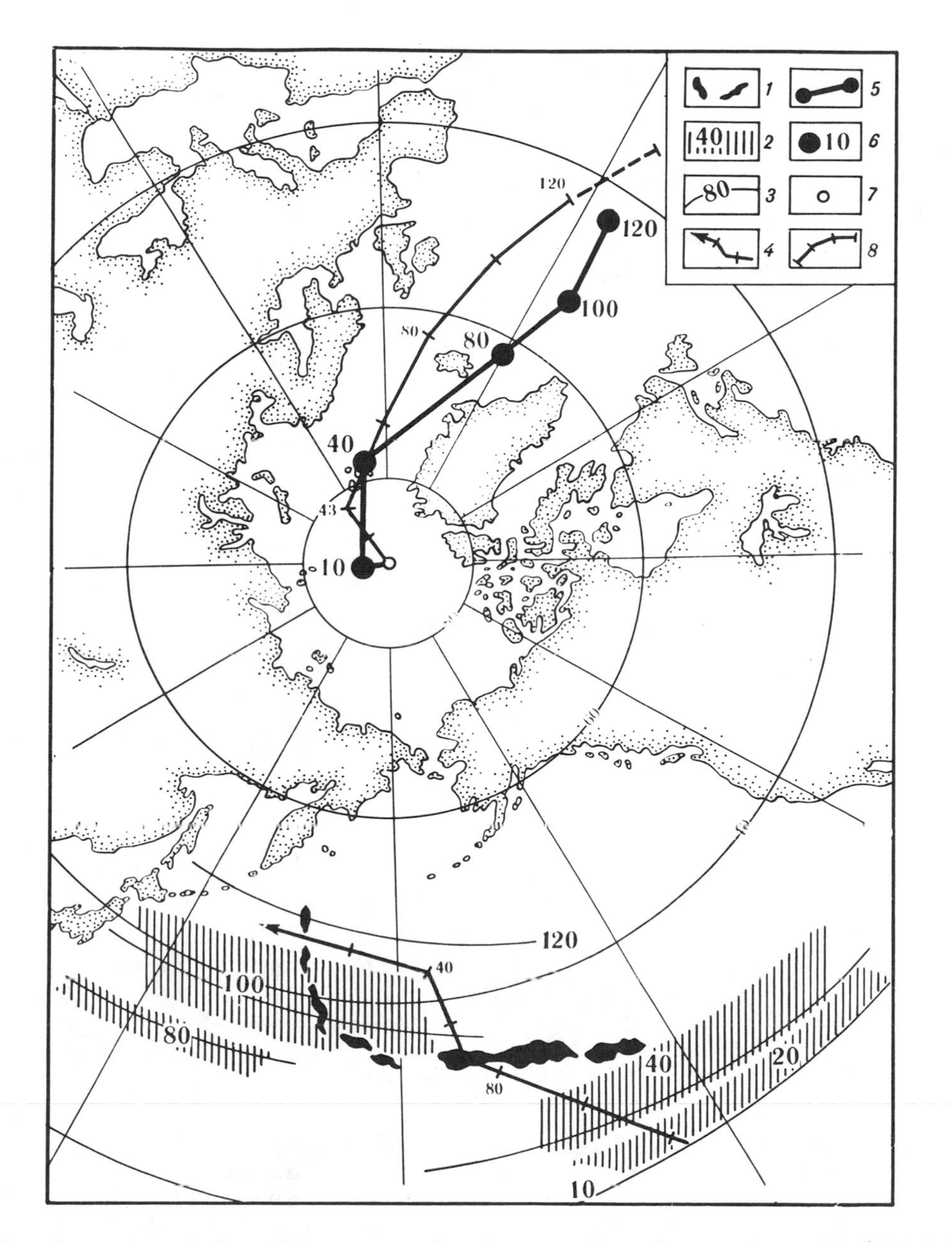

Fig. 10. Comparison between the apparent paleomagnetic polar wander path (thick line) and the migration path (thin line) of the geographical North Pole based on hotspot trajectories data. Stereographic projection centered on the North Pole. 1-Hawaiian-Emperor volcanic chain; 2-Paleoclimatic equatorial zone and its age; 3-Paleomagnetic equator and its age; 4-Trajectory of the Pacific Plate motion from 120 Ma to the present, with 20 Ma increments, according to the hotspot data in Table 3; 5-Apparent polar wander path of the Pacific Plate; 6-Paleomagnetic pole position; 7-Present North Pole; 8-Migration path of the geographical North Pole according to hotspot data (Table 3) with 20 Ma increments.

fossil spreading center (Erickson et al., 1969) and the Emperor Fracture Zone as a transform fault. Woods and Davies (1982) suggested that the Chinook Trough belongs to the same transform fault system as the Surveyor and Mendocino Fracture zones, despite the apparent difference of at least 10° between the strike of the Surveyor and Mendocino Fracture zones, and the Chinook Trough.

Fig. 11. Comparison between Pacific/Eurasia motion paths inferred from closure of relative movements (lines with solid circles) and from closure of absolute movements (lines with open circles). As far back as 56 Ma both curves coincide well; between 56-84 Ma they diverge. The difference reaches 3000 km, Base: Lambert projection centered 0,150°W. The figures are the following numbers of magnetic anomalies: 1-A5, 2-A6, 3-A13, 4-A16, 5-A18, 6-A19, 7-A21, 8-A22, 9-A24, 10-A25, 11-A27, 12-A29, 13-A31, 14-A32, 15-A33, 16-A34, 17-M0, 18-M1, 19-M15.

In this case it is very difficult to understand the nature of the Emperor Fracture zone, which is strictly orthogonal to the Chinook Trough. We took the strike of the Emperor Fracture Zone and also the long axis of the Hess Rise as flow directions with which to compute, conveniently, the Kula/Pacific pole position for the no-reversal period. The half-spreading rate is estimated as 2.5 cm/year.

The third spreading event is coincident with formation of the A34-A23 anomaly set in the west-east direction. The pole position for the Kula/Pacific rotation was obtained using the azimuths of the Adak, Amlia and other transform faults cutting the magnetic lineation. The distance between anomalies gives a half-spreading rate of 3-4 cm/year from 80 to 50 Ma.

The Pacific Plate Absolute Motion

Three independent ways are used for estimation of absolute motion: paleoclimatic, paleomagnetic and hotspot trajectory data.

Figure 3 shows the DSDP sites in the North

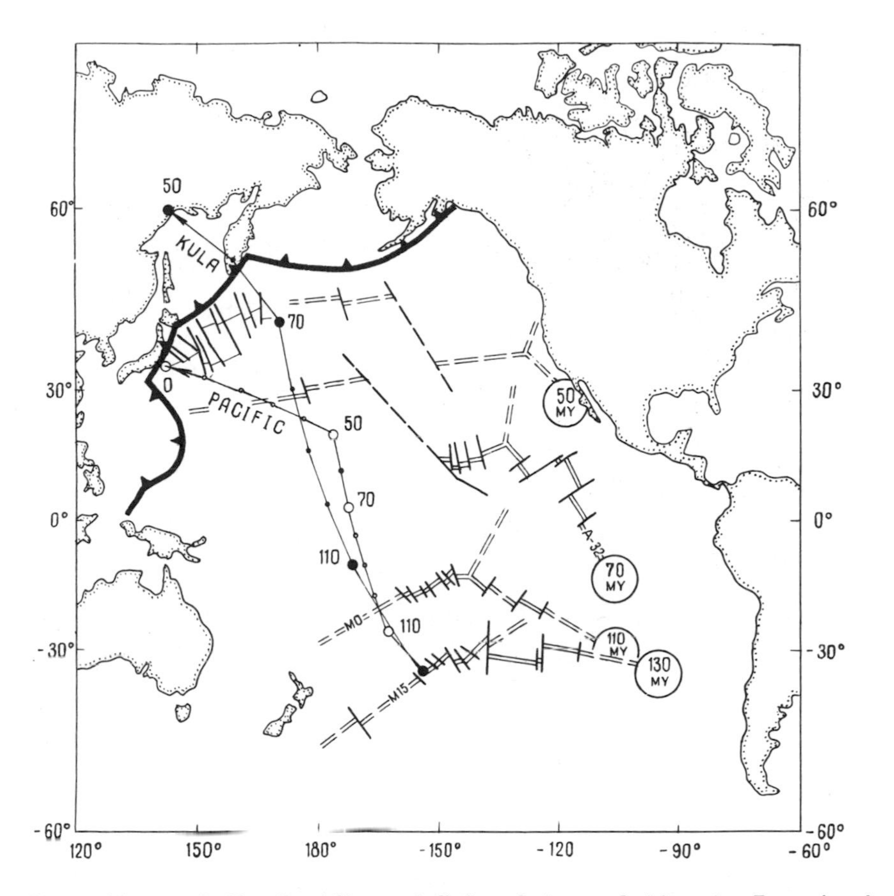

Fig. 12. The motions of the Pacific and Kula plates relative to Eurasia during the last 130 Ma. The line with open circles corresponds to the Pacific/Eurasian motions, and the line with solid circles to the Kula/Eurasian motions, through 10 Ma increments. In addition, successive positions of the triple junction and Kula/Pacific and Farallon/Pacific boundaries for 130, 110, 70 and 50 Ma are shown as double lines. The thick toothed line indicates the present northwestern Pacific subduction zone. Base: Mercator projection centered 0,150°W.

Pacific where the time of crossing of the high productivity equatorial zone is recorded by the thickened sedimentary sequence and by increased sedimentation rates. The large figures indicate the time when a site crossed the equatorial high productivity zone (from 100 Ma in the north to 10 Ma in the south). From these data, paleo-equatorial positions are readily inferred (Fig. 4). According to paleoclimatic data the movement of the Pacific plate was almost 4000 km northwards during the last 100 million years.

Paleomagnetic data were obtained from the following sources:

1. from magnetism of seamounts (Francheteau et al., 1970; Harrison et al., 1975);
2. from skewness of the magnetic lineation (Larson and Chase, 1972);
3. from paleoinclinations measured in DSDP cores (Sclater and Cox, 1970).

Available paleomagnetic poles obtained from different sources are unfortunately scattered over a broad area (Table 2). Therefore we obtained new paleomagnetic poles for 80,100,120 Ma using DSDP paleoinclination data. The procedure is a simple one: small circles corresponding to measured paleolatitudes are drawn with centers in DSDP sites (Fig. 5-7) and the intersection cloud gives the statistical pole position shown as a dot. In this manner we obtained poles for 80 Ma, 100 Ma and 120 Ma. All three poles are shown on Figure 8 with 95% confidence intervals, and coincide relatively well with poles recently published by Cox and Gordon (1984).

The next step is to compare paleoclimatic and paleomagnetic data to choose the best pole positions. Shown in hachured patterns in Figure 9 are paleoequator positions for 20, 30, 45, 70, 80, 90, 100 Ma, obtained from paleoclimatic data, and shown in thin lines are paleoequator positions derived from different paleomagnetic poles. Some

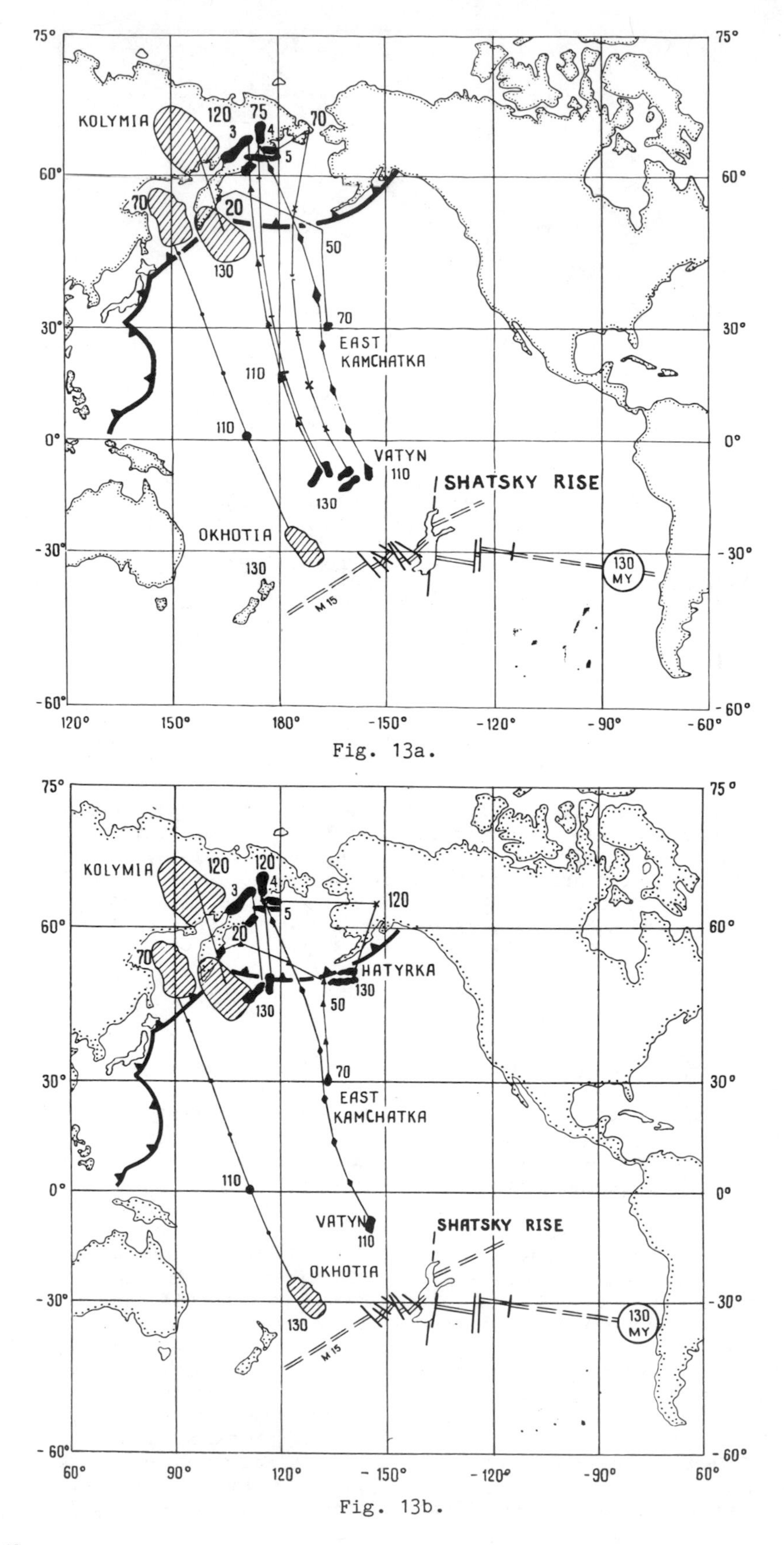

Fig. 13a.

Fig. 13b.

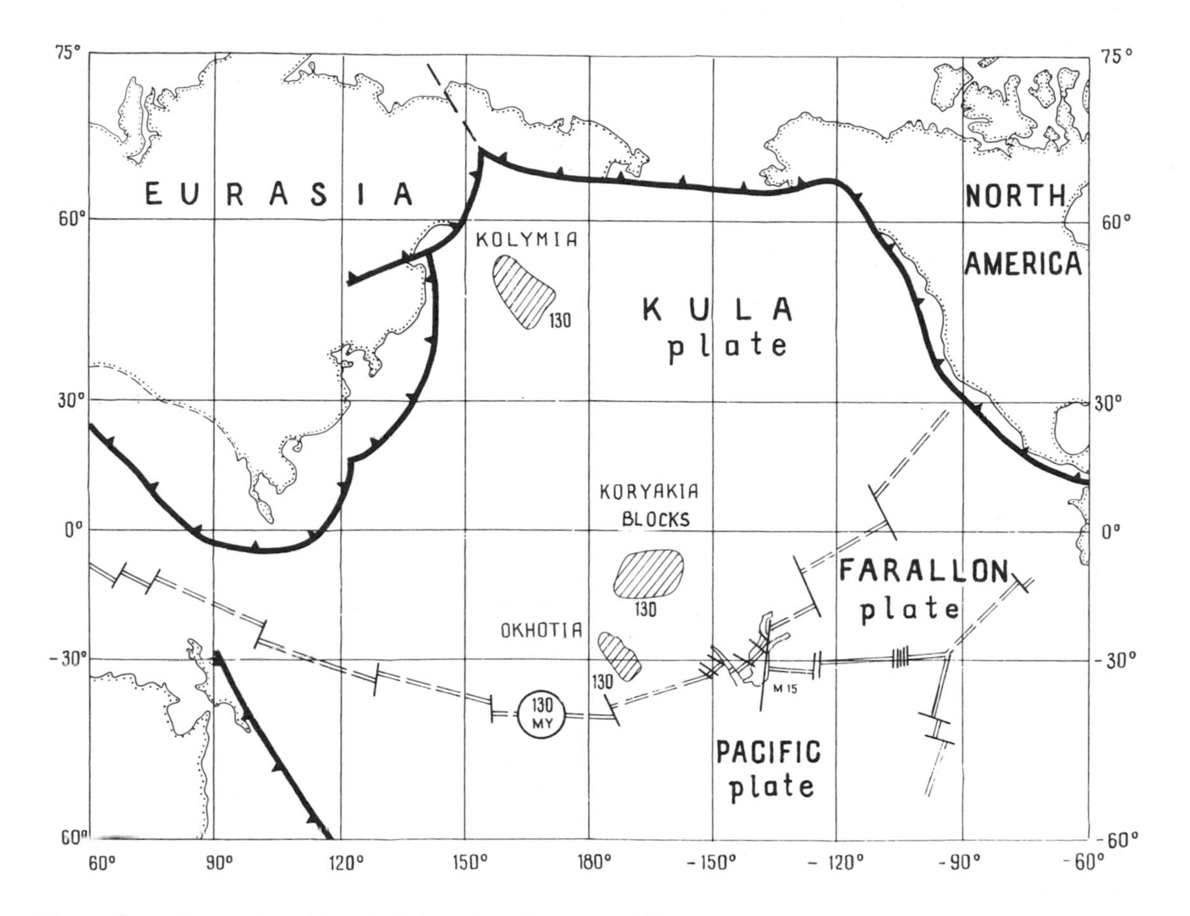

Fig. 14. Reconstruction of the Pacific for 130 Ma. Eurasia is fixed. Subduction zones (thick toothed lines), spreading centers (double lines) are shown. Exotic blocks are hachured. The Shatsky Rise is outlined near the Kula/Pacific/Farallon triple junction. Base: Mercator projection centered 0,180°.

of these lines coincide well with the paleoclimatic equators whereas others do not. We use herein paleomagnetic equators numbered 1, 4, 5, 9, 13, 16 on Figure 9 as these coincide well with the paleoclimatic equators. The paleomagnetic poles corresponding to these equators are listed in Table 2.

These preferred paleomagnetic poles and the apparent polar wander path for the Pacific plate are compared with the computed path of the geographical North Pole defined from hotspot trajectories data in Figure 10. This figure also shows the Hawaiian-Emperor volcanic chain as a hotspot track. Hotspot trajectory data are taken from Clague and Jarrard (1973) and Lancelot (1978). It is seen in Figure 10 that polar wandering and hotspot paths coincide relatively well up to 40 Ma, but diverge slightly in the interval from 40 to 80 Ma. Along a distance of 50 degrees on the great circle, the maximum difference between the two paths is 9 degrees, which may be due either to poor quality of data or to hotspot motions

---

Fig. 13. Travelling paths for exotic terranes of northeast Asia. Large terranes with a continental core are hachured and small terranes of oceanic and island arc origin are black. Successive positions of the terranes are shown in 10 or 20 Ma increments. The positions of the 130 Ma spreading center (=M15) (double line) and of the Shatsky Rise (thin line) are also shown. a) Shows the "long way" option if the Hatyrka (5 on the map), Pekulnei (4) and Talovo-Mainsky (3) terranes collided with the Eurasian margin in the Late Cretaceous. b) Shows the "short way" option if these terranes collided with the margin at Early Cretaceous. Large figures indicate the time of collisional events and small figures give the time calibration of the travelling paths. Bases: Mercator projection centered 0,150°W.

TABLE 4. Pole Parameters for Computation of Pacific/Eurasian Motions Through the Absolute Motion Closure

Magnetic Anomaly	Time span	Pacific/Hotspot Inferred From Table 3			Hotspot/Africa After Duncan (1981)			Eurasia/Africa			Pacific/Eurasia		
		North Latitude	East Longitude	Angle	North Latitude	East Longitude	Angle	North Latitude	East Longitude	Angle	North Latitude	East Longitude	Angle
5	0-10	69.0	-68.0	7.64	61.0	-45.0	2.1	30.95	-20.42	1.71	70.77	- 77.73	8.67
6	0-20	69.	-68.0	15.98	61.0	-45.0	4.37	37.78	-19.02	3.11	70.92	- 75.94	18.22
13	0-37	69.	-68.	29.7	41.9	-25.75	9.9	20.83	-16.45	4.09	66.56	- 60.93	35.59
16	0-39	69.	-68.	31.48	41.2	-26.5	10.25	24.8	-16.75	5.0	66.73	- 62.64	36.95
18	0-42	69.	-68.	33.25	40.65	-27.16	10.63	27.4	-16.9	5.8	66.88	- 64.22	38.29
21	0-48	66.23	-71.6	35.66	39.15	-28.81	11.77	32.0	-16.6	8.4	64.6	- 71.41	39.51
24	0-53	63.56	-74.4	37.04	38.14	-29.86	12.67	33.55	-18.05	9.88	62.05	- 76.78	40.07
27	0-63	59.0	-78.	39.8	32.0	-44.10	17.5	33.8	-15.8	12.2	53.7	- 83.58	46.52
31	0-68	55.6	-79.6	41.55	31.1	-49.1	19.5	33.75	-15.4	13.2	51.03	- 86.52	50.0
32	0-71	55.7	-80.2	42.4	31.5	-51.3	20.6	33.73	-15.3	13.6	50.	- 87.67	51.96
33	0-78	53.4	-83.0	47.05	30.9	-57.3	24.8	35.72	-14.7	15.	47.42	- 91.45	60.68
34	0-84	51.8	-84.7	50.9	30.0	-55.4	27.0	33.77	-14.1	16.	45.73	- 91.57	65.8
M0	0-111	46.6	-89.5	68.7	28.0	-41.1	32.5*	45.1	- 5.79	30.04	38.04	- 99.78	79.87
M1	0-115	46.1	-89.8	70.9	28.3	-40.3	32.9*	45.246	- 5.43	31.15	37.83	-100.23	81.81
M15	0-130	44.1	-91.3	81.2	29.3	-35.7	34.5*	45.8	- 3.9	35.6	37.08	-102.19	91.07

* After Crough et al., 1980

TABLE 5. Differential Pacific/Eurasia and Kula/Eurasia Poles of Rotation

Magnetic Anomaly	Time span	Pole		
		North Latitude	East Longitude	Angle
		Pacific/Eurasia*		
5-6	10-20	71.1	- 74.92	9.56
6-13	20-37	62.52	- 47.08	17.64
13-16	27-39	60.04	- 98.92	1.43
16-18	39-42	60-24	- 99.36	1.41
18-21	42-48	5.31	- 97.2	2.78
21-24	48-53	11.34	82.55	- 2.5
24-27	53-63	11.8	- 81.7	9.4
27-31	63-68	15.6	- 91.2	4.4
31-32	68-71	21.73	- 92.2	2.24
32-33	71-78	28.12	- 96.08	9.36
33-34	78-84	29.29	- 81.09	5.43
34-M0	84-111	2.37	- 97.35	18.24
M0-M1	111-115	23.3	-105.28	2.02
MM1-M15	115-130	23.97	-106.79	9.56
		Kula/Eurasia**		
23-32B	52-71	6.81	- 64.73	25.28
32B-M0	71-111	21.17	-108.37	60.35
M0-M15	111-130	22.56	-104.08	29.7

* Used data of the Table 4
** Used data of Tables 1 and 4

relative to the spin axis. In any case, for a first approximation, the data for absolute plate motions are in good agreement. Parameters of the absolute motion for the Pacific plate are given in the Table 3.

Pacific/Eurasian and Kula/Eurasian Relative Motion

To obtain the relative motions of Pacific and Eurasian plates we took Pacific hotspot parameters and African hotspot parameters (Duncan, 1981; Crough et al., 1980), and Africa/Eurasian relative motion parameters (Olivet et al., 1984). The rotational parameters for the Pacific/Eurasian and Kula/Eurasian motions are listed in Tables 4 and 5.

The Pacific/Eurasian motion paths obtained from the closure of relative motions through Antarctic and Africa and from closure through the absolute Pacific and Africa motions coincide well for the last 56 Ma (anomaly 25), but strongly diverge for the 84-56 Ma (anomalies 34-25) interval (Fig. 11). The difference is inferred to be due to movement between the eastern and western parts of Antarctica. From this difference, it can be easily inferred that the position of the

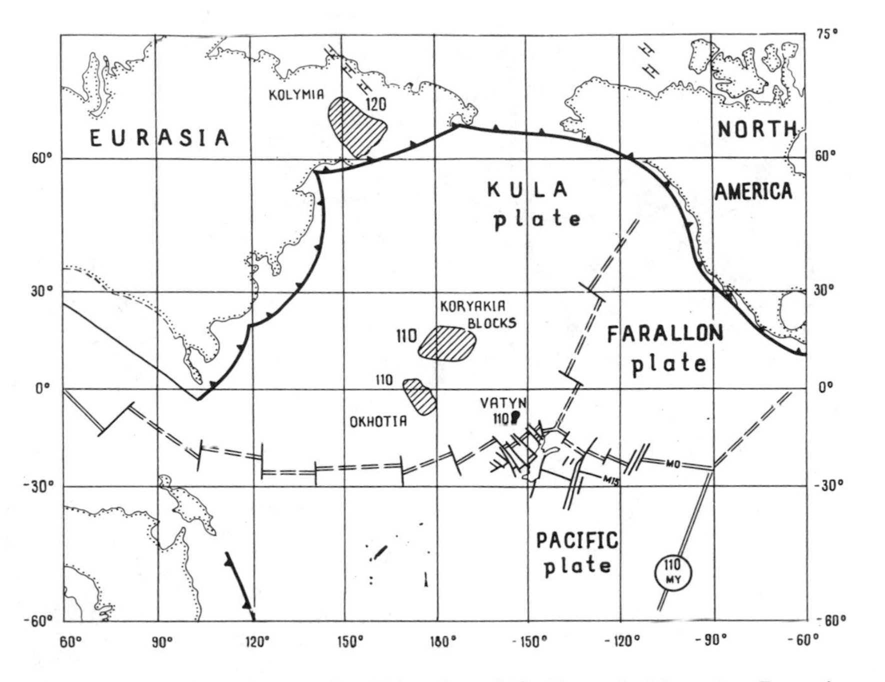

Fig. 15. Reconstruction of the Pacific for 110 Ma relative to Eurasia. Oceanic exotic blocks are black. For other explanations, see Figure 14.

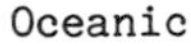

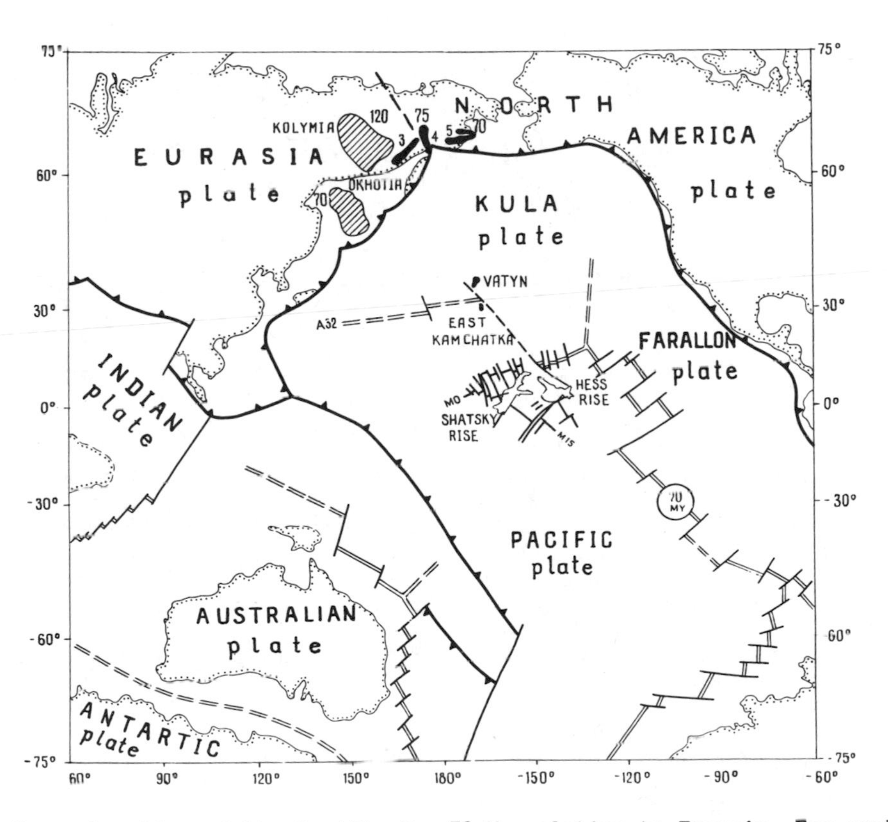

Fig. 16. Reconstruction of the Pacific for 70 Ma relative to Eurasia. For explanation see Figures 14 and 15. Koryakya exotic blocks numbered as on Figure 1.

Fig. 17. Absolute reconstruction of the Pacific plate for 130 Ma. For explanation see Figure 14. Lambert projection centered 0,180°.

eastern and western Antarctic rotational pole is 47.2°S, 107.8°W, with an angle of rotation equal to 21° during the interval between 84 and 56 Ma. This data gives about 2000 km of left-lateral displacement between the eastern and western Antarctic plates with a mean rate of 7 cm/year in a northeasterly direction.

Figure 12 shows the Pacific and Kula plate motions relative to Eurasia. A point on anomaly M15 at the 130 Ma spreading center was taken as the starting point. This point remains in the Pacific basin until present and is part of the Japanese anomaly set, whereas the same point travelling with the Kula plate was subducted under the Eurasian margin 60 million years ago.

During the time of its existence, from 130 to 50 Ma, the Kula plate moved relative to Eurasia almost strictly northwards with a mean rate 16.4 cm/year. If the Engebretson et al. (1984) parameters of the Izanagi/Pacific rotation were taken the mean rate of the Izanagi/Eurasia convergence would be 12 cm/year. Meanwhile the Pacific plate moved, during the interval 130 to 110 Ma, northwestwards with a rate of 6 cm/year, from 110 to 43 Ma, northwards with a rate of 7 cm/year, and from 43 Ma to the present, in a northwestwards direction with a rate of 9.4 cm/year. The Kula/Pacific spreading center shifted with respect to the Eurasian margin at a rate of 11.5 cm/year and passed it about 40 million years ago.

Possible Initial Positions of Exotic Terranes

The Kula and Pacific trajectories relative to Eurasia are used as flow lines to show the motions of exotic terranes in the northeastern USSR and to estimate their positions at 130 Ma, or, if they were formed later, at the time of their formation. Kolymia undoubtedly has continental type basement and probably was torn off

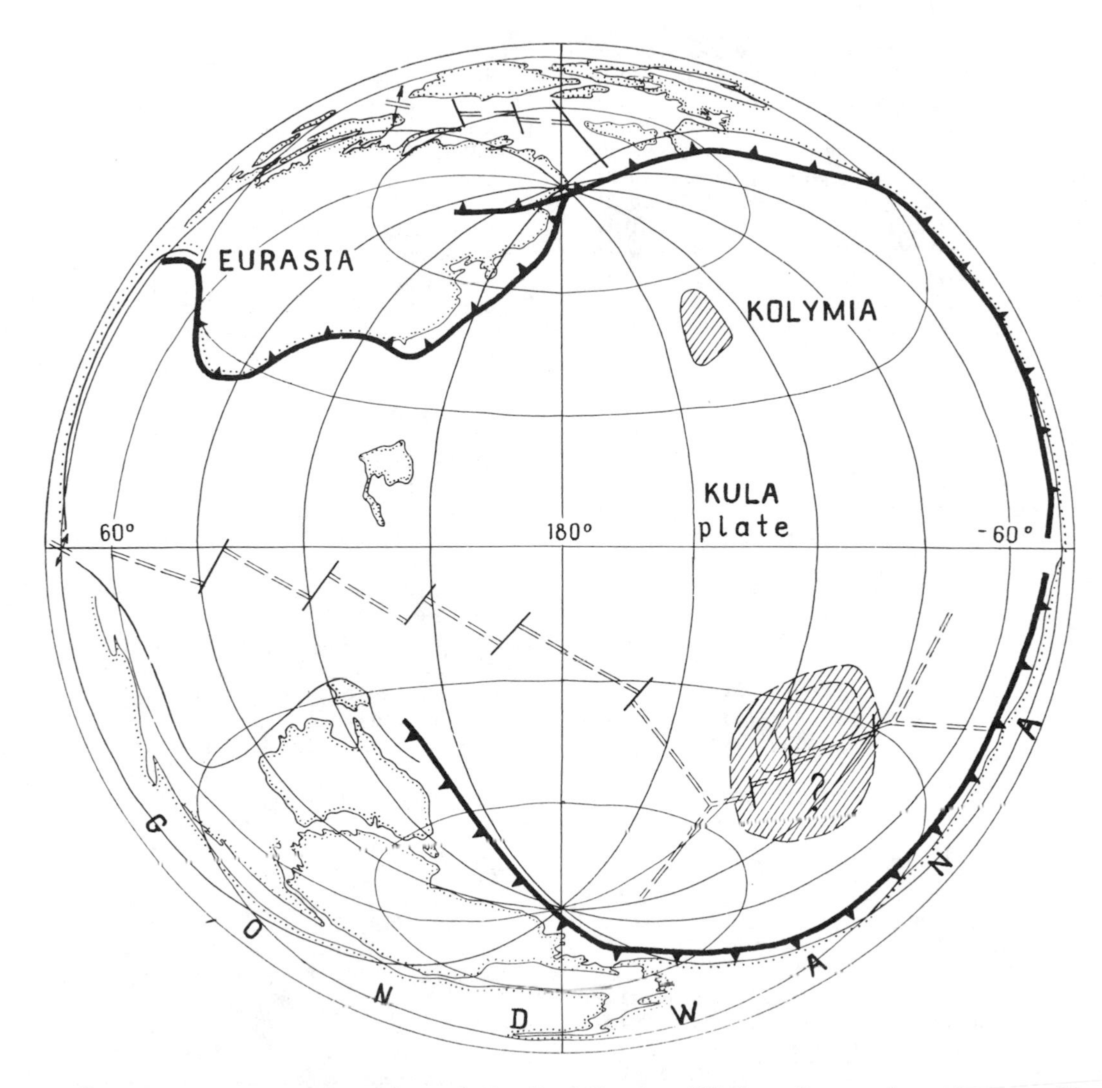

Fig. 18. Absolute reconstruction of the Pacific for 190 Ma. For explanation see Figure 14. Lambert projection centered 0,180°.

southeast Eurasia, whereas Okhotia and Koryakia include partly continental and partly oceanic and island arc assemblages. We have paleomagnetic data for Kolymia only, that are consistent with a southern origin for the block (Khramov, 1982). We consider two options for the small Talovo-Mainsky, Pekulnei, and Hatyrka terranes which constitute parts of Koryakia, namely (1) long travelling paths if these terranes collided with the continental margin in Maastrichtian time (70 Ma) (Fig. 13a), and (2) short travelling paths if these terranes collided with the margin in Hauterivian time (120 Ma) (Fig. 13b). In both cases we see three major blocks with different characteristics within the ancestral Pacific Ocean basin. Small blocks like the Vatyn terrane in Olutorovsk Peninsula and the Eastern Kamchatka terrane originated in mid- and Late Cretaceous time, respectively, and have an intra-oceanic origin.

Reconstructions of the Pacific and Kula plates relative to Eurasia at 130, 110 and 70 Ma are shown on Figures 14, 15 and 16 respectively. We used the long travel path of Koryakia. The position of the Shatsky Rise on the Pacific plate is also shown. It is possible to speculate that the Vatyn terrane was a part of Shatsky Rise that split off and moved with the Kula plate, leaving behind the Shatsky Rise on the Pacific plate. Koryakia and Okhotia collided with the Eurasian margin 70 Ma ago. The position of the Emperor Fracture Zone is restored on the reconstruction (Fig. 16) and the dashed line shows the possible continuation of the zone northwestwards with offset of the spreading ridge. It separates Vatyn and East Kamchatka terranes.

The positions of continents, restored within the absolute frame at 130 Ma, are shown in Figure 17. The three terranes, Kolymia, Koryakia (shown according to the long way version) and Okhotia, were situated at this time in the center of the Pacific at a distance of 3000 km from the mid-oceanic ridge. Speculatively, if we accept the same half rate for plate motion as from 130 to 110 Ma ago, that is about 5 cm/year, it takes 60 million years to close the space between these terranes and the spreading ridge. Possible positions of the terranes are shown on the absolute reconstruction for 190 Ma (Fig. 18). What were their origins and provenances? Possibly they were derived from the Antarctic part of Gondwanaland. Regardless, it appears that this landmass travelled more than 10,000 km during the last 130 Ma.

Acknowledgments. This work was made within the framework of the International Lithosphere Program WG-2. We thank very much R. Price, R. Van der Voo and J. Monger for fruitful discussions, and E. Irving for helpful criticism.

References

Aleksandrov, A. A., Nappe structures in the Koryakyaupland, M., Nauka, 1978 (in Russian).

Aleksandrov, A. A., N. A. Bogdanov, S. A. Palandjan et al., Tectonics of the north part of the Koryakya upland, Olutorovsk zone, Geotectonics, 2, 111-122, 1980 (in Russian).

Alekseev, Ed. S., Main features of the evolution and the structure of the south part of Koryakya upland, Geotectonics, 1, 85-95, 1979 (in Russian).

Ben-Avraham, Z., A. Nur, D. Jones, and A. Cox, Continental accretion orogeny: from oceanic plateaus to allochthonous terranes, Science, 213, 47-54, 1981.

Churkin, M. Jr., Paleozoic marginal ocean basin-volcanic arc systems in the Cordilleran foldbelt, in: Modern and Ancient Geosynclinal Sedimentation, Dott, R. H., and R. H. Shaver (eds.), Soc. Econom. Paleontologists Mineralogists, Spec. Pub., 19, 174-192, 1974.

Clague, D. A., and R. D. Jarrard, Tertiary Pacific plate motions deduced from the Hawaiian - Emperor Chain, Geol. Soc. Amer. Bull. 84, 1135-1154, 1973.

Coney, P. J., D. L. Jones, and J. W. H. Monger, Cordilleran suspect terranes, Nature, 288, 329-333, 1980.

Cooper, A., D. W. Scholl, and M. S. Marlow, Plate tectonic model for the evolution of the eastern Bering Sea basin, Geol. Soc. Amer. Bull., 87, 1119-1126, 1976.

Cox, A., A new method for obtaining poles from heterogeneous paleomagnetic data from oceanic plates, EOS AGU, 56, 1110, 1974.

Cox, A., and R. G. Gordon, Paleomagnetic evidence for large Tertiary motion of Marie Byrd Land (West Antarctica) relative to East Antarctica, EOS AGU, 59, 1059, 1978.

Cox, A., and R. G. Gordon, Paleolatitudes determined from paleomagnetic data from vertical cores, Rev. Geophys. Space Phys., 22, 47-72, 1984.

Crough, S. T., M. J. Morgan, and R. B. Hargraves, Kimberlites: their relation to mantle hotspots, Earth Planet. Sci. Lett., 50, 260-274, 1980.

Duncan, R. A., Hotspots in the southern oceans - an absolute frame of reference for motion of the Gondwana continents, Tectonophysics 74, 29-42, 1981.

Engebretson, D. C., A. Cox, and R. G. Gordon, Relative motions between oceanic plates of the Pacific Basin, J. Geophys. Res. 89, B12, 10291-10310, 1984.

Erickson, B. H., D. K. Rea, and F. P. Naugler, Chinook trough: a probable consequence of north-south sea floor spreading, EOS AGU, 50, 633, 1969.

Francheteau, J., C. G. Harrison, J. G. Sclater, and M. Richards, Magnetization of Pacific seamounts: A preliminary polar curve for the Northeastern Pacific, J. Geophys. Res., 75, 2035-2061, 1970.

Geodekjan, A. A., G. B. Udintsev, B. V. Baranov et al., Bedrock within the central Sea of Okhotsk region, Soviet Geology, 6, 12-31, 1976 (in Russian).

Gordon, R. G., The Late Maastrichtian paleomagnetic pole of the Pacific plate, Geophys. J. Roy. Astr. Soc., 70, 129-140, 1982.

Gordon, R. G., Late Cretaceous apparent polar wander of the Pacific plate: evidence for a rapid shift of the Pacific hotspots with respect to the spin axis, Geophys. Res. Lett., 10, 709-712, 1983.

Gordon, R. G., and A. Cox, Calculating paleomagnetic poles from oceanic plates, Geophys. J. Roy. Astr. Soc., 63, 619-640, 1980.

Harrison, C. G., R. D. Jarrard, V. Vacquier, and R. L. Larson, Paleomagnetism of Cretaceous Pacific seamounts, Geophys. J. Roy. Astr. Soc., 42, 859-882, 1975.

Hilde, T. W. C., N. Isezaki, and J. M. Wageman, Mesozoic sea-floor spreading in the North Pacific, Am. Geophys. Union, Geophys. Monogr., 19, 205-226, 1976.

Jones, D. L., N. J. Silberling, and J. W. Hillhouse, Wrangellia: A displaced terrane in northwestern North America, Can. J. Earth Sci., 14, 2565-2577, 1977.

Jones, D. L., N. J. Silberling, W. Gilbert, and P. J. Coney, Character, distribution and tectonic significance of accretional terranes in the central Alaska Range, J. Geophys. Res., 87, 3709-3717, 1982.

Khramov, A. N., (ed.), Paleomagnetology, Leningrad, Nedra, 312, 1982 (in Russian).

Kononov, M. V., Pacific plate absolute motion for the last 120 m.y., Oceanology, 293, 484-492, 1984a (in Russian).

Kononov, M. V., Absolute Pacific plate motion according to paleoclimatic, paleomagnetic data and hot spot trajectories for the last 120 m.y., Doklady Acad. Sci. USSR, 276, 6, 1442-1445, 1984b (in Russian).

Lancelot, Y., Relations entre evolution sédimentraire et tectonique de la plaque pacifique depuis le Crétacé inférieur, Mém. Sociéte Géologique France, 57, 134, 40, 1978.

Larson, R. L., and C. G. Chase, Late Mesozoic evolution of the western Pacific ocean, Geol. Soc. Amer. Bull., 83, 3627-3644, 1972.

Markov, M. S., G. E. Nekrasov, and S. A. Palandjan, Ophiolites and melanocratic basement of Koryakya upland, in: Precis of the Koryakia upland tectonics, Pyjarovskiy, Yu. M., and S. M. Tilman (eds.), M., Nauka, 30-69, 1982 (in Russian).

Molnar, P., T. Atwater, J. Mammerickx, and S. M. Smith, Magnetic anomalies, bathymetry and the tectonic evolution of the South Pacific since the Late Cretaceous, Geophys. J. Roy. Astr. Soc., 40, 383-420, 1975.

Monger, J. W. H., Upper Paleozoic rocks of the Western Canadian Cordillera and their bearing on Cordilleran evolution, Can. J. Earth Sci., 14, 1832-1859, 1977.

Nekrasov, G. E., Role of ultramafics, basic rocks and radiolarites in the evolution of Taigonos Peninsula and Penjima Ridge, Geotectonics, 5, 43-59, 1971 (in Russian).

Norton, I. O., and J. G. Sclater, A model for the evolution of the Indian Ocean and the break-up of Gondwanaland, J. Geophys. Res. 84, 12, 6803-6830, 1979.

Olivet, J. L., J. Bonnin, P. Beuzart, and MJ. M. Auzende, Cinématique de L'Atlantique Nord et Central, Brest, Rapports Scientifiques et Techniques Centre National pour L'Exploitation des Océans, 54, 110, 1984.

Parfenov, L. M., Mesozoic continental margins and island arcs of the North-East Asia, Novosibirsk Nauka, 192, 1984 (in Russian).

Peirce, J. W., Assessing the reliability of DSDP paleolatitudes, J. Geophys. Res., 81, 4173-4177 1976.

Raznitsyn, Yu. N., S. D. Sokolov, N. V. Tsukanov, and V. S. Vishnevskaya, Serpentinitic melange position in the structure of the east Kronotsky Peninsula (Kamchatka), Doklady Acad. Sci. USSR, 260, 6, 1437-1438, 1981 (in Russian).

Sager, W. W., An Eocene paleomagnetic pole for the Pacific plate, EOS AGU, 63, 445, 1982.

Sager, W. W., A late Eocene paleomagnetic pole for the Pacific plate, Earth Planet. Sci. Lett. 63, 408-422, 1983.

Sclater, J. G., and A. Cox, Paleolatitudes from JOIDES deep-sea sediment cores, Nature, 226, 934, 1970.

Stock, J., and P. Molnar, Uncertainties in the relative positions of the Australia, Antarctic, Lord Howe and Pacific plates since the Late Cretaceous, J. Geophys. Res., 1987, 4697-4714, 1982.

Woods, M. T., and G. F. Davies, Late Cretaceous genesis of the Kula plate, Earth Planet. Sci. Lett., 58, 161-166, 1982.

Zinkevich, V. P., Formations and periods of the tectonic evolution of the North Koryakia, M., Nauka, 107, 1981 (in Russian).

Zinkevich, V. P., O. V. Lyashenko, and V. M. Basmanov, Ophiolites of the Ozerny Peninsula (East Kamchatka), Koklady Acad. Sci., USSR, 270, 6, 1429-1432, 1983 (in Russian).

Zonenshain, L. P., Tectonics of the inner continental fold belts, in: 27 Inter. Geol. Congress Proc., Tectonics, M., Nauka, 48-59, 1984.

DEVELOPMENT OF THE CIRCUM-PACIFIC PANTHALLASSIC OCEAN DURING THE EARLY PALEOZOIC

Christopher R. Scotese

Institute for Geophysics, University of Texas, Austin, Texas 78751

Abstract. Though the Pacific plate is less than 200 million years old, the Circum-Pacific ocean basin (Panthallassic ocean basin) has probably been in existence since Precambrian times. During the early Paleozoic the Tasman, trans-Antarctic, and southern Andean margins of the Panthallassic ocean basin appear to have been the site of active subduction. This convergent system may have continued north into southeast Asia, northern China, and southern Siberia. Continental reconstructions for the Late Cambrian, Late Ordovician, Late Silurian, and Late Devonian times are presented from a "Panthallassic" point-of-view. Paleomagnetic, biogeographic, and paleoclimatic evidence supporting three different Late Devonian reconstructions is reviewed.

Introduction

Panthallassa, the "universal sea," was the name given by the ancient Greeks to describe the vast oceanic expanse surrounding the known world. In Wegner's (1929) scheme, Panthallassa became the primordial ocean, just as Pangea was the primordial continent. Our present understanding of Paleozoic plate motions, however, requires us to abandon the concept of Pangea as the primordial continent for we now know that Pangea did not exist during the Paleozoic. Pangea formed as a result of continental collisions that began in the middle Paleozoic, continued through the late Paleozoic, but were not completed until the Early Jurassic.

Panthallassa, whose modern descendant is the Pacific Ocean, certainly existed before Pangea, and may justifiably be called "primordial." As far back as we can see into the Paleozoic, the Circum-Pacific region appears to have been a large ocean basin bordered by a rim of continents. We do not have a clear picture, however, of the exact geometry of this ocean basin or the plates that composed it, for less attention has been given to Panthallassa, than has been given to Pangea.

This is due, in part, to the fact that there is no extensive oceanic crust of Paleozoic age, and it has been difficult to rigorously estimate both the width of oceans and the nature of oceanic plate boundaries during the Paleozoic. Indeed, the life histories of Paleozoic oceans have often been summarized by the simple epitaph, "It was born during rifting, and was consumed through continental collision." This ignorance arises both from the lack of direct evidence and from the fact that the tools that are required to resolve the pattern of Paleozoic plate evolution have not yet been sufficiently developed.

Ideally, these tools would include: 1) a complete paleomagnetic record for the major continental cratons, 2) a geologic data base comprising the age and distribution of subduction-related and rift-related volcanics and tectonism, 3) a comprehensive biogeographic data base, 4) a global summary of climatically sensitive lithofacies and finally, 5) an understanding of the forces that drive the plates that would allow us to predict plate motions given the changing geometry of plate boundaries.

At present, though the early Paleozoic paleomagnetic record of North America, Siberia, and Gondwana is relatively complete (Scotese et al., 1979), there are still large gaps in the apparent polar wander paths of Baltica, China, and southeast Asia. Biogeographic data, though useful, are often incomplete and anecdotal. The geologic and paleoclimatic databases have not been compiled, and though we have an increasingly accurate picture of the history of plate motions, we still do not understand the complex interplay of forces that drive the plates.

Despite the fact that the data and our understanding of the plate tectonic process are incomplete, the problem of early Paleozoic plate evolution is not completely unconstrained. This paper summaries what we do know, and highlights the existing areas of uncertainty and controversy. However, rather than present this information from the usual perspective of a "Pangea-centered" world, the figures and discussion will be given from a "Panthallassic" point-of-view.

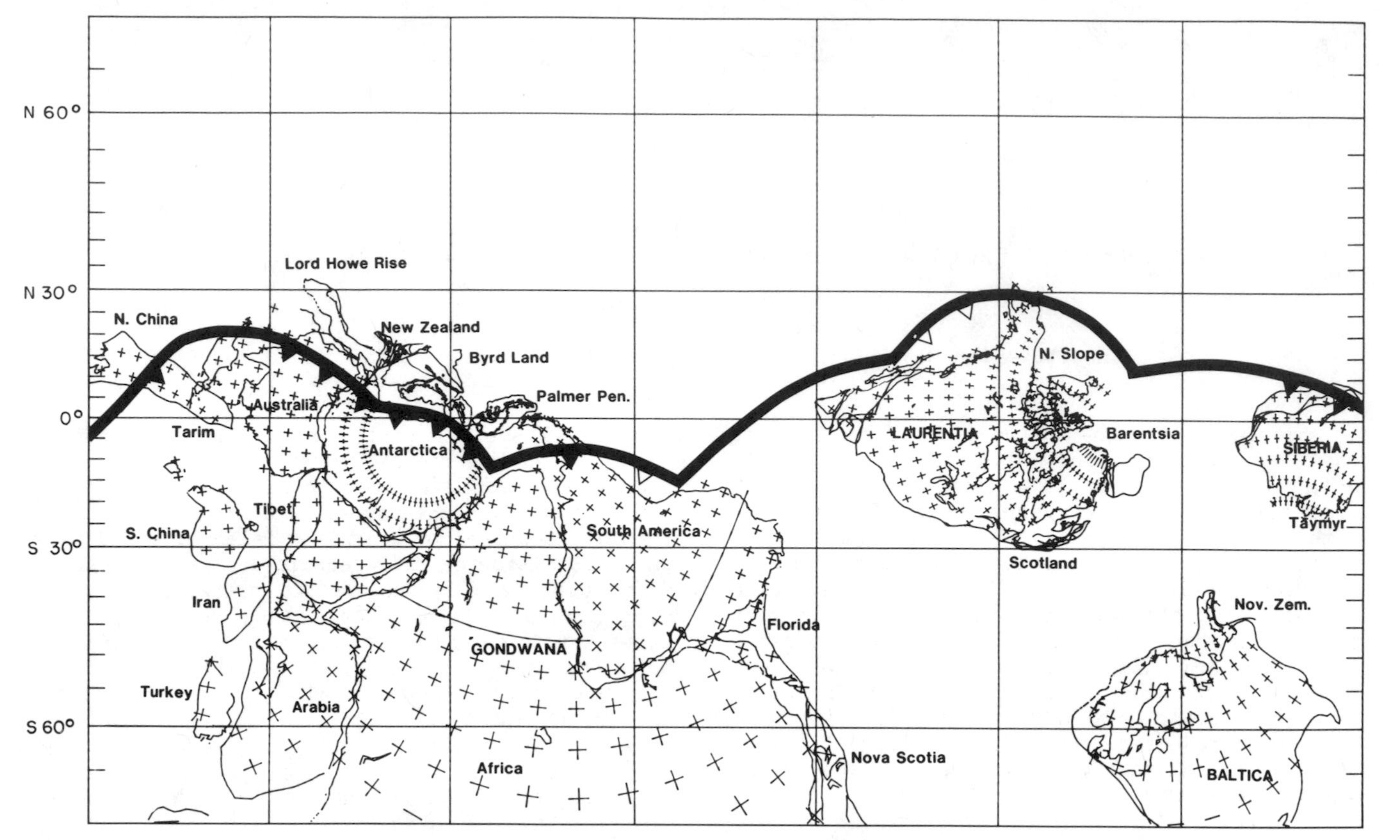

Fig. 1. Late Cambrian reconstruction showing Panthallassic Ocean occupying much of Northern Hemisphere. Thick black line is position of subduction zone from evidence of subduction related volcanics and plutonism (solid triangles) and probable island arc volcanics (open triangles) of early and middle Paleozoic age. Areas outside of subduction zone boundary represent terranes accreted after the early Paleozoic. Present-day geographic areas are labelled. Lines of latitude spaced at 30° intervals.

Defining the Extent of the Circum-Pacific Panthallassa

From a plate tectonic perspective, most oceans are defined by the lithospheric plates that carry them. In the case of Panthallassa, however, the oceanic plates comprising this ocean basin have been completely subducted, so an alternate tack must be taken. We choose to define the Panthallassic ocean basin by the continents that formed its perimeter. These continents, in most cases, are the same as those that border the modern Pacific Ocean. Proceeding clockwise, they include North America, the Gondwanan continents of South America, Antarctica, and Australia; probably southeast Asia and the two Chinese cratons, and completing the circuit, Siberia, which lay adjacent to North America.

As illustrated in Figures 1-6, during the early Paleozoic the Panthallassic Ocean was relatively constant in size. Spanning 200 degrees of longitude, it was over 20,000 km wide and covered nearly twice the area of the modern Pacific Ocean. Though Panthallassa remained relatively constant in size, its orientation with respect to the geographic pole changed gradually during the Paleozoic. During the early Paleozoic, the Panthallassic Ocean covered the entire northern hemisphere. Through the middle Paleozoic this hemispherical cap rotated southward, eventually occupying a meridionial orientation similar to the orientation of the modern Pacific Ocean.

Cambrian and Ordovician Configuration of Panthallassa (Figs. 1 and 2)

Figures 1 and 2 illustrate the relative positions of the continents during the Late Cambrian and Late Ordovician, respectively. The reconstructions have been rotated so that the Panthallassic Ocean occupies much of the Northern Hemisphere.

The position of Gondwana during the Cambrian and Ordovician is fairly well constrained by

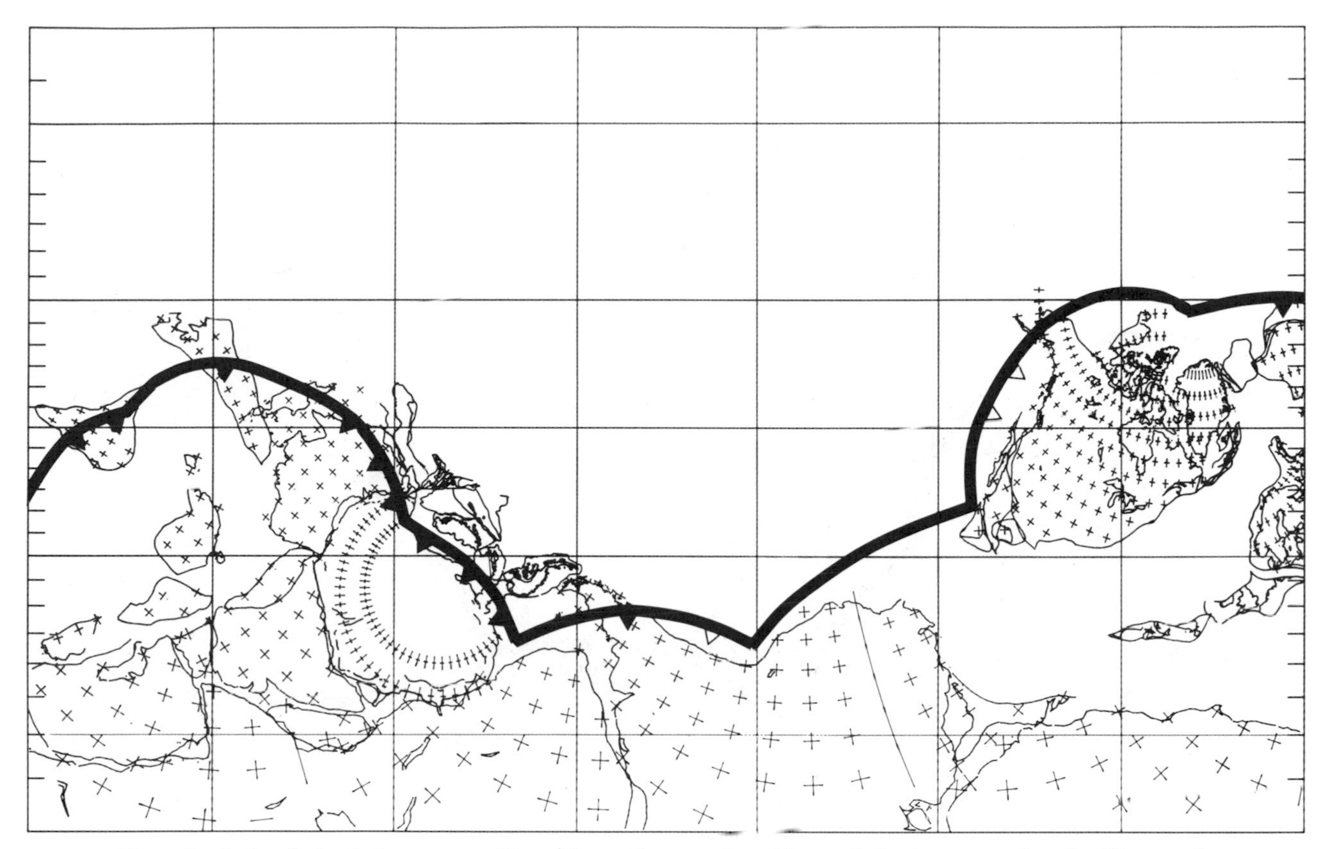

Fig. 2. Late Ordovician reconstruction. For explanation of features refer to Figure 1.

paleomagnetic and paleoclimatic data. Both lines of evidence indicate that the South Pole was in the vicinity of North Africa during the Late Cambrian, and moved steadily southward during the Ordovician (Scotese, 1984).

The best paleomagnetic data for this interval are from northwest Africa, South America, and Australia (Hailwood, 1974; Thompson and Clark, 1982). Paleomagnetic determinations place the South Pole at approximately 40°N, 13°E, during the Late Cambrian and at 30°N, 5°E during the Late Ordovician. (Pole positions in African coordinates). These results from Gondwana-proper are confirmed by recent paleomagnetic results from western France (Armorica) that indicate these areas were located at near polar latitudes during the Late Ordovician (450 Ma; Perroud and Van der Voo, 1985).

During the Cambrian and Ordovician, carbonates and evaporites occur throughout a broad belt that stretched from northern India, across Australia and Antarctica, and into southernmost South America (Ronov et al., 1984). In Australia, an east to west progression of evaporite basins can be seen as the continent appears to have rotated across the equator from the dry northern subtropics to the dry southern subtropics. Probably the most significant paleoclimatic indicators, however, are the Late Ordovician tillites that occur throughout the northern half of Gondwana, from the Sahara (Biju-Duval et al., 1981) north to south-central Europe (Dore, 1981; Hamaumi, 1981; Robardet, 1981), and eastwards to Arabia (McClure, 1978).

Extensive carbonate platforms covering much of North America, Siberia, and the South China platform confirm paleomagnetic evidence that indicates that these continents occupied equatorial latitudes during the Cambrian and Ordovician (Van der Voo, 1986; Lin et al., 1983). North China and southeast Asia have been placed adjacent to northeastern Australia, though the exact position of these areas is not well constrained. Ordovician paleomagnetic data from the North China platform places it at latitudes of 10-30 degrees. However, because the polarity of this magnetic signature is not known, it is not clear whether North China lay north or south of the equator. The northern position of the North China craton that we have chosen, is consistent with our interpretation that the Chilian-Shan mobile belt may be a continuation of the Trans-Antarctic-Tasman subduction zone. Similarly, following the reconstruction of Audley-Charles (1983), I prefer

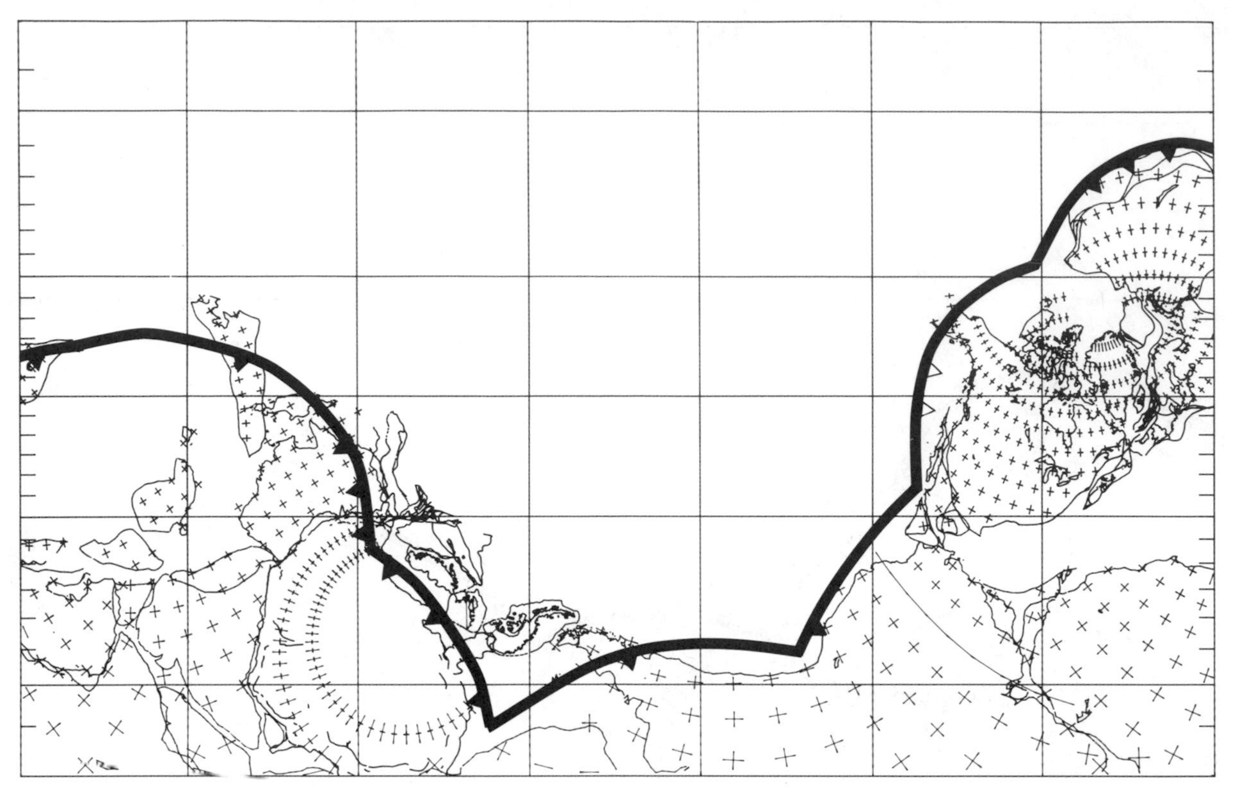

Fig. 3. Late Silurian reconstruction. For explanation of features refer to Figure 1.

to keep parts of southeast Asia adjacent to northern Australia during the early Paleozoic, despite the fact that paleolatitudes of 45 degrees have been determined from Ordovician limestones of the Malay peninsula (Haile, 1980).

Silurian and Devonian Configuration of Panthallassa (Figs. 3-6)

Figures 3 through 6 illustrate the changing configuration of the Panthallassic ocean basin during the Late Silurian and Late Devonian. Because of the uncertainty in the orientation of Gondwana, three alternate Late Devonian reconstructions are presented (Figs. 4,5,6). In all of these reconstructions, the extent of Panthallassa is approximately the same.

The uncertainty in the position of Gondwana is due to the fact that Siluro-Devonian paleomagnetic poles for Gondwana are widely scattered (Scotese et al., 1985) and that the better established paleomagnetic directions make paleo-latitudinal predictions that do not agree with available biogeographic and paleoclimatic evidence (Heckel and Witzke, 1979; Barrett, 1985). The South Pole, which was in the vicinity of North Africa during the Late Ordovician, had moved to a position adjacent to the southeastern coast of Africa by the end of the Carboniferous. Though the end points of this apparent polar wander path are well established, controversy continues to rage regarding the timing and trajectory of this transition.

Three independent lines of evidence may be used to constrain the orientation of Gondwana during the middle Paleozoic: 1) paleomagnetism, 2) biogeographic affinities, and 3) the distribution of climatically sensitive lithofacies. Unfortunately, for the middle Paleozoic, these criteria are not in agreement. In each of the reconstructions shown in Figures 4 through 6, one of these lines of evidence has been given precedence.

The position of Gondwana shown in Figure 4 is the orientation that has the best paleomagnetic support (Hurley et al., 1985). A paleomagnetic pole from an undated Morroccan intrusive, truncated by Famennian red beds (Hailwood, 1974) when combined with recent results from Late Devonian rocks of the Canning Basin (Hurley et al., 1985) and preliminary results from the Table Mountain Group, South Africa (Bachtadse and Van der Voo, 1984), indicates that during the Late Devonian the South Pole was located in equatorial Africa (3S, 15E; African coordinates).

This paleomagnetic pole places the northern margin of Gondwana at latitudes of 45-60 degrees south during the Late Devonian. This paleolatitudinal prediction is significant because by the Early Devonian, carbonates had reappeared along

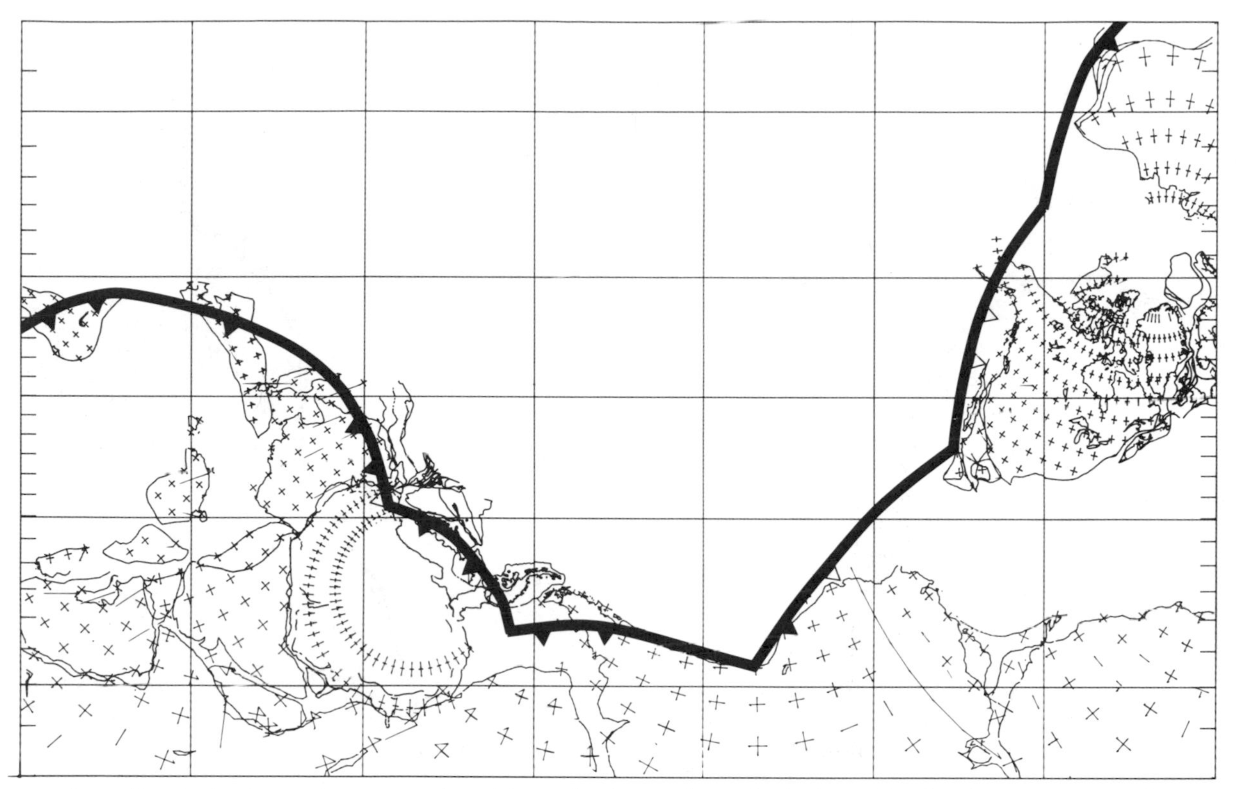

Fig. 4. Late Devonian reconstruction. Gondwana is oriented on the basis of a recently determined Late Devonian pole (Hurley et al., 1985). For explanation of features refer to Figure 1.

the northern margins of Gondwana, and by the Middle Devonian reefs were also present. When compared with the Mesozoic and Cenozoic distribution of carbonates and reefs, we find that carbonates rarely occur at such high latitudes. Only 7% of Mesozoic and Cenozoic carbonate localities occur above latitudes of 45 degrees; less than 1% occur above latitudes of 55 degrees (Ziegler et al., 1984).

Even more problematic is the occurrence of biostromes and carbonate buildups at these high latitudes. Due to the decrease in isolation and the increasing degree of seasonality, Mesozoic and Cenozoic reef building organisms do not range beyond 35 degrees of the equator (Ziegler et al., 1984; Johannes et al., 1983). Evidence of isolated carbonate buildups and coral wackestones along the northern margin of Gondwana during the early Middle Devonian (Wendt, 1985; Oliver 1980), and the extensive occurrence of reefs, stromatoporoids, and receptaculids (Heckel and Witzke, 1979) during Late Devonian, argues against a high latitudinal position for this margin.

Another important aspect of the orientation of Gondwana shown in Figure 4 is that it requires a wide seaway separating Gondwana from North America and Europe. Strong Early Devonian biogeographic affinities between Gondwana and North America (Eastern Americas Realm) and between Europe and Gondwana (Old World Realm) argue against a wide oceanic barrier (Boucot et al., 1969; Barrett, 1985).

The orientation of Gondwana shown in Figure 5 best fits the distribution of climatically sensitive lithofacies data (Scotese and Barrett, 1985; Caputo and Crowell, 1985). In this reconstruction the South Pole does not lie along the trans-African apparent polar wander path, but rather, is located in central Argentina (22S, 10E, African coordinates). This configuration brings the wide carbonate platform that bordered the northern and northeastern margin of Gondwana into tropical latitudes. Probable Devonian tillites described from Brazil (Rocha-Campos, 1981a,b) are positioned at latitudes of 60 degrees south.

The biogeographic data within Gondwana also matches this orientation. The low diversity, cold-water Malvinokaffric Realm occurs at high temperate and circum-polar latitudes on this reconstruction, and the high diversity Old World and Americas Realm faunas are located at low latitudes. A problem arises, however, because this orientation of Gondwana, as in the case of the orientation shown in Figure 4, does not permit a match between equivalent faunal realms in Gondwana and Laurasia (North America and

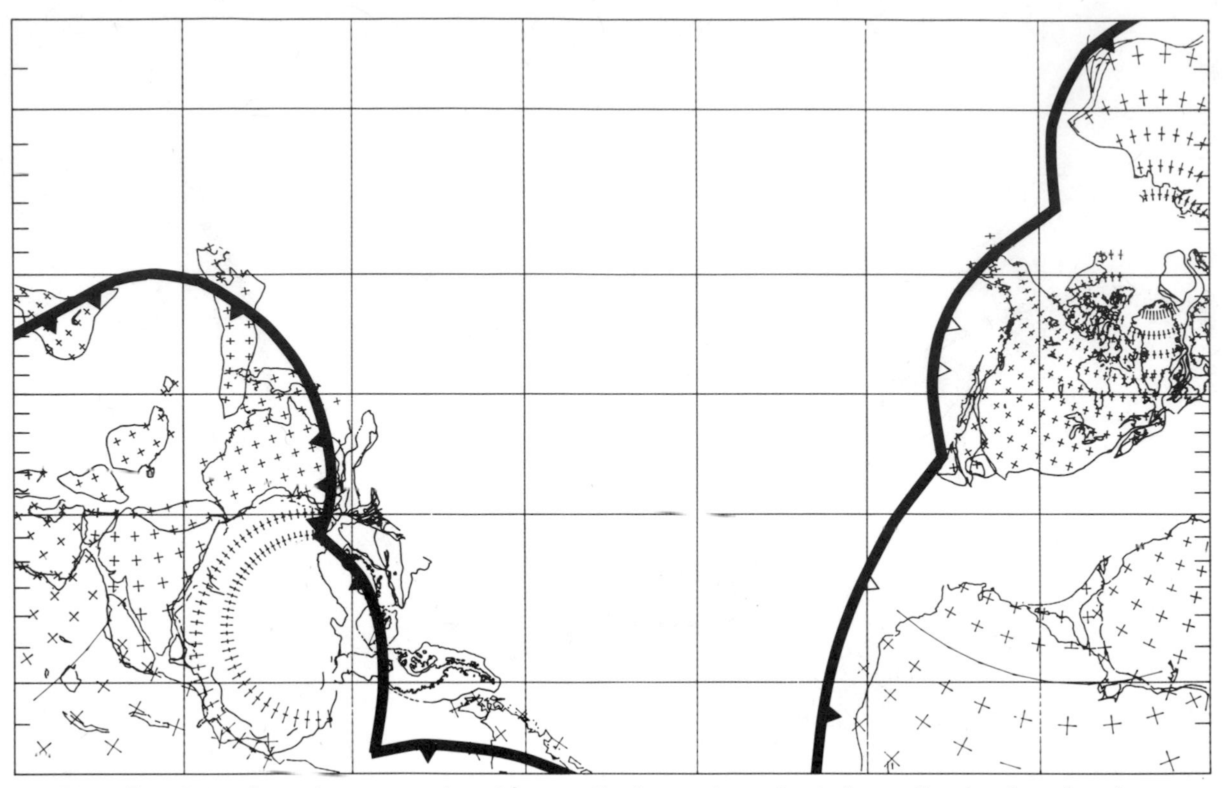

Fig. 5. Late Devonian reconstruction. Gondwana is oriented on the basis of paleoclimatic indicators (Scotese and Barrett, 1985). For explanation of features refer to Figure 1.

Europe). This problem may be resolved if North America and Europe are positioned slightly further south. At present, the Late Devonian position of North America is uncertain. The paleomagnetic results from the Late Devonian red beds of the Catskill Mountains (Kent and Opdyke, 1978; Van der Voo et al., 1979), which previously have been used to orient North America (Scotese et al., 1979, 1985, Scotese, 1984) are now thought to represent a late Paleozoic remagnetization (Van der Voo, 1986). The position of North America shown on the Late Devonian reconstructions (Fig. 4 and 5) is an interpolation between control points in the Early Devonian and Early Carboniferous, and therefore, may be in error.

The paleomagnetic support for the orientation of Gondwana shown in Figure 5 is equivocal. There are several paleomagnetic poles that plot in the vicinity of south-central Argentina, however they have been regarded by most authors as unreliable due to the fact that they are based on results from tectonically unstable areas (Tasman Orogen). A recent result from Middle Devonian volcanics of the Tasman orogenic belt (Comerong volcanics, New South Wales; Schmidt et al., 1986), however, confirms these earlier determinations, and places the South Pole in the vicinity of east-central Argentina (34S, 2E, African coordinates). The agreement between these paleomagnetic poles, and an estimation of the position of the geographic pole calculated from the distribution of paleoclimatic data (Scotese and Barrett, 1985), suggests that the earlier paleomagnetic results from the Tasman orogenic belt may be more accurate than previously thought.

The position of Gondwana shown in the final Figure (6) is based on a questionable paleomagnetic pole from Moroccan red beds (Kent et al., 1984) and a reasonably reliable pole from eastern Australia (Mulga Downs; Embleton, 1977). The Moroccan pole, as in the case of the Catskill red beds, is now thought to be a secondary remagnetization. This reassembly produces a "premature" Pangea-like configuration that best explains the disjunct distribution of biogeographic data, but is only in fair agreement with the paleoclimatic indicators.

Summary and Conclusions

In this review, I have attempted to present a framework from which we can begin to consider the evolution of the Circum-Pacific region from a plate tectonic perspective. By mapping the positions of the continents that bordered Panthallassa, we can get a rough idea of the changing shape and size of this ocean basin during the early and middle Paleozoic. It is clear, however,

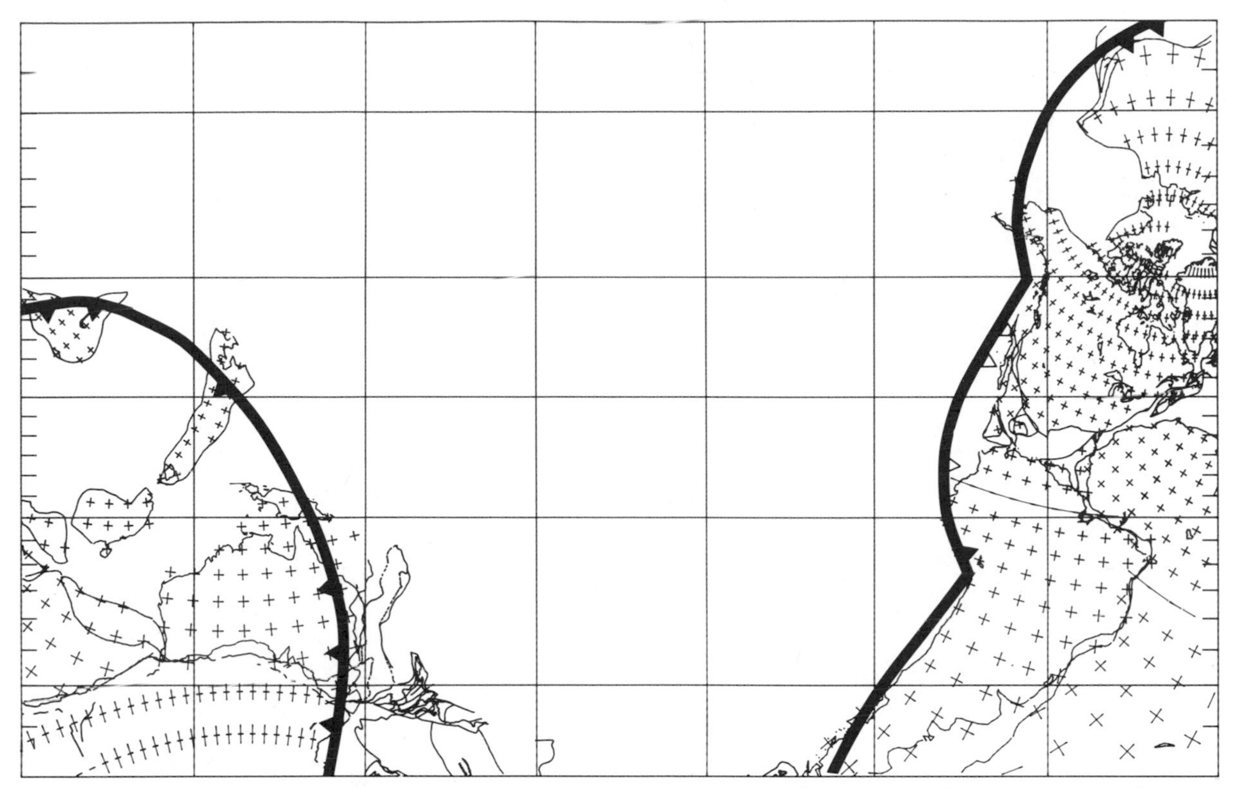

Fig. 6. Late Devonian reconstruction. The orientation of Gondwana is consistent with biogeographic data that indicates the ocean between Gondwana and North America was nearly closed. For explanation of features refer to Figure 1.

that the size and shape of the Panthallassic ocean basin ultimately must have been controlled by the changing geometry of the plate boundaries within it, and along its perimeter. In order to understand the early and middle Paleozoic plate tectonic development of Panthallassa, several important questions remain to be answered: 1) Where were the early Paleozoic margins of Panthallassa? 2) When did these margins form? and 3) What is the evidence for subduction along these margins? At present we can only begin to answer these questions.

The most unusual aspect of the Panthallassic/Pacific ocean basin is its size and stability through Phanerozoic history. For approximately the past billion years there has been an ocean basin, roughly twice the size of the modern Pacific, bounded by a ring of subduction zones. The apparent stability of this configuration may be telling us something important about the dynamics of the plate tectonic process. To maintain an ocean basin of this extent, there must have been delicate balance between the forces of subduction and the forces of sea floor spreading. How was this equilibrium maintained? What special plate geometries were required? At present we can only speculate. However, by carefully building an accurate model of plate evolution during the past 600 million years we may be able to recognize the broader patterns of plate evolution and arrive at a new understanding of the processes that drive the plate tectonic system.

Acknowledgments. I would like to thank Rob Van der Voo for suggesting this paper, and Jim Monger for his support and encouragement.

References

Audley-Charles, M. G., Reconstruction of eastern Gondwanaland, Nature, 306, 48-50, 1983.

Bachtadse, V., and R. Van der Voo, Paleomagnetic results from the Lower Devonian part of the Table Mountain Group, South Africa, (abs.), EOS, Trans. AGU, 65, 863, 1984.

Barrett, S. F., Early Devonian continental positions and climate: a framework for paleophytogeography, in: Geological Factors and the Evolution of Plants, Tiffany, B., ed., Yale Univ. Press, New Haven, 1985.

Biju-Duval, B., M. Deynoux, P. Rognon, Late Ordovician tillites of the Central Sahara, in: Earth's pre-Pleistocene Glacial Record, Hambrey M. J., and W. B. Harland, eds., 99-107, Cambridge Univ. Press, London, 1981.

Boucot, A. J., J. G. Johnson, and J. A. Talent, Early Devonian brachiopod zoogeography, Geol. Soc. Amer. Spec. Pap., 119, 197 p., 1969.

Caputo, M. V., and J. C. Crowell, Migration of glacial centers across Gondwana during the Paleozoic Era, Geol. Soc. Amer. Bull. 96, 1020-1036, 1985.

Dore, F. The Late Ordovician tillites in Normandy (Armoricain Massif) in: Earth's pre-Pleistocene Glacial Record, Hambrey, M. J., and W. B. Harland, eds., 99-107, Cambridge Univ. Press, London, 1981.

Embleton, B. J. J., A Late Devonian paleomagnetic pole for the Mulga Downs group, western New South Wales, J. and Proc. Roy. Soc. New South Wales, 110, 25-27, 1977.

Haile, N. S., Paleomagnetic evidence from Ordovician and Silurian rocks of northwest peninsular Malaysia, Earth Planet. Sci. Lett., 48, 233-236 1980.

Hailwood, E. A., Paleomagnetism of the Msissi norite (Morocco) and the Paleozoic reconstruction of Gondwanland, Earth Planet. Sci. Lett., 23, 376-386, 1974.

Hamaumi, N., Analyse sedimentologique des formations de l"Ordovicien Superieur en Presqu'ile de Crozon (Massif Armoricain), Ph.D. thesis, Brest, 1981.

Heckel, P. H., and B. J. Witzke, Devonian world paleogeography determined from the distribution of carbonates and related lithic paleoclimatic indicators, in: The Devonian System, House, M. R., C. T. Scrutton, and M. G. Bassett, eds., Spec. Pap. Palaeon., 23, 1-353, 1979.

Hurley, N. F., V. Bachtadse, M. Ballard, and R. Van der Voo, Early and Mid-Paleozoic paleomagnetism of South Africa and Australia - new constraints for the Gondwana apparent polar wander path (abs.), Sixth Gondwana Sym., 19-23 Aug., 1985, Ohio State Univ., Misc. Pub. 231, Instit. Polar Studies, Columbus, Ohio, 1985.

Johannes, R. E., W. J. Wiebe, C. J. Crossland, D. W. Rimmer, and S. V. Smith, Latitudinal limits of coral reef growth, Mar. Ecol. Prog. Serv., 11, 105-111, 1983.

Kent, D. V., O. Dia, and J. M. A. Sougy, Paleomagnetism of Lower-Middle Devonian and Upper Proterozoic-Cambrian (?) rocks from Mejeria (Mauritania, West Africa), in: Plate Reconstruction from Paleozoic Paleomagnetism, Van der Voo, R., C. R. Scotese, and N. Bonhommet, eds., Geodynam. Series, 12, 1-10, 1984.

Kent, D. V., and N. D. Opdyke, Paleomagnetism of the Devonian Catskill red beds: evidence for motion of coastal New England-Canadian Maritime region relative to cratonic North America, J. Geophys. Res., 83, 4441-4450, 1978.

Lin, J.-L., M. Fuller, and W.-Y. Chang, Position of the South China block on the Cambrian world map (abs.), EOS. Trans. AGU., 64, 320, 1983.

McClure, H. A., Early Paleozoic glaciation in Arabia, Palaeogeog., Palaeoclim., Palaeoecol., 25, 315-326, 1978.

Oliver, W. A. Corals in the Malvinokaffric Realm, Munster. Forsch. Geol. Palaont., 52, 13-27, 1980.

Perroud, H., and R. Van der Voo, Paleomagnetism of the late Ordovician Thouars massif, Vendee province, France, J. Geophys. Res., 90, 4611-4625, 1985.

Robardet, M., Late Ordovician Tillites in the Iberian peninsula, in: Earth's pre-Pleistocene Glacial Record, Hambrey, M. J., and W. B. Harland, eds., 99-107, Cambridge Univ. Press, London, 1981.

Rocha-Campos, A. C., Late Devonian Curua formation, Amazon Basin, Brazil, in: Earth's pre-Pleistocene Glacial Record, Hambrey, M. J., and W. B. Harland, eds, 99-107, Cambridge Univ. Press, London, 1981a.

Rocha-Campos, A. C., Middle-Late Devonian Cabecas formation, Parnaiba basin, Brazil, in: Earth's pre-Pleistocene Glacial Record, Hambrey, M. J., and W. B. Harland, eds, 99-107, Cambridge Univ. Press, London, 1981b.

Ronov, A., V. Khain, and K. Seslavinsky, Atlas of Lithological-Paleogeographical Maps of the World, Late Precambrian and Paleozoic Continents, 1-70, Leningrad, 1984.

Schmidt, P. W., B. J. J. Embleton, T. J. Cudahy, and C. McA. Powell, Prefolding and premegakinking magnetizations from the Devonian Comerong volcanics, New South Wales, Australia, and their bearing on the Gondwana pole path, Tectonics, 5, 135-150, 1986.

Scotese, C. R., Paleozoic paleomagnetism and the assembly of Pangea, in: Plate Reconstruction from Paleozoic Paleomagnetism, Van der Voo, R., C. R. Scotese, and N. Bonhommet, eds., Geodynam. Series, 12, 1-10, 1984.

Scotese, C. R., R. K. Bambach, C. Barton, R. Van der Voo, and A. M. Ziegler, Paleozoic Basemaps, J. Geology, 87, 217-277, 1979.

Scotese, C. R., and S. F. Barrett, Paleoclimatic constraints on the motion of Gondwana during the Paleozoic (abs.), Sixth Gondwana Sym., 19-23 Aug., 1985, Ohio State Univ., Misc. Pub. 231, Instit. Polar Studies, Columbus, Ohio, 1985.

Scotese, C. R., R. Van der Voo and S. F. Barrett, Silurian and Devonian base maps, Phil. Trans. Roy. Soc. Lond., B309, 57-77, 1985.

Thompson, R., and R. M. Clark, A robust least-squares Gondwanan apparent polar wander path and the question of paleomagnetic assessment of Gondwanan reconstructions, Earth Planet Sci. Lett., 57, 152-158, 1982.

Van der Voo, R., Paleomagnetism of Continental North America: the craton, its margins, and the Appalachian Belt, in: Geophysical Framework of the Continental United States, Pakiser, L. C., and W. D. Monney, eds., Geol. Soc. Amer. Mem. (in press), 1986.

Van der Voo, R., A. N. French, and R. B. French, A paleomagnetic pole position from the folded Upper Devonian Catskill red beds, and its tectonic implications, Geology, 7, 345-348, 1979.

Wegner, A., The Origin of the Continents and Oceans, John Biram, Trans. 1929, Dover, New York, (1966 reprint).

Wendt, J., Distribution of the continental margin of northwestern Gondwana: Late Devonian of the eastern Anti-Atlas (Morocco), Geology, 13, 815-818, 1985.

Ziegler, A. M., M. L. Hulver, A. L. Lottes, and W. F. Schmachtenberg, Uniformitarianism and Paleoclimates: Inferences from the distribution of carbonate rocks, in: Fossils and Climates, Benchley, P. J., ed., John Wiley and Sons, Chichester, 3-25, 1984.

CIRCUM-PACIFIC TECTOGENESIS IN THE NORTH AMERICAN CORDILLERA

Peter J. Coney

Department of Geosciences, Laboratory of Geotectonics,
The University of Arizona, Tucson, Arizona 85721

Abstract. Periods of prolonged compressive and transpressive deformation, magmatic patterns, and a certain measure of Cordilleran-wide synchroneity in major tectonic response, can be approximately correlated with what appear to be major near-global reorganizations and regimens of plate motion. It is also clear that almost 70% of the North American Cordillera is made up of suspect terranes, many of which are certainly allochthonous and somehow collided with and accreted against North America's Pacific margin. Best estimate times of collisions correlate in a general way with times of initiation of major periods of tectonism in large parts of the Cordillera. There is also evidence that "absolute" motion of the North American plate affected a high-stress regimen within an accretionary Late Jurassic through Eocene North American Cordillera Thus far, the only correlation identified with near Cordilleran-wide termination of Laramide orogeny and the unparalleled extension that followed it, was a marked change in the Pacific Plate motion vector near 40 Ma, and a slowing of the North American plate. Types of orogenic systems are a field continuum in which along-strike and temporal changes reflect differences of absolute and relative plate velocity vectors, size and nature of colliding and accreting masses, and relative lithospheric-crustal densities resulting from thermal disturbance, origin, and age. Collisions are accidents, but post-collisional consolidation includes the massive telescoping, rotations and large-scale translations that characterize Cordilleran tectonic evolution and are largely a poorly understood intraplate response.

Introduction

The North American Cordillera has long been recognized as part of the complex system of circum-Pacific orogenic belts. Since the beginnings of the plate tectonic concept it has seemed obvious that if there were any profound implications in the concept there should exist a demonstrable relationship between events recorded on the floor of the Pacific Ocean and the timing of tectonic events in the Cordilleran orogen. The idea originally was quite simple. Circum-Pacific orogeny was viewed mainly as a process of subduction of oceanic lithosphere beneath continental lithosphere with minor collisions of intra-oceanic arcs against the continental margins, occasional formation of small-ocean basins behind arcs as they migrated off-shore from those margins, and collapse of these small ocean basins, as possible variations. Actualistic examples from around the present-day Pacific Ocean seemed to be abundant and most of the tectonic history of the North American Cordillera was reconstructed using such models. There has been a strong effort to capitalize on the kinematic implications of plate tectonics and seek correlations between directions and magnitudes of plate interactions and the tectonic response on-land, and thus derive interpretative models of Cordilleran tectonic evolution based on perceived plate motions.

The quest for correlations between events recorded on the floor of the Pacific Ocean and events in the North American Cordillera (Atwater, 1970; Coney, 1971, 1972, 1976, 1978; Dickinson and Snyder, 1978; Engebretson et al., 1982) has not been entirely satisfying. There is a lack of precision in the oceanic sector (Stock and Molnar, 1982), and the difficulty in dating tectonic events within orogenic belts is well known. In spite of this there have been some insights, particularly regarding the Laramide orogeny and late Tertiary evolution of the San Andreas fault. On the other hand, when good correlations can be demonstrated, as in the above cases, it is not always clear what the correlations mean in terms of orogenic dynamics. Most recently the entire issue has been clouded by the notion that much of the North American Cordillera is made up of "suspect terranes" (Coney et al., 1980) which, besides modifying Cordilleran tectonic thinking, has questioned (Ben-Avraham et al., 1981) the basic ideas of Cordilleran mountain building as derived through classic plate tectonic theory (Dewey and Bird, 1970).

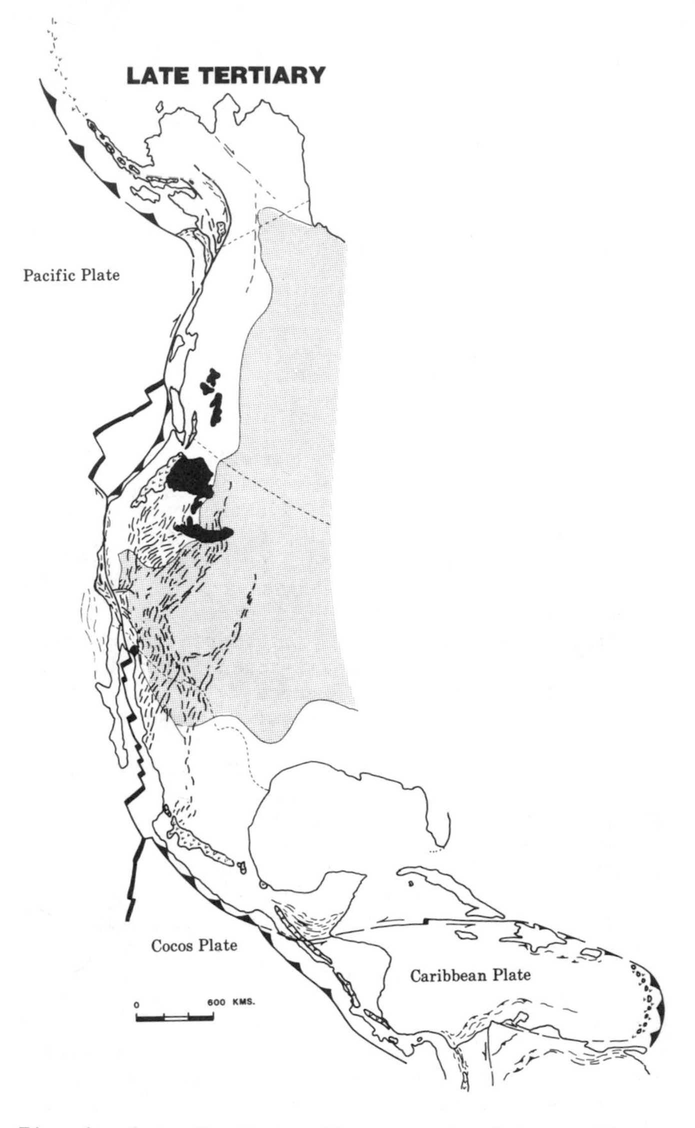

Fig. 1. Late Tertiary: The present plate configuration including structural and magmatic features representative of the last 17 million years. Heavy barbed lines show subduction zones, heavy dark lines show spreading centers. Finer dark lines indicate faults. V-pattern shows magmatic arcs. Dark areas are late Tertiary basalt provinces.

I will first review the broader aspects of Cordilleran tectonic evolution working backwards in time to the Jurassic. The review goes no farther back since identification of possible relationships between tectonics on land and marine data from the Pacific Ocean floor is not possible. In fact, little correlation is possible before Late Cretaceous time and north-east Pacific plate geometry is very uncertain before Eocene time. I will then review the record preserved on the cratonic foreland, inboard of the accreted terranes, since that geology, at least, is reasonably secure in terms of paleogeography and should record events taking place further west in the accreting ground. I will conclude with some general observations on Cordilleran tectonic evolution. The main point to be made is that the Cordillera is not a collision orogen, nor is it a simple Andean-type subduction orogen. Instead it is an accretionary orogen which has been strongly effected by large-scale intraplate tectonics.

In the following discussion a series of very generalized reconstructions are used which are more palinspastic than paleogeographic (Figs. 1-5). The positions shown for the various "suspect terranes" are simply representative of where and when they can first, with certainty, be accounted for with respect to the evolving North American margin, aftertaking into consideration existing paleomagnetic (May et al., 1982) and geologic data. The ultimate origins of the terranes and to what degree they are, or are not, far travelled are not discussed here.

Broad Outline of Cordilleran Tectonic Evolution Since the Jurassic

Late Tertiary (Middle Miocene and Younger)

It is appropriate to begin with the late Tertiary (Fig. 1) for it was the study by Atwater (1970) of the marine magnetic anomalies of the northeastern Pacific Ocean that was the first quantified step toward correlation of marine data with Cordilleran tectonic evolution. Her work remains convincing as she showed that the progressive extinction of the East Pacific rise replaced a trench regimen with a transform one as variably moving triple junctions migrated along the Cordilleran coast. Furthermore, the timing she mastered seemed to correlate with and explain the history of the San Andreas fault and the last 17 million years or so of intraplate extension (Smith, 1978), rotation, and translation that created the Basin and Range structural province (Stewart, 1978). Both volcanic geochemistry and tectonic timing and style (Eberly and Stanley, 1978) seemingly separate this period of extension from the middle Tertiary extension that preceded it. The Basin and Range pattern may have migrated northward following the shrinking Cascade arc trend (Dickinson and Snyder, 1979) and it may have waned in the south following the recent opening of the Gulf of California. In the far north no such widespread extension is recognized inboard of the Queen Charlotte transform fault off western Canada, but massive translation and rotation continues outboard in the Gulf of Alaska (Plafker et al., 1978; Lahr and Plafker, 1980). Intraplate deformation mainly south and west of, and along, the Denali fault system continues to transpressively wrench southern Alaska (Lanphere, 1978).

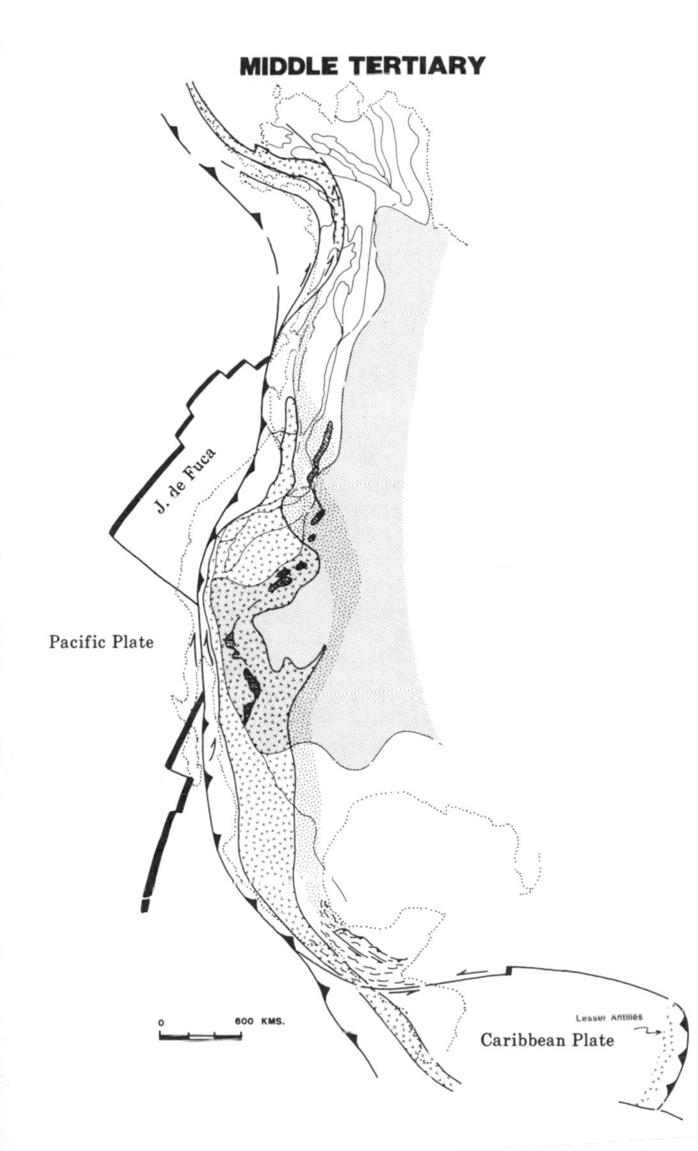

Fig. 2. Middle Tertiary: V-pattern is the middle Tertiary ignimbrite province and similarly aged rocks. The stippled pattern to the east represents similar eruptions of the Eocene. The dark-rayed areas locate the metamorphic core complexes. The dotted line on this and subsequent maps shows the present-day outline of the North American coastline for reference.

Middle Tertiary (Middle Eocene to Middle Miocene)

The middle Tertiary (Fig. 2) was a most peculiar time. It was characterized by intra-plate extension as recorded in the metamorphic core complexes and associated listric normal faults (Davis and Coney, 1979; Coney, 1979, 1980; Coney and Harms, 1984; Armstrong, 1982) and massive outbursts of generally caldera-centered ignimbrites (Elston, 1976). The extension-related phenomena spread southward and westward across the entire central and southern Cordillera, commencing in the Eocene in the Pacific Northwest (Armstrong, 1974) and continuing during the Oligocene-Miocene further south (Coney and Reynolds, 1977).

The direction of extension varies throughout the central Cordillera from northwest by west in the north, to southwest in the south, as recorded by the strike of listric normal faults and by the direction of a penetrative cataclastic mineral "stretching" lineation that is found in mylonitic gneiss of basement terranes of the metamorphic core complexes (see contents of Crittenden et al., 1980). The direction of extension usually closely parallels directions of assumed pre-middle Tertiary compressive slip-lines and thus reverses them, giving the impression that extensional middle Tertiary flow was toward the plate margin in the direction of least resistance. Gravitational collapse of a previously tectonically overthickened crustal welt is implied, which is probably aided by a reduction in crustal viscosity resulting from profound thermal disturbance recorded by extensive coeval ignimbrites (Coney and Harms, 1984). Plate convergence continued into the early part of the period, at least, and there is no reason why even deeper seated compression, beneath the extension, could not have occurred.

Nuclear Central America, by then part of the Caribbean plate, shifted eastward along the Mexican coast (Coney, 1978), a process that continues to the present day. An unstable triple junction between the Caribbean, Cocos, and North American plates, which probably formed in Eocene time, caused left transpressive deformation in the Chiapas highlands of southern Mexico.

Late Cretaceous to Early Eocene

From Late Cretaceous to early Tertiary time (Figs. 3,4) the entire Cordillera was gripped in a siege of compressive and transpressive intraplate deformation known as the Laramide orogeny (Coney, 1976; Dickinson and Snyder, 1978). The deformation reached over 1000 km inboard of the western edge of North America and, particularly in the central Cordillera, affected large areas of the foreland underlain by Precambrian cratonic basement. In Alaska and the northern Canadian Cordillera there was mainly massive translation and deep-seated transpressive deformation within a system of right-shear, which was associated with intraplate strike-slip faults such as the Denali, Tintina, and Kaltag fault systems. Further south in Canada and in the northern conterminous United States there was major low-angle thrust telescoping in the Rocky Mountains. In the United States a large expanse of cratonic shelf was compressed and sheared on crustal scales to produce the basement-cored fault-bounded uplifts of the classic Laramide Rocky Mountains and the enormous monoclines of the Colorado Plateau. In Mexico, a mainly upper Mesozoic carbonate and

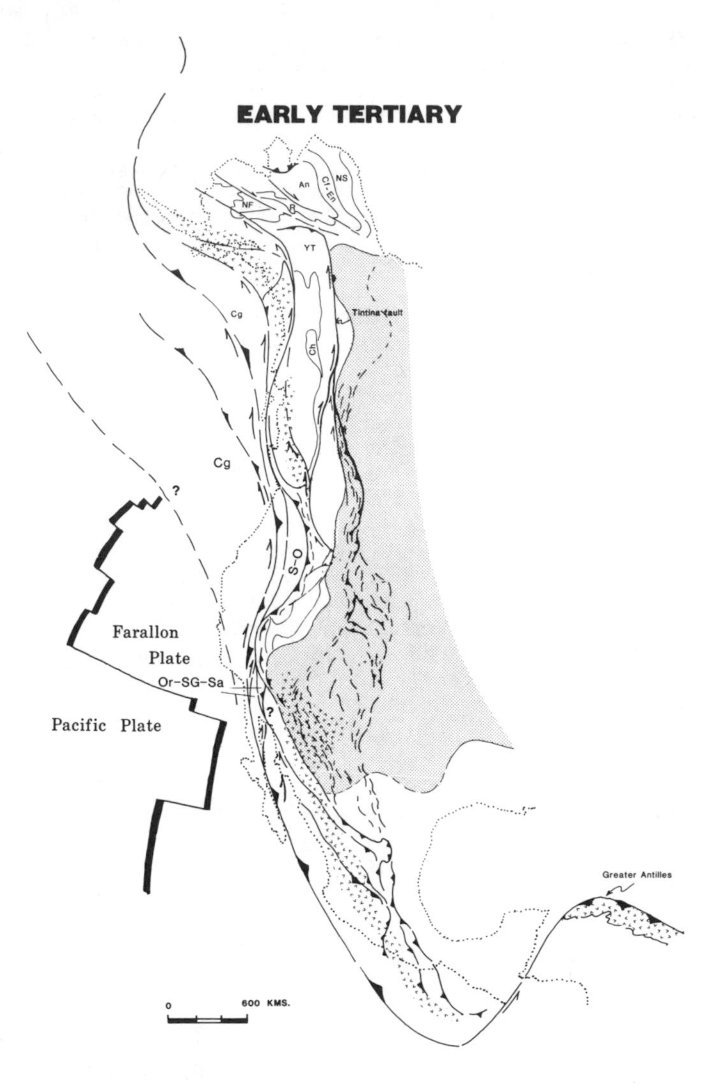

Fig. 3. Early Tertiary: The Laramide orogeny. Features shown are representative of the period from Late Cretaceous to Early Eocene. For terrane names see Figure 6.

clastic overlap assemblage, which had transgressed out of the opening Gulf of Mexico, was severely deformed by northeast-verging low-angle thrust faulting and décollement-style folding to produce the Sierra Madre Oriental.

Timing of the Laramide orogeny neatly corresponds with that of formation of the Emperor leg of the Hawaii-Emperor seamount chain, and the orogeny has long been inferred to correlate with vector circuit-derived very high rates (12-15 cm/yr) of northeast-southwest Farallon-North America relative plate convergence (Coney, 1978; Engebretson et al. 1982). The end of the Laramide orogeny and significant shifts in magmatic patterns (Coney and Reynolds, 1977), among other things, roughly correlate with the Hawaii-Emperor "elbow" dated near 45 Ma (Clague and Jarrard, 1973). The abrupt change in direction of Pacific plate motion implied by the "elbow" can be argued to have reduced Farallon-North America convergence rates and thus presumably reduced the stress-coupling. Probably more than anything else, all these collective correlations between events recorded on the Pacific Ocean floor and events recorded deep within, and along, the entire Cordillera have encouraged faith in the concept of linkage between plate kinematics and orogenic response in mountain chains.

The prolonged nature of deformation within the Cordilleran foreland during Laramide time is seemingly well recorded by subsidence in the intra-orogen Laramide basins within the Rocky Mountains of central western United States. An envelope above the various basin histories is similar to an envelope of Late Cretaceous-early Tertiary relative convergence rates between Farallon and North America plates suggesting, of course, a coupling of some sort.

Northward there was significant northeast vergent telescoping along listric low-angle thrust faults in the northern Rocky Mountains (Price, 1981), which possibly transforms northward into mainly right lateral strike-slip along the Tintina and related faults. Movement on these faults shifted large segments of western Canada and southern Alaska 450 to 1000 km or more northwest-

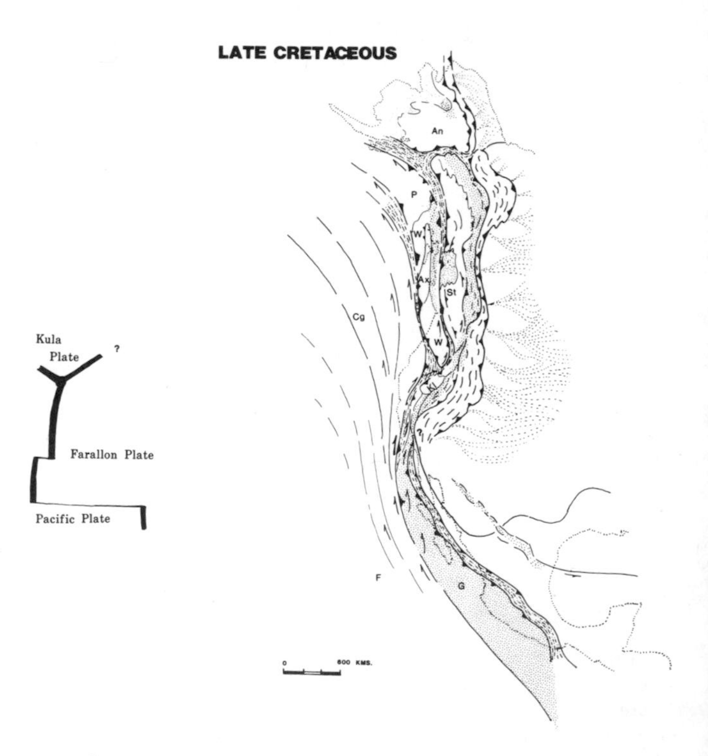

Fig. 4. Late Cretaceous: Accretion of Wrangellia-Guererro terranes. Heavy stipple shows magmatic belts. Dot-dash pattern shows flysch basins. Various finer dot-dash patterns on the foreland indicate the foreland basins.

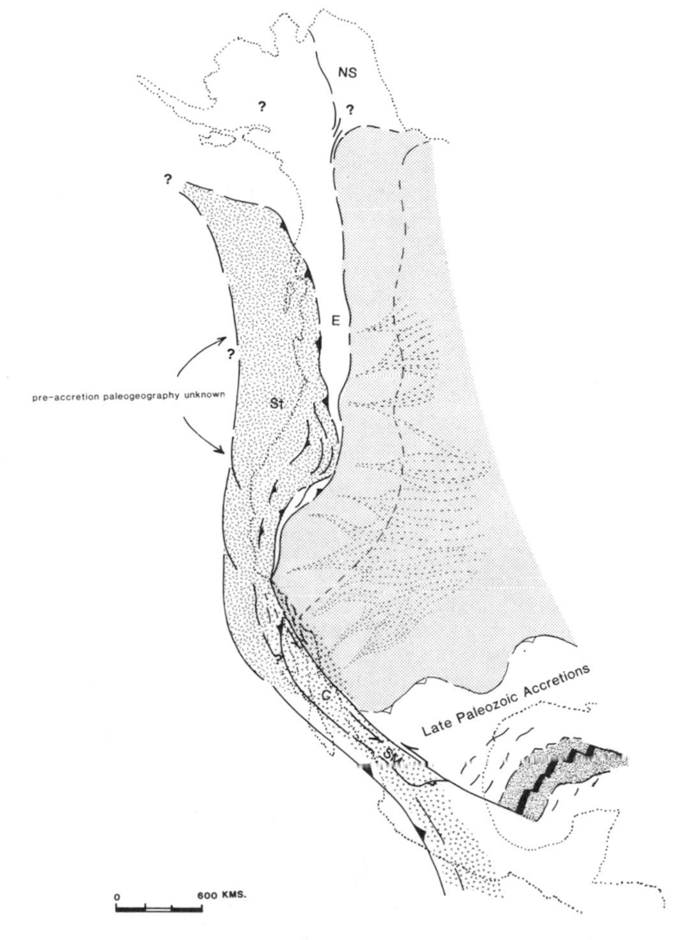

Fig. 5. Late Jurassic-Early Cretaceous: V-pattern shows Early to Middle Jurassic magmatic belts constructed upon North America while the dash pattern shows the Jurassic "arcs" on the amalgam of Stikinia, Cache Creek, Quesnellia and other terranes shown in Figure 6. The fine stipple and heavy dark lines represent Jurassic seafloor spreading in the Gulf of Mexico.

ward (Monger and Price, 1979; Tempelman-Kluit, 1979; Gabrielse, 1985).

It is in Cretaceous time that the presence of "Greater Wrangellia" can be recognized within the Cordilleran system. This is seemingly recorded in the north by Late Cretaceous closure of deep-marine flysch basins on its inboard side in southern and southeastern Alaska (Fig. 4; Coney et al., 1980, 1981; Jones et al., 1982) although closure to the south in southern British Columbia and northwestern United States may have been slightly earlier. Similarly, in Mexico the Guerrero terrane has definitively "emerged" by Late Cretaceous time (Campa and Coney, 1981, 1983). In fact, much of Laramide deformation seems to have been an intraplate consolidation of the Cordilleran foreland and various late Mesozoic "accretions" into the Cordillera over a period of 30 to 40 million years under a regimen of high rates of oblique convergence and high-stress coupling.

The Chugach terrane of southern Alaska was apparently far to the south as late as early Tertiary time (Plumley et al., 1982).

Jurassic-Early Cretaceous

Beginning in late Early Jurassic time, but mainly from Late Jurassic to mid-Cretaceous time, telescoping was widespread in central and northern Alaska and western Canada and perhaps only slightly less so in the western United States (Fig. 5). In Alaska, vast sheets of sedimentary and volcanic sequences of deep-marine oceanic affinity, and bearing fossils ranging in age from Late Devonian through Early Jurassic, were stacked as nappes over tectonic elements of continental affinity (Coney and Jones, in press). The continental "basement" terranes, which include the North Slope basement of the Brooks Range, are of obscure origins and cannot as yet be tied definitively to North America. In Canada, a vast amalgam or collage of late Paleozoic to early Mesozoic oceanic arcs and/or transtensional deep-marine rifts, pieces of ocean floor with Tethyan faunas, and other fragments of various sorts appears to be tied to North America by late Early Jurassic to Late Jurassic-Early Cretaceous time (Fig. 6; Monger et al., 1982; Saleeby, 1983). It has been proposed that the emplacement was by obduction over the ancient North American margin (Tempelman-Kluit, 1979). The original paleogeography of all this, particularly in pre- and syn-accretion time, is very obscure. Once accreted the collage was subjected to prolonged telescoping and major strike-slip dispersion which has continued up to the present time.

It has been suggested (e.g. see Monger et al., 1982) that much in the same way as Laramide deformation may have been initiated by the accretion of a "Greater Wrangellia", the deformation in the Middle Jurassic to Cretaceous was initiated by the accretion of what might be termed a "Greater Stikinia". In fact, precise control on exactly where or when either accretions first took place is poorly constrained. Large-scale, post-accretionary telescoping, translation, and rotation has obscured original relationships.

The projection of this belt of major accretion southward into western United States is still much discussed. The conservative choice is to place the suture west of the Klamath-Sierran basement terranes along the so-called Foothills suture zone (Saleeby, 1983). A more controversial option, as yet undocumented, would be to place the suture, now cryptic, in northwestern Nevada, east of the Sierra Nevada.

Nothing of this nature is presently known to the south in Mexico. In contrast, most of this region was suffering the effects of the Jurassic

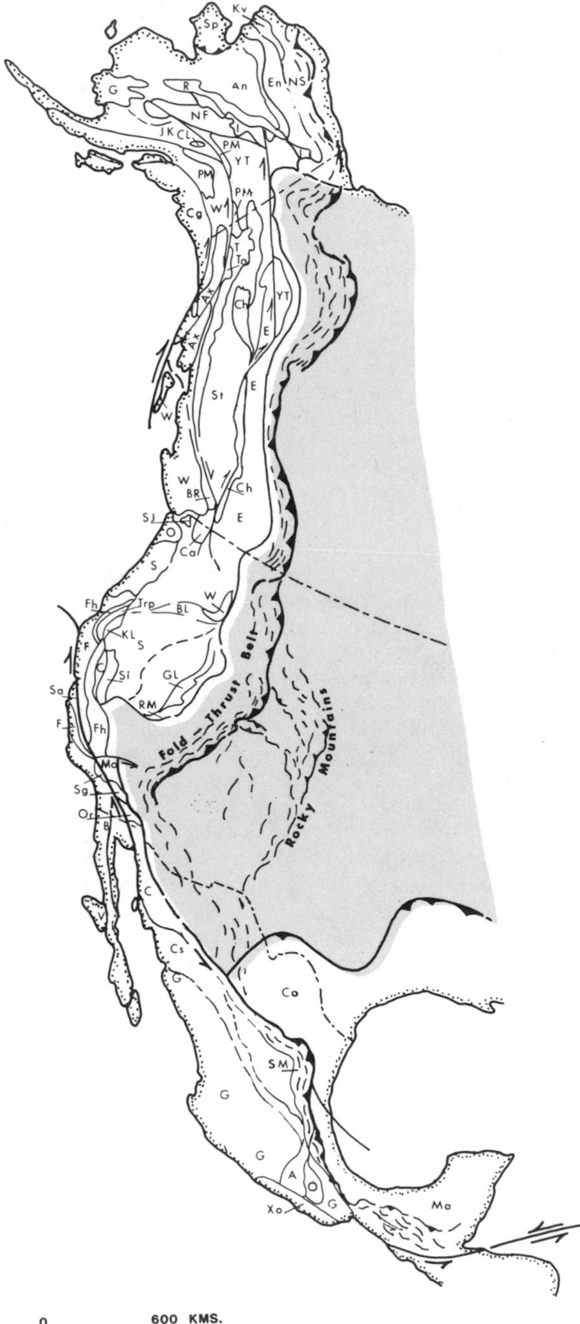

opening of the Gulf of Mexico. This opening may have transformed across Mexico as the Mojave-Sonora megashear (Silver and Anderson, 1974). The exact paleogeographic tie between the several Early to Middle Jurassic submarine "arc-like" magmatic belts which form part of the various suspect terranes (Stikinia, Quesnellia, etc. Fig. 6) in the north, with roughly similarly aged belts apparently constructed upon North American cratonic crust in Arizona and Mexico is still not known. Paleomagnetic data suggests North America was moving northwestward very rapidly during Middle to Late Jurassic time (O'Hare et al., 1982).

Accretion and the Cordilleran Foreland

Since we know most of the Cordilleran foreland is anchored to North America, it gives one some sense of security to examine the record of deformation there to see what evidence it holds for collision and accretion in the Cordilleran "hinterland".

The Cordilleran foreland consists of a belt of folds and thrust faults and associated foreland depositional basins that extends from Alaska to southern Mexico. This belt for the most part formed in miogeoclinal or shelf sequences of late Precambrian to early Mesozoic age that, except in Alaska, were deposited upon North American cratonic basement. This is not the arena for a detailed analysis of the belt but a few general observations are germane to this discussion. The principal point to be made is that deformation on the foreland apparently was characterized by prolonged periods of intraplate strain.

From the Alaska-Canada border south through Canada, down the northern Rocky Mountains into western Wyoming, and southwestward across western Utah and eastern Nevada, the belt is made up of what is usually termed "thin-skinned" eastward vergent folds and low-angle thrust faults (Price

Fig. 6. Cordilleran suspect terranes (from Coney, 1981). For terrane descriptions see Coney et al., 1980; Coney, 1981. Terrane names as follows: Alaska: NS-North Slope, En-Endicott, An-Angayuchum, R-Ruby, Kv-Kagvik, Sp-Seward Peninsula, G-Goodnews, NF-Nixon Fork, I-Innoko, JK-Upper Jurassic-Late Cretaceous flysch, Cl-Chulitna, PM-Pingston-McKinley, W-Wrangellia, P-Peninsular, Cg-Chugach, YT-Yukon-Tanana. Canada and southeastern Alaska: W-Wrangellia, Ax-Alexander, T-Taku, Ta-Tracey Arm, St-Stikinia, Ch-Cache Creek, BR-Bridge River, E-Quesnellia-Eastern assemblages. United States: O-Olympia, SJ-San Juan, S-Siletzia, Ca-Cascade, RM-Roberts Mountain GL-Golconda, BL-Blue Mountains, KL-Eastern Klamath, Trp-Triassic-Paleozoic, Fh-Foothills, F-Franciscan, C-Calaveras, Si-Sierran, S-Sonomia, Sa-Salinia, Or-Sg-Orocopia and San Gabriel, Mo-Mojave. Mexico: B-Baja, V-Vizcaino, Cs-Cortes, C-Caborca, G-Guererro, Co-Cohuilla, Ma-Maya, SM-Sierra Madre, A-Acatlan, O-Oaxaca, Xo-Xolapa.

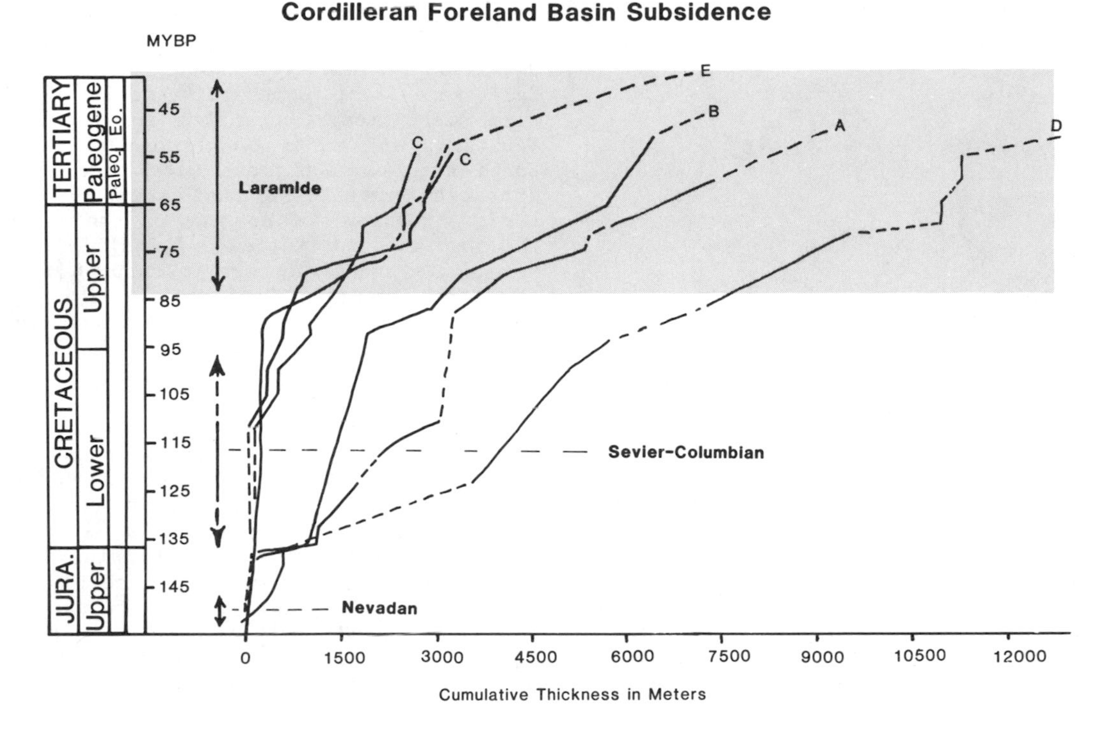

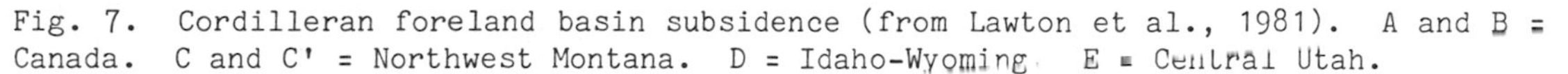

Fig. 7. Cordilleran foreland basin subsidence (from Lawton et al., 1981). A and B = Canada. C and C' = Northwest Montana. D = Idaho-Wyoming. E = Central Utah.

and Mountjoy, 1970; Armstrong, 1968). These structures preferentially evolved in the thick, late Precambrian through early Mesozoic, Cordilleran miogeocline. The deformation endured from Middle to Late Jurassic to Late Eocene time. West of the foreland folds and thrust faults is a series of crystalline culminations of uplifted poly-deformed, east-vergent metamorphic core complexes (Coney, 1979, 1980; Armstrong, 1982). Lying east of the thin-skinned belt in central and southwestern United States are the distinctive provinces of the Late Cretaceous to early Tertiary Laramide central and southern Rocky Mountains and the Colorado Plateau. These are characterized by generally arcuate basement-cored thrust-bound uplifts and monoclinal flexures surrounded by intermontane basins. Southward in Mexico, the Sierra Madre Oriental is northeast vergent and thin-skinned in structural style (Tardy, 1980). The Sierra Madre Oriental evolved during Laramide time and mainly deformed a moderately thick, Upper Jurassic to Late Cretaceous, carbonate shelf overlap assemblage which had transgressed out of the opening Gulf of Mexico (Campa and Coney, 1983). Characteristic of all the foreland is a striking set of sedimentary basins into which was shed debris eroded from rocks to the west, with their evolving thrust faults (Dickinson, 1976).

The prolonged thrusting on the Cordilleran foreland is seemingly recorded by cumulative subsidence in the foreland basins (Fig. 7). If subsidence in foreland basins is due to stacking of thrusts on their margins, as has been proposed (Price, 1973), thrust faulting in one place or another on the foreland endured for over 100 million years between Late Jurassic and Late Eocene time.

The Late Jurassic marks a profound turn-around in Cordilleran tectonic history. At that time the polarity of sediment dispersal in the central and northern Cordillera reversed from source areas generally to the east on the craton before Late Jurassic time, to source areas to the west in the Cordilleran hinterland after Late Jurassic time. This fundamental change apparently marks the time when crustal telescoping was sufficient to bring large areas of the accreted hinterland and the edge of the continental margin high enough above sea level to form source areas to the west. Thus, the most inboard of the "suspect" terranes must have either accreted or emerged by that time. The next 100 million years or so of Cordilleran tectonic history saw what appears to be nearly continuous intraplate telescoping on the Cordilleran foreland. It should be noted that this reversal of sedimentary polarity in Mexico is in the Late Cretaceous, and thus much younger. It also marks, however, the emergence of the first major accretion there, in this case of the Guerrero terrane. The important point, however, is that in both regions the telescoping

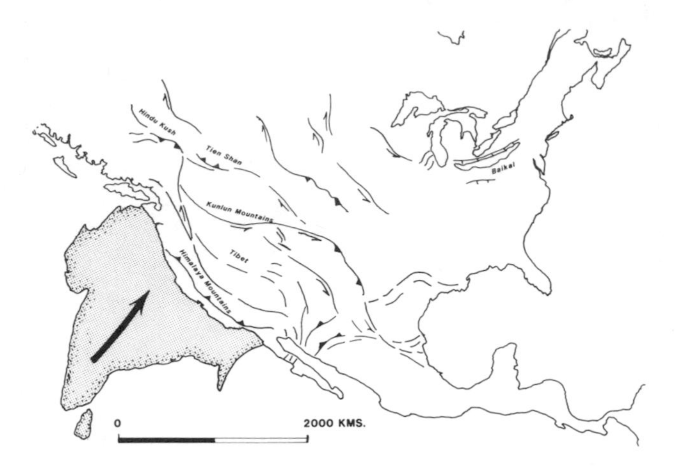

Fig. 8. The Himalayas and the Cordillera. The Indian subcontinent (shaded) is shown impinging on the North American coast. Both India and North America are at the same scale. The structural features shown and named are those north of India deployed at scale. Note that the Himalayas stretch from Seattle to San Diego. Note that, for the analogy to be complete, after 40 million years, India would have "indented" North America at least 400-800 km, placing its leading edge nearly to Salt Lake City.

recorded on the foreland and within the accreted hinterland seems to endure for millions of years after the time of initial accretion. In the central and northern Cordillera, deformation lasted 100 million years (Nevadan, Sevier-Columbian, Laramide), while in Mexico it lasted nearly 40 million years (Laramide).

The Cordillera as a Collision Orogen

Since development of the plate tectonic model, circum-Pacific orogeny has been viewed as a special type where subduction of oceanic lithosphere beneath a continental margin takes place, and is often contrasted with Alpine-Himalaya collision orogenies. At times, subduction beneath the continental margin was interrupted by brief collisions of intra-oceanic island arcs and/or collapse of back arc basins. The more recent concepts of "suspect" terranes and accretion tectonics in the circum-Pacific region in general, and in the North American Cordillera in particular, has clouded these perceptions and introduced the notion that all mountain systems are collisional in origin. This then raises the very important question as to whether the North American Cordillera has been produced by collision tectonics. Let us consider this from two points of view.

Figure 8 is a composite cartoon of the North American Cordillera with the present-day tectonics of the India-Asia collision superimposed on it at the same scale. India is shown impinging on the coast of California. Notice that the Himalayas stretch from about Seattle to southern California. More importantly, notice that strike-slip disruption, intraplate folding and thrust faulting, and rifting extend clear across the North American continent with Lake Baikal, for instance, positioned just south of the Great Lakes. Nothing, of course, on the scale, or of the crustal constitution, of India has ever been recognized among the myriad suspect terranes in the North American Cordillera. This in itself suggests that we are dealing with something other than classic collision orogeny. But let us look at this further by examining some kinematic matters.

The collision of India with Asia took place about 40 million years ago (Molnar and Tapponier, 1975), but the collision process has continued to the present day. Another way of phrasing this is that throughout the past 40 million years of collision between India and Asia, the Indian subcontinent has remained attached to, and a part of, the Australia-India plate, and subduction has not yet stepped south of India. As a result, buoyant India has indented Asia 400 to 800 km or more and has pushed much of Asia aside to the east causing massive disruption across thousands of square kilometres north of the Himalayas, far into and across China.

Figure 9 is a cartoon of the North American continent with the North American Cordillera outlined. Shown along the northwest coast of the United States is Wrangellia in a position where it might have first collided with the North American margin in Cretaceous (?) time. Also shown are the trajectories Wrangellia would have taken if it had remained attached to the Farallon or Kula plates (as India has remained attached to the Australia-India plate) based on reconstruction of the relative motions of these plates with respect to North America during the 40 million years or so of the Late Cretaceous to early Tertiary Laramide orogeny. The actual locations of the several fragments of Wrangellia are shown scattered up the Cordilleran margin from Idaho to southern Alaska. This suggests Wrangellia became detached from whatever oceanic plate it was a passenger on almost immediately upon collision. Once detached, with no inertia, it could have done no damage and certainly cannot explain 40 million years or so of progressive telescoping in the thrust belts of the Cordilleran foreland. These combined arguments suggest that calling the North American Cordillera a collision orogen (Mattauer et al., 1983) is an oversimplification of its tectonic history. Instead, the prolonged deformation in western North America must result from some combination of brief initial accretions followed by prolonged post-accretionary consolidation by telescoping, translation, and rotation resulting from what is perhaps best referred to as intraplate tectonics.

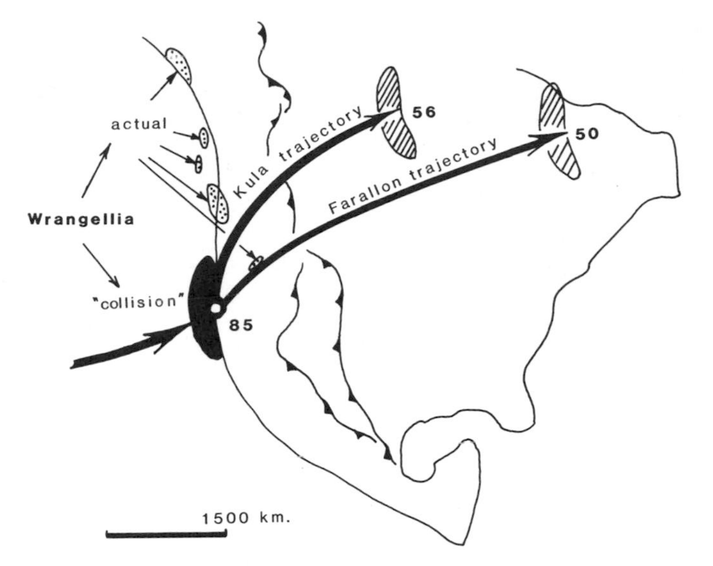

Fig. 9. The "collision" of Wrangellia. Wrangellia (solid object) is shown "colliding" with the Cordillera in Late Cretaceous time. The two heavy arrows show the trajectory and (lined objects) positions Wrangellia would have attained if it had remained part of the Farallon or Kula plates (as India has remained part of the Australia-India plate) during the time span of Laramide orogeny (±20-40 million years). The dotted objects scattered northward to Alaska are the present day locations of Wrangellia. Numbers in Ma.

Conclusions

If the telescoping and translations seen on the Cordilleran foreland are added to those seen within the accreted terranes, significant intraplate velocity gradients (Richardson, 1982) must have accumulated across the Cordillera during the prolonged periods of deformation that have been proposed. These intraplate velocity gradients imply that significant percentages of interplate relative motion are actually absorbed as intraplate strain rather than in normal subduction. They also imply disposal of commensurate lower crust and sub-crustal lithosphere by processes still unclear, but not accounted for by usual plate tectonic theory.

It seems that compressional orogenic systems fall in a field continuum (Coney, 1981) between two end-members represented by the Himalayas at one extreme and parts of the Andes at the other. In the case of the Himalayas, and much of China north of them, all the relative motion between the India-Australia and Asian plates is absorbed as intraplate telescoping and translation (Molnar and Tapponier, 1975). There is no plate boundary here in the classic sense. In the case of the Andes, the percentage of interplate relative motion between the Nazca and South America plates absorbed as intraplate strain is obviously much less, even though the telescoping there is not insignificant. In other words, intraplate telescoping and translation typical of mountain chains is due to a partial, to total, fusion of plate boundaries causing high-stress systems and widespread, prolonged intraplate compressive tectonic response. Some significant percentage of interplate motion is taken up as intraplate strain rather than subduction, which may be focussed in regions of lower plate viscosity.

The cause of the stress coupling, either high or low, has been debated for years. Some combination of the absolute motion of the overriding plate (Coney, 1973; Uyeda and Kanamori, 1979), rates of relative convergence, size and nature of accreting masses, and/or buoyancy of the subducting slab (Molnar and Atwater, 1978) seem to be most important. At a further extreme, across what might be termed a catastrophe-type transition, are the apparently low-stress environments typical of some intra-oceanic arc-trench systems where intra-arc and/or arc-rear extension occurs. The extensional intraplate velocity gradients typical of the central part of the middle and late Tertiary Cordillera were perhaps due to stress drops at the North American edge, permitting equilibration through spreading of the thickened and thermally disturbed "hinterland" (Coney and Harms, 1984). These ideas differ somewhat from normal plate tectonic notions of rigid plates and how mountains are formed, and that plates must always everywhere have boundaries. It is quite remarkable that the North American Cordillera has ranged across all of a spectrum of convergent, divergent, and strike-slip intraplate tectonics at various times and places in its post-Middle Jurassic history.

References

Armstrong, R. L., Sevier orogenic belt in Nevada and Utah, Geol. Soc. Am. Bull., 79, 429-458, 1968.

Armstrong, R. L., Geochronology of the Eocene volcanic-plutonic episode in Idaho, Northwest Geology, 3, 1-14, 1974.

Armstrong, R. L., Cordilleran metamorphic core complexes - from Arizona to southern Canada, Ann. Rev. Earth Planet. Sci., 10, 129-154, 1982.

Atwater, T., Implication of plate tectonics for the Cenozoic tectonic evolution of western North America, Geol. Soc. Am. Bull., 81, 3513-3536, 1970.

Ben-Avraham, Z., A. Nur, D. L. Jones, and A. Cox, Continental accretion: From oceanic plateaus to allochthonous terranes, Science, 213, 47-54, 1981.

Campa, M. F., and P. J. Coney, Tectonostratigraphic terranes and related metallogeny of Mexico, Geol. Assoc. Can. Ann. Mtg. Abs., 6, 8, 1981.

Campa, M. F., and P. J. Coney, Tectonostrati-

graphic terranes and mineral resource distributions in Mexico, Can. J. Earth Sci., 20, 1040-1051, 1983.
Clague, D. A., and R. D. Jarrard, Tertiary Pacific plate motion deduced from the Hawaiian-Emperor chain, Geol. Soc. Am. Bull., 84, 1135-1154, 1973.
Coney, P. J., Cordilleran tectonic transitions and motion of the North American plate, Nature, 233, 462-465, 1971.
Coney, P. J., Cordilleran tectonics and North American plate motion, Am. J. Sci., 272, 603-628, 1972.
Coney, P. J., Non-collision tectogenesis in western North America, in: Tarling, D. H., and S. H. Runcan, eds., Implications of continental drift to the earth sciences, New York, Academic Press Inc., 713-722, 1973.
Coney, P. J., Plate tectonics and the Laramide orogeny, New Mexico Geol. Soc. Sp. Pap., 6, 5-10, 1976.
Coney, P. J., Mesozoic-Cenozoic Cordilleran plate tectonics, Geol. Soc. Am. Mem., 152, 1978.
Coney, P. J., Tertiary evolution of Cordilleran metamorphic core complexes, in: Cenozoic Paleogeography of Western United States, Armentrout, J. W., M. R. Cole, and H. Terbest (eds.), Soc. Econ. Paleontologists and Mineralogists, Pac. Sec. Symp. III, 1979.
Coney, P. J., Cordilleran metamorphic core complexes: An overview, in: Cordilleran metamorphic core complexes, Crittenden, M. C., P. J. Coney, and G. H. Davis (eds.), Geol. Soc. Am. Mem., 153, 7-34, 1980.
Coney, P. J., Accretionary tectonics in western North America, in: Relations of tectonics to ore deposits in the southern Cordillera, Dickinson, W. R., and W. D. Payne (eds.), Ariz. Geol. Soc. Digest, 14, 23-37, 1981.
Coney, P. J., and S. J. Reynolds, Cordilleran Benioff zones, Nature, 270, 403-406, 1977.
Coney, P. J., D. L. Jones, and J. W. H. Monger, Cordilleran suspect terranes, Nature, 188, 329-333, 1980.
Coney, P. J., N. J. Silberling, D. L. Jones, and D. H. Richter, Structural relations along the leading edge of Wrangellia terrane in the Clearwater Mountains, Alaska, U.S. Geol. Surv. Circ. 823-B, 56-58, 1981.
Coney, P. J., and T. A. Harms, Cordilleran metamorphic core complexes: Cenozoic extensional relics of Mesozoic compression, Geology, 12, 550-554, 1984.
Coney, P. J., and D. L. Jones, Accretion tectonics and crustal structure in Alaska, Tectonophysics, in press.
Crittenden, Jr., M. D., P. J. Coney, and G. H. Davis, eds., Cordilleran metamorphic core complexes, Geol. Soc. Am. Mem., 153, 490 p., 1980.
Davis, G. H., and P. J. Coney, Geologic development of the Cordilleran metamorphic core complexes, Geology, 7, 120-124, 1979.
Dewey, J. F., and J. M. Bird, Mountain belts and the new global tectonics, J. Geophys. Res., 75, No. 14, 2625-2647, 1970.
Dickinson, W. R., Sedimentary basins developed during evolution of Mesozoic-Cenozoic arc-trench systems in western North America, Can. J. Earth Sci., 13, 1268-1287, 1976.
Dickinson, W. R., and W. S. Snyder, Plate tectonics of the Laramide orogeny, Geol. Soc. Am. Mem., 151, 355-366, 1978.
Dickinson, W. R., and W. S. Snyder, Geometry of subducted slabs related to San Andreas transform, J. Geology, 87, 609-627, 1979.
Eberly, L. D., and T. J. Stanley, Jr., Cenozoic stratigraphy and geologic history of southwestern Arizona, Geol. Soc. Am. Bull., 89, 921-940, 1978.
Elston, W. E., Tectonic significance of mid-Tertiary volcanism in the Basin and Range province, in: Cenozoic volcanism in southwestern New Mexico, Elston, W. E., and S. A. Northrop (eds.), New Mexico Geol. Soc. Sp. Pub., 5, 93-151, 1976.
Engebretson, D. C., A. V. Cox and G. A. Thompson, Convergence and tectonics: Laramide to Basin and Range, EOS, 63, 911, 1982.
Gabrielse, H., Major dextral transcurrent displacements along the northern Rocky Mountain trench and related lineaments in north-central British Columbia, Geol. Soc. Am. Bull., 96, 1-14, 1985.
Jones, D. L., N. J. Silberling, W. Gilbert, and P. J. Coney, Character, distribution, and tectonic significance of accretionary terranes in the central Alaska Range, J. Geophys. Res., 87, B-5, 3709-3717, 1982.
Lahr, J. C., and G. Plafker, Holocene Pacific-North America plate interaction in southern Alaska: Implications for the Yakataga seismic gap, Geology, 8, 483-486, 1980.
Lanphere, M. A., Displacement history of the Denali fault system, Alaska and Canada, Can. J. Earth Sci., 15, 817-822, 1978.
Lawton, T. F., M. Crespi, D. A. Currier, L. D. Ditullio, D. A. Hambrick, A. B. Kauffman, R. W. Krantz, P. T. Ryberg, J. R. Spencer, and P. J. Coney, Compilations for the Laboratory of Geotectonics, Dept. of Geosciences, Univ. of Arizona, 1981.
Mattauer, M., B. Collot and J. Van den Driessohe, Alpine model for the internal metamorphic zones of the North American Cordillera, Geology, 11, 11-15, 1983.
May, S. R., P. J. Coney, and M. E. Beck, Jr., Paleomagnetism and suspect terranes of the North America Cordillera, Open File Map and Report, Laboratory of Geotectonics, Dept. of Geosciences, Univ. of Arizona, Tucson, AZ. 1982.
Molnar, P., and T. Atwater, Interarc spreading and Cordilleran tectonics as alternates related to the age of subducted oceanic lithosphere, Earth Planet. Sci. Lett., 41, 330-340, 1978.
Molnar, P., and P. Tapponnier, Cenozoic tectonics

of Asia: Effects of continental collision, Science, 189, 419-426, 1975.

Monger, J. W. H., and R. A. Price, Geodynamic evolution of the Canadian Cordillera - progress and problems, Can. J. Earth Sci., 16, 770-791, 1979.

Monger, J. W. H., R. A. Price and D. J. Tempelman Kluit, Tectonic accretion and the origin of the two major metamorphic and plutonic welts in the Canadian Cordillera, Geology, 10, 70-75, 1982.

O'Hare, S., R. G. Gordon, and A. Cox, Absolute motion of North America during the late Paleozoic and Mesozoic as determined from paleomagnetic Euler poles (abs.), EOS, 63, No. 45, 912, 1982.

Plafker, G., T. Hudson, T. Bruns, and M. Rubin, Late Quaternary offsets along the Fairweather fault and crustal plate interactions in southern Alaska, Can. J. Earth Sci., 15, 805-816, 1978.

Plumley, P. W., R. S. Coe, T. Byrne, M. R. Ried, and J. C. Moore, Paleomagnetism of volcanic rocks of the Kodiak Islands indicates northward latitudinal displacement, Nature, 300, 50-52, 1982.

Price, R. A., Large-scale gravitational flow of supracrustal rocks, southern Canadian Rockies, in: Gravity and Tectonics, Dejong K. A., and R. Scholten (eds.), 491-502, 1973.

Price, R. A., The Cordilleran foreland thrust and fold belt in the southern Canadian Rocky Mountains, in: Thrust and nappe tectonics, Geol. Soc. London Sp. Pub., 9, 427-448, 1981.

Price, R. A., and E. W. Mountjoy, Geologic structure of the Canadian Rocky Mountains between Bow and Athabasca Rivers - a progress report, Geol. Assoc. Can. Sp. Pap., 6, 7-25, 1970.

Richardson, R. M., Viscous finite element modelling of plate driving forces (abs.), EOS, 63, No. 45, 1105, 1982.

Saleeby, J. B., Accretionary tectonics of the North American Cordillera, Ann. Rev., Inc., 1983.

Silver, L. T., and T. H. Anderson, Possible left-lateral early to middle Mesozoic disruption of the southwestern North American craton margin, Geol. Soc. Am. Abs. with Programs, 6, 955-956, 1974.

Smith, R. B., Seismicity, crustal structure, and intraplate tectonics of the interior of the western Cordillera, in: Cenozoic tectonics and regional geophysics of the western Cordillera, Smith, R. B., and G. E. Eaton (eds.), Geol. Soc. Am. Mem., 152, 111-144, 1978.

Stewart, J. H., Basin-Range structure in western North America: A review, in: Cenozoic tectonics and regional geophysics of the western Cordillera, Smith, R. B. and G. E. Eaton (eds), Geol. Soc. Am. Mem., 152, 1-31, 1978.

Stock, J., and P. Molnar, Uncertainties in the relative positions of Australia, Antarctica, Lord Howe, and Pacific plates since the late Cretaceous, J. Geophys. Res., 87, B6, 4697-4714, 1982.

Tardy, M., Contribution of L'étude geologique de la Sierra Madre oriental du Mexique, These du Doctorat d'Etat, Univ. de Paris, 445 p., 1980.

Tempelman-Kluit, D. J., Transported cataclasite, ophiolite and granodiorite in Yukon Geol. Surv. Can. Pap., 79-14, 1-27, 1979.

Uyeda, S., and H. Kanamori, Back-arc opening and the mode of subduction, J. Geophys. Res., 84, B3, 1049-1062, 1979.

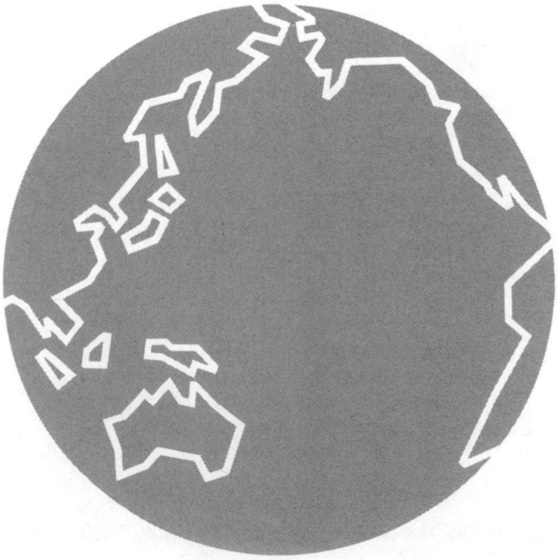

CORDILLERAN ANDES AND MARGINAL ANDES: A REVIEW OF ANDEAN GEOLOGY NORTH OF THE ARICA ELBOW (18°S)

Francois Mégard

Centre Géologique et Géophysique, Université des Sciences et Techniques du Languedoc
Place Eugène Bataillon - 34060 Montpellier Cedex, France

Abstract. The Andes display similar mid-Tertiary to Quaternary features, such as the classical Andean magmatic belt, the east-vergent Subandean thrust and fold belt, and cordilleras trending parallel with the western South American trench. These features arose in response to subduction of Pacific oceanic lithosphere beneath the South American plate (SOAM), and overprint earlier structures that vary markedly with latitude. This variation permits subdivision of the Andes into Northern Andes from the Caribbean to 3°S, Huancabamba Andes from 3° to 8°S, and Central Andes further south. The Northern Andes are orogens related in a part to the collision and accretion to SOAM of terranes that may have formed far from the continent. In the Northern Andes, the coastal lowlands and the western parts of the mountain ranges consist of accreted terranes of oceanic origin. In Colombia, the terranes comprise mostly dismembered ophiolitic complexes and their sedimentary cover, and form two discrete sets of thrust sheets that were obducted eastward over the edge of SOAM in, respectively, late Early Cretaceous and latest Cretaceous times. These accreted terranes make up the Western Cordillera, parts of the western flank of the Central Cordillera, and klippen on its top. A younger terrane forming the Serranía de Baudó was accreted in early Tertiary time as a west-vergent structural slice synthetic to the present oceanic subduction zone. In Ecuador north of 3°S, a Cretaceous and early Tertiary island arc together with its contiguous fore-arc was accreted to SOAM in Eocene time. This terrane comprises the Coastal area and the Western Cordillera, and is overridden along a steep west-vergent thrust by the edge of SOAM. One of the major features in the evolution of the Huancabamba Andes is the westward jump of the continental volcanic arc in Early Cretaceous time. The jump may be related to accretion of a continental block that was obducted eastward over SOAM at this time. In contrast with the Northern and Huancabamba Andes, the Central Andes seem to be an orogen exclusively related to subduction of oceanic lithosphere from the Early Jurassic to the Present. No terranes accreted during this interval have been identified, and the stable location or migration of the continental magmatic arcs away from the Pacific border seem to disprove a collisional origin for the Central Andes. Mountain building appears to be caused both by compression occurring in discrete short episodes and by accession of magma from the mantle into the crust that forms a thick sialic root. Compressive episodes may have been related to subduction above a flat or shallowly-dipping oceanic plate, whereas periods of active magmatism may have taken place at times when the oceanic plate dipped 30° or more.

Introduction

The late Tertiary evolution of the Andes displays a common tectonic pattern along its entire length, despite the role played by a possible segmented Benioff zone, acting in a manner similar to today. During this time, the major linear physiographic features that border western South America continuously for 9000 km, and the typical Andean Cenozoic magmatic belt, evolved in response to subduction of Pacific Ocean lithosphere beneath the South American plate (SOAM). A widespread pulse of compression in latest Miocene-Early Pliocene time was exclusively related to subduction of oceanic lithosphere, and created a belt along which the Andes from eastern Venezuela to mid-Argentina were thrust eastward over pericratonic foreland basins.

These prominent mid-Tertiary to Quaternary features are overprinted upon earlier structures that vary markedly along the length of the Andes. In the northern half of the Andes, north of latitude 18°S, this structural variation allows the Andes to be subdivided into discrete segments which basically depend on the presence or absence of terranes accreted to SOAM during Mesozoic or early Tertiary time (Fig. 1). Each terrane is defined by a specific geological evolution and a distinct boundary with adjacent terranes. Commonly the boundaries contain slivers of mafic and/or ultramafic rocks.

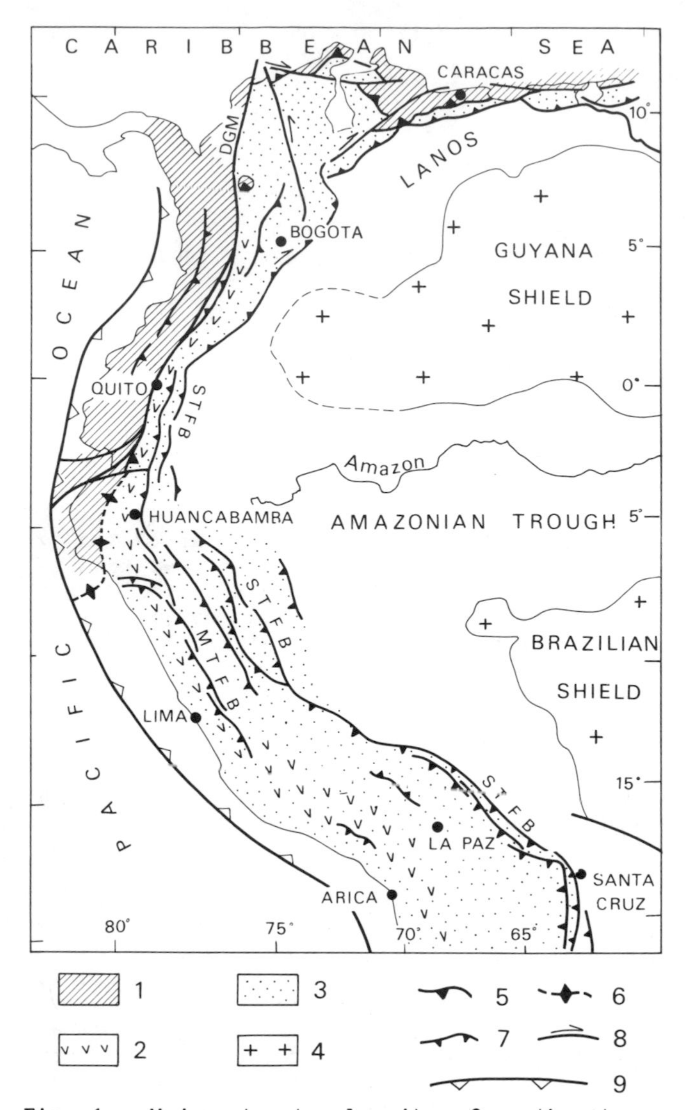

Fig. 1. Major structural units of northwestern South America. DGM=Dolores-Guayaquil megashear, MTFB=Marañon thrust and fold belt, STFB=Subandean thrust and fold belt. 1) accreted terranes; 2) Andean volcanic belt of Cenozoic age; 3) integral Andes; 4) Precambrian shields; 5) sutures between SOAM and accreted terranes; 6) sutures, when concealed and/or uncertain; 7) intra-continental overthrusts; 8) strike-slip faults; 9) oceanic trenches.

In this paper I shall restrict myself to description of the Mesozoic and early Tertiary evolution of the two large segments north of the Arica-Santa Cruz deflection at latitude 18°S (Fig. 1). As emphasized by Aubouin (1972), Gansser (1973), Zeil (1979) and Mégard (1984a), among others, the Northern Andes, north of latitude 3°S, and the Central Andes, south of 7°S, are extremely different in their tectonic evolution. The Northern Andes are characterized by the presence in their western parts of exotic terranes of oceanic and/or island arc origin, while the Central Andes are an integral part of the edge of SOAM, at least for their post-Paleozoic history. The intermediate segment between latitudes 3° and 8°S seems to be transitional as it possibly includes an exotic block mostly made of continental crust. This intermediate segment is also the site of the large Huancabamba deflection (Fig. 1), in which the mean strike of the Andes varies from nearly north-south, north of latitude 3°S, to northwest, south of latitude 8°S (Gansser, 1973), and is called the Huancabamba Andes. The deflection is characterized by very complex structures, with anomalous northeast and east-west structural trends, which are possibly related to the interaction of north-south and northwest Andean trends with pre-Andean, east-west, Amazonian trends.

Northern Andes

The similarity between modern oceanic crust and island arc rocks and the Mesozoic and early Tertiary basic igneous complex of the coastal area and Western Cordillera of Ecuador and Colombia, as well as from Central America, was first stressed by Goossens (1968), Goossens and Rose (1973), and Pichler et al. (1974). Gravimetric studies led Case et al. (1971, 1973) to propose that the crust of western parts of the Andes of Colombia was oceanic. Feininger and Sequin (1983) made a similar interpretation of the Bouguer anomaly map of Ecuador (Feininger, 1977). The basic igneous complex and associated sedimentary rocks comprise the accreted terranes of Colombia and Ecuador, but accretionary processes seem to have been different in both countries. For this reason, I shall review separately the evolution of the Andes in Colombia and Ecuador.

In both countries, the limit between Mesozoic continental SOAM and terranes accreted to it from the west is commonly taken to be the north-south orientated Dolores-Guayaquil megashear (DGM) (Fig. 1; Case et al., 1971, 1973; Campbell, 1974a,b; Meissner et al., 1976; Flüh, 1983). In Colombia, this limit follows the probably still-active eastern boundary of the Cauca-Patía graben (Fig. 2). In Ecuador, the boundary follows the inter-Andean graben (Fig. 2) until latitude 2°S, where it bends southwest towards the Gulf of Guayaquil. This boundary is represented by a dextral strike-slip fault zone that may end at a triple point where it meets the Peru-Ecuador trench near latitude 3°30'S.

The Andes of Colombia

I use herein the traditional subdivision of Colombia into Occidente and Oriente (Fig. 2). The Occidente extends over the coastal hills (Serranía de Baudó) and lowlands, the Western Cordillera, and westernmost parts of the Central Cordillera. It corresponds, geographically, with

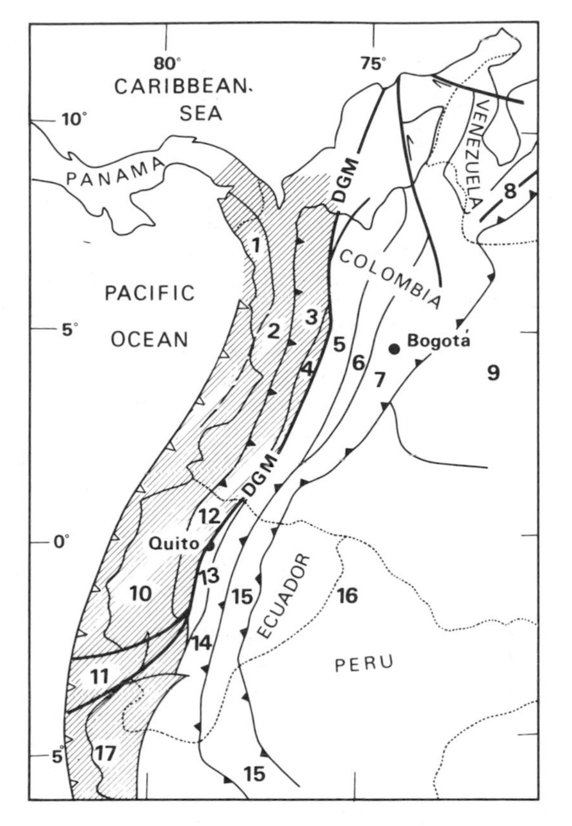

Fig. 2. Principal physiographic and structural provinces of northwestern South America. Colombia, eastern Panama and northwestern Venezuela: (1) Serranîa de Baudó, (2) Atrato-San Juan basin/ coastal lowlands, (3) Western Cordillera, (4) Cauca-Patîa graben, (5) Central Cordillera, (6) Magdalena valley, (7) Eastern Cordillera, (8) Cordillera de Mérida, (9) Llanos. 1+2+3+4 = "Occidente" of Colombia; 5+6+7+9 = "Oriente" of Colombia. Ecuador and northern Peru: (10) coastal lowlands, (11) Gulf of Guayaquil pull-apart basin, (12) Western Cordillera, (13) inter-Andean graben, (14) Eastern Cordillera, (15) Subandean hills, (16) Amazonian lowlands, (17) Amotape-Tahuin probable accreted terrane. 15+16 correspond to the Ecuadorian and Peruvian "Oriente". Hachured pattern denotes accreted terranes. Line with open triangles indicates the oceanic trench and lines with full triangles indicate overthrusts. DGM=Dolores-Guayaquil megashear zone.

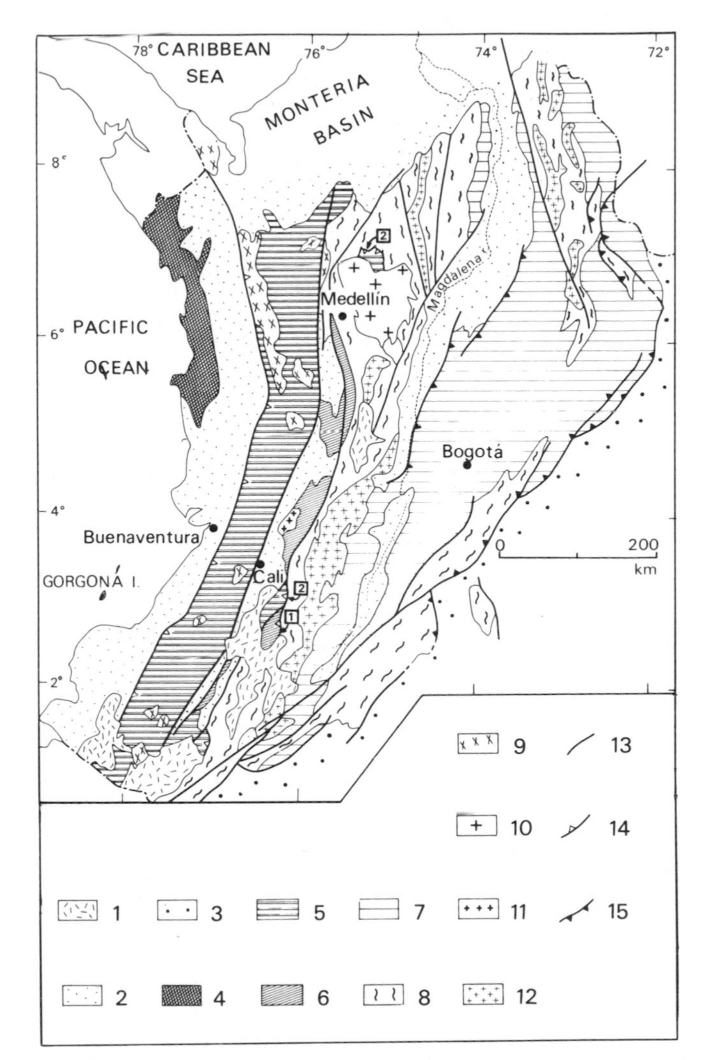

Fig. 3. Schematic geologic map of Colombia. Based on the geological map of Colombia, scale 1:1.500,000; (Arango et al., 1976; modified partly after Bourgois et al., 1985). Legend: (1) Plio-Quaternary volcanic cover, (2) Cenozoic infilling of the Atrato = San Juan basin, (3) undeformed Meso-Cenozoic sequence of the Llanos, (4) Serranîa de Baudó oceanic terrane, of mainly Tertiary age, (5) Western Cordillera oceanic terranes and related klippen on the Central Cordillera, of Early and mid-Cretaceous age, (6) Pre-Early Cretaceous oceanic terrains, (7) Mesozoic and Tertiary autochthonous units of the Central and Eastern Cordilleras, (8) Precambrian and Paleozoic basement of the Oriente, (9) Eocene to Miocene granitoids, (10) Late Cretaceous to Paleocene granitoids, (11) Early Cretaceous granitoids, (12) pre-Cretaceous intrusives, (13) major vertical faults, mostly with strike-slip displacement, (14) thrusts related to obduction; numbers 1 & 2 in boxes along the thrusts indicate respectively the first (Early Cretaceous) and the second (Late Cretaceous) episodes of obduction, (15) late Cenozoic intracontinental thrusts.

the location of the accreted terranes and their post-accretionary sedimentary and volcanic cover.

Accreted terranes of western Colombia. Many authors (McCourt et al., 1984; De Souza et al., 1984; Bourgois et al., 1985) agree that there is evidence for two discrete episodes of accretion, in Early and Late Cretaceous times, in the Western Cordillera, the Cauca-Patîa graben and the western part of the Central Cordillera (Fig. 3), and a third episode of early Tertiary accretion of rocks in the Serranîa de Baudó.

During Early Cretaceous time, variably dismembered ophiolitic complexes, including local komatiitic basalt (Espinosa in McCourt et al., 1984), and associated sediments, were thrust eastwards onto Paleozoic metamorphic rocks in the western part of the Central Cordillera (Bourgois et al., 1985). The allochthonous rocks are commonly metamorphosed in greenschist and, rarely, amphibolite facies. High-pressure metamorphic rocks, which include glaucophane schists, have been reported from two places along the Dolores-Guayaquil megashear (DGM) in southern Colombia (Orrego et al., 1980; Feininger, 1982a; De Souza et al., 1984). They form isolated fault-bounded blocks and the degree of metamorphism drops off markedly across the bounding faults. At Jambaló, according to Bourgois et al. (1985), the protolith of these high-pressure rocks is of Paleozoic age and the K-Ar whole-rock age of the metamorphic assemblage is of Cretaceous age (125 $\pm$ 15 Ma) (Orrego et al., 1980). Glaucophane from a nearby area (De Souza et al., 1984) yielded a 104 $\pm$ 14 Ma date. The conditions under which the glaucophane formed, namely pressures corresponding to 15 to 20 km burial and temperatures less than 400°C (Feininger, 1982a), suggest rapid burial of Paleozoic rocks from the ancient continental edge, and subsequent rapid uplift. McCourt et al. (1984) list 16 reset ages ranging from 120 $\pm$ 5 to 107 $\pm$ 4 Ma from schists of the Central Cordillera. A K-Ar date on amphibole from schist in the footwall of a thrust beneath dunite, east of Medellin, is 108 $\pm$ 12 Ma (Restrepo and Toussaint, 1975). Despite their lack of precision, these radiometric data provide evidence for a pulse of Early Cretaceous metamorphism that is reasonably ascribed to emplacement, or obduction, of ophiolites over rocks in the Central Cordillera which represent the former continental edge. An upper age limit to the emplacement episode is indicated by the K-Ar date of 113 $\pm$ 10 Ma from crosscutting granitoids intruding the Cauca ophiolite (Toussaint et al., 1978).

The Western Cordillera, and some klippen in the Central Cordillera, comprise a second set of tectonic units, that were emplaced in Late Cretaceous time (Fig. 4; Bourgois et al., 1985). These units have been described by several authors, including Nelson (1957), Barrero (1979) and most recently Bourgois et al. (1985). The thrust sheets are 1.5 to 5.5 km thick, and are separated by shear surfaces that were originally flat but later were refolded with an open antiform. The base of the thrust sheets comprises massive and/or pillowed basalts, in places greater than 1 km thick, overlain by a predominantly sedimentary sequence that in places contains interlayered basalts. It has been suggested by Barrero (1979) and Espinosa (1980) that the volcanics belong at least in part to immature tholeiitic arc assemblages. However, analyses of REE and other incompatible elements in tholeiitic rocks from part of the Western Cordillera, show that these are typical MORBs (Marriner and Millward, 1984; Millward et al., 1984). According to Lebrat (1985), they are similar to the Piñon tholeiites from coastal Ecuador, that are interpreted by him to be N-type MORBs. Island arc suites, including differentiated rocks similar to the andesites commonly found further south in the Macuchi arc of Ecuador, have not been documented yet. In the uppermost thrust sheet, the basalts are commonly overlain by a thick agglomerate and tuff sequence that in turn underlies a flysch. The lower part of the flysch sequence consists of distal, thin layers of fine to very fine grained sandstone and siltstone with volcanic clasts suggesting an oceanic or island arc source. The upper part seems to have formed in a setting proximal to a continental source, since it contains a high-proportion of coarse grained sandstones with abundant subangular quartz grains, probably derived from a sialic source, as are conglomerate lenses containing quartz, gneiss and schist clasts in the upper part of the flysch. Fossils from the thrust sheets are sparse but range in age from Barremian to Coniacian, with the latter providing a lower age limit to time of emplacement. An upper limit is given by 80-60 Ma K-Ar dates from the Antioquian batholith (Pérez in Toussaint and Restrepo, 1982), which near latitude 7°N, intrudes both Central Cordilleran basement and the remnant of a flysch thrust sheet that is part of the upper tectonic unit (Bourgois et al., 1985).

The Tertiary evolution of the region is related to underthrusting, beneath terranes accreted in the Cretaceous and beneath SOAM, of the Serranía de Baudó volcanic arc (Figs. 3,4). A deep seismic profile at latitude 4°N shows a flat surface at a depth of 3.5 km beneath the Upper Cretaceous to Eocene Atrato-San Juan basin, which links up with a thrust extending for 20 km beneath the Western Cordillera to a depth of 12 km. The large, flat, west-vergent thrust is related by Flüh (1982, 1983) to Eocene underthrusting. The Serranía comprises basalts, dated at 70 $\pm$ 3.5 and 41 $\pm$ 3 Ma, overlain by volcaniclastic mass flow deposits that contain Late Paleocene to Early Miocene microfossils, on top of which are more basalts (Bourgois et al., 1982). The Serranía de Baudó assemblage is submerged beneath the Pacific Ocean south of latitude 7°N (Fig. 3) but extends southwards offshore under the thick sedimentary cover of the trench-coast gap (Flüh, 1983). The Gorgona and Gorgonilla islands near latitude 3°N may well represent outcrops of the Serranía assemblage (Pérez Téllez, 1980). According to the model of Flüh (1982, 1983), the Serranía de Baudó is a slightly emergent bulge that belongs to a large slab of oceanic and/or island arc rocks beneath the present trench-coast gap, the Atrato-San Juan basin, the Western Cordillera and the Cauca-Patía graben, as far east as the Dolores-Guayaquil megashear. Most authors, (e.g. Case et al., 1971; Toussaint and Restrepo, 1982; Flüh, 1983) agree that accretion of the Baudó-Atrato slab occurred

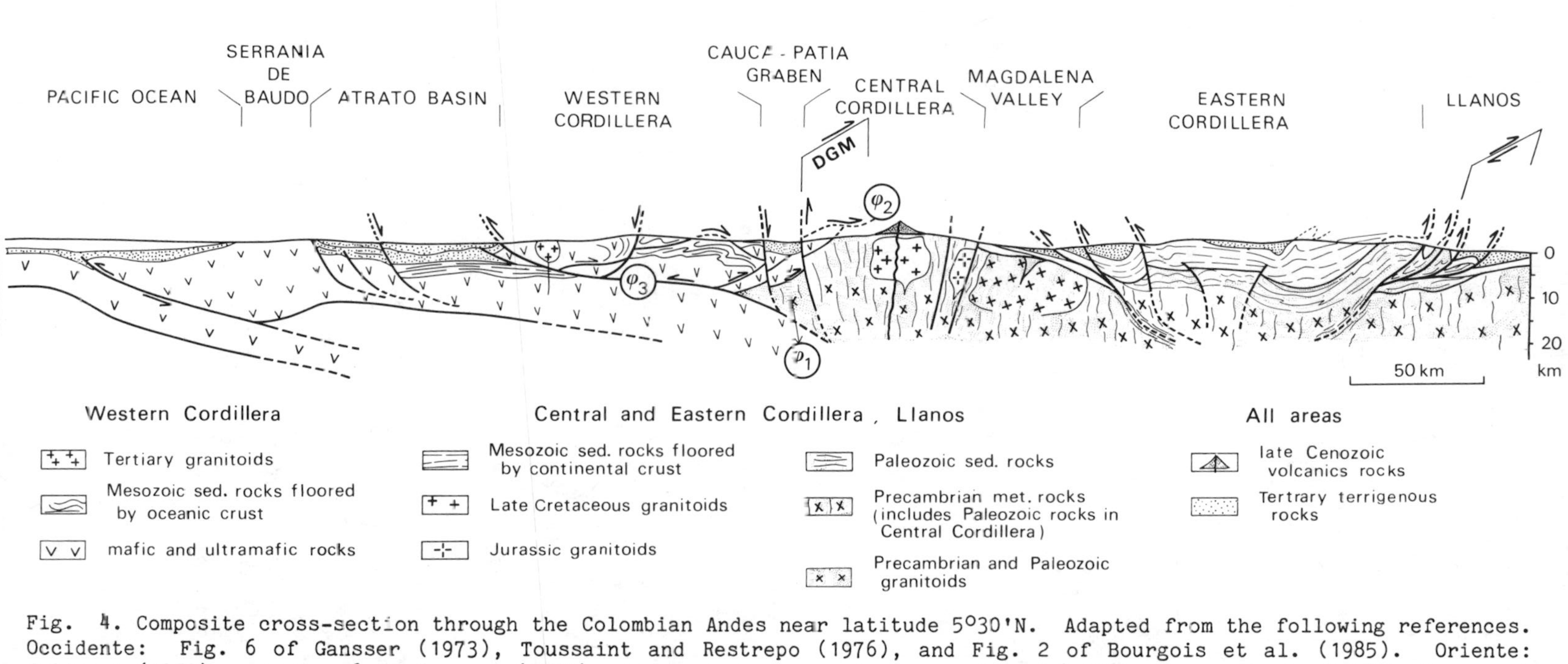

Fig. 4. Composite cross-section through the Colombian Andes near latitude 5°30'N. Adapted from the following references. Occidente: Fig. 6 of Gansser (1973), Toussaint and Restrepo (1976), and Fig. 2 of Bourgois et al. (1985). Oriente: Julivert (1970), and Fig. 6 of Gansser (1973). Deep structures: Meissner et al. (1976), Meyer et al. (1976), Mooney et al. (1979), Flüh (1983). Thrusts related to: $\phi1$, Early Cretaceous obduction; $\phi2$, Late Cretaceous obduction; $\phi3$, Tertiary subduction/accretion.

when the Baudó volcanic bulge clogged the Atra subduction zone. The change in depositional style from distal fine-grained to proximal conglomeratic turbidites in the Middle Eocene in the Atrato-San Juan basin (Pérez Téllez, 1980), as well as calc-alkaline plutonism with 47 Ma K-Ar dates in the Western Cordillera (Göbel and Stibane, 1979), indicate that accretion probably occurred in Middle Eocene time.

Cretaceous and Tertiary accretion processes and the overall structure of the Columbian Andes. The overall interpretation of the Western Cordillera of Colombia is conjectural. McCourt et al. (1984) consider that two oceanic arc sequences were successively accreted to the western edge of SOAM in the Early and Late Cretaceous times as steep east-vergent tectonic slices. Mooney (1980) proposed both an "accretionary prism" and a "westward stepping" model, the latter being similar to that of McCourt et al. (1984). The presence of klippen of the accreted western terranes on top of rocks of the integral Andes in the Central Cordillera, together with the eastward vergence of most large recumbent folds within accreted terranes of the Western Cordillera, is interpreted by both Bourgois et al. (1985) and the writer to show that at least some of the thrust sheets of the Occidente moved eastward or southeastward, and were thrust directly, or obducted, on to the ancient continental margin. The obducted terranes cover the western edge of Cretaceous SOAM, which may explain the apparent lack of a Cretaceous arc-trench gap assemblage west of the Central Cordillera. Alternatively, or in addition, the absence of such an assemblage could be accounted for by truncation of the continental margin by large, dextral movements along the precursor of the Dolores-Guayaquil megashear (DGM), that were possibly related to oblique subduction of the Phoenix and Farallon plates (e.g. Duncan and Hargraves, 1984). Following this, in Eocene time, the Serranía de Baudó, was accreted by underthrusting of the earlier accreted terranes and SOAM.

In summary, the evolution of the Occidente of Colombia comprises (i) two early phases of eastward obduction along shear surfaces apparently antithetic to internal earlier east-dipping subduction zones, and (ii) a late phase of accretion along west-vergent shear surfaces synthetic to the subduction zone. Oblique subduction was common at other times and caused dextral movements along north-south trending fault-zones.

Integral SOAM in Colombia: the Oriente. Isolated outcrops of Precambrian metamorphic rocks can be found in the Central and Eastern Cordilleras of Colombia, some of which belong to a 1200-1300 Ma granulite belt (review in Kroonenberg, 1982). Upon this basement were deposited Cambrian and Ordovician terrigenous sediments with interbedded basic tuffs. These rocks were deformed and metamorphosed in a Silurian "Taconic" event. At the end of Early Devonian time, mainly terrigenous marine sedimentation resumed in a north northeast-trending elongate basin located in the Eastern Cordillera, but sedimentation was interrupted by uplift in Late Devonian time. According to McCourt et al. (1984), many metamorphic rocks of the Central Cordillera belong to a paired metamorphic belt of probable Late Devonian/Early Carboniferous age. The formation of the belt and the uplift can be roughly correlated with the Late Devonian to Early Carboniferous early Hercynian compressive pulse of the Central and Southern Andes. A basin similar to that which formed in the Middle Devonian lay in the Eastern Cordillera in Late Mississippian to Early Permian times. The siliciclastic and carbonate rocks deposited in this basin were folded together with Middle Devonian sandstones and shales during a Late Permian pulse comparable to the late Hercynian pulse of the Central Andes. Reviews on pre-Mesozoic evolution of Colombia can be found in Stibane (1967), Irving (1975), Gansser (1973) and Théry (1982).

Red fluviatile conglomerates and sandstones are typical of the Late Permian and/or Early Triassic interval and locally conformably underlie shelf carbonates and/or clay-sandstones bearing Norian marine fossils (Fig. 5). Local marine transgressions of Early and Late Jurassic age are also recorded both in the Central and Eastern Cordilleras. Elsewhere, the whole Late Permian to Late Jurassic interval is represented by red beds, that in places are rich in volcanic intercalations, and are locally more than 3 km thick. The volcanic intercalations range in composition from lower rhyolite to higher andesite and seem to be coeval with plutonism in the Central Cordillera. There, a suite of plutons with K-Ar dates ranging from 248 $\pm$ 10 to 214 $\pm$ 7 Ma is succeeded by another suite that includes large calc-alkaline batholiths with K-Ar dates between 174 $\pm$ 10 and 142 $\pm$ 6 Ma (review in McCourt et al., 1984). The calc-alkaline granitoids of the latter suite and the contemporaneous volcanic rocks in the Jurassic red beds are thought to belong to a Jurassic continental magmatic arc located in the eastern flank of the Central Cordillera (Mooney, 1980; Toussaint and Restrepo, 1982; McCourt et al., 1984). The red bed basin east of the magmatic arc may be interpreted as a back-arc basin in which tensional deformation is evidenced by pre-Cretaceous normal faults (Fig. 5). More information on the Triassic-Jurassic evolution of the Colombian Oriente can be found in Julivert (1968), Irving (1975), Geyer (1980), Théry (1982) and Mojica (1984).

The stratigraphy of the Cretaceous of the Oriente is well known. Thick sequences of fossiliferous marine shale and sandstone were deposited, which range in age from Neocomian, and locally Tithonian, to Maastrichtian (review in Julivert, 1968; Irving, 1975; Théry, 1982). The Cretaceous basin in eastern Colombia had a mean

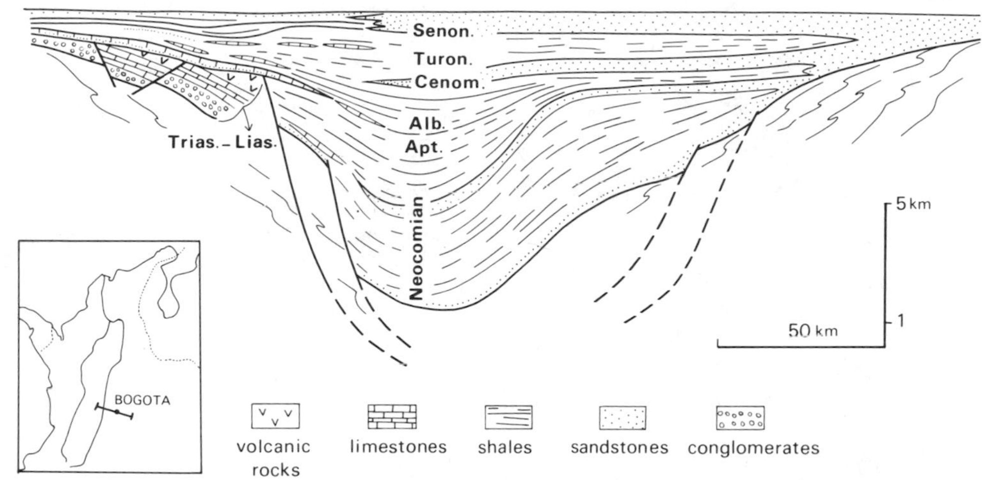

Fig. 5. Restored section across the Bogotá trough in Late Cretaceous time, modified after Julivert (1973).

north-northeasterly-trend, underwent rapid subsidence in Neocomian to Cenomanian times, and accumulated about 10 km of sandstones and dark shales. This great sedimentary thickness suggests that the sialic basin floor was thinned, a suggestion which is supported by contemporaneous normal faulting. The Cretaceous sedimentary sequence thins towards the Guyana shield to the east, and towards the Central Cordillera to the west. The latter formed a barrier between the intracontinental eastern basin and the open sea to the west. The many calc-alkalic plutons with K-Ar dates ranging from 119 to 68 Ma (review in McCourt et al., 1984), provide good evidence for the presence of a Cretaceous plutonic arc located in the partly emergent Central Cordillera, but volcanic equivalents are represented only by a few calc-alkaline lavas and tuffs (e.g. Feininger et al., 1972). Accretion of part of the western Cordilleran exotic terranes to SOAM may account for the end of this plutonic episode in latest Cretaceous times, for accretion is accompanied by a westward shift of plutonic activity, which may reflect a similar shift in position of the subduction zone. Such a shift might also explain the marked decrease in subsidence rates recorded in the eastern Colombian basin in Senonian times (e.g. Julivert, 1973). Late Cretaceous instability with local uplifts is recognized in the eastern Colombian basin, but marine sedimentation prevailed till the Paleocene. The seas then withdrew from most of the Oriente region except for a few places in northern Colombia, and subsidence rates decreased markedly. Basal sandstones and/or conglomerates in Tertiary formations and local or regional unconformities in the Magdalena valley and the Eastern Cordillera, indicate minor pulses of deformation and/or uplift in these areas and the Central Cordillera. The only major episode of folding and thrusting in the Eastern Cordillera and the Magdalena valley (Fig. 4) occurred in Late Miocene and/or Early Pliocene time (data in Van der Hammen, 1961; Campbell, 1974a; and Irving, 1975). This phase is contemporaneous with formation of the Subandean thrust and fold belt of Ecuador, Peru, Bolivia and northern Argentina.

The Andes of Ecuador

Accreted terranes of western Ecuador. Much of the Western Cordillera of Ecuador consists of a typical island arc assemblage called the Macuchi Formation (Fig. 6), that includes rocks ranging in composition from tholeiites to quartz-andesites (Pichler et al., 1974; Henderson, 1979; Lebrat, 1985), with interbedded volcaniclastic and, typically silicified, fine grained sedimentary rocks with Senonian faunas. In most places, the arc assemblage is overlain by flysch of Maastrichtian to Paleocene age (Faucher and Savoyat, 1973). Locally, the flysch passes upwards and laterally into conglomerate that contains quartz pebbles clearly derived from erosion of the continental basement of the Eastern Cordillera of Ecuador (Tschopp, 1948; Faucher and Savoyat, 1973). The presence of these pebbles and the increased content of quartz grains and mica flakes in the upper part of the flysch indicate that the Macuchi arc was close to the former continental margin when these rocks were deposited. In one restricted area at latitude 1°S in the Western Cordillera, Macuchi-type arc assemblages are interspersed with, and overlain by, Eocene limestones (Henderson, 1979), suggesting that the island arc was still active, at least locally, by this time.

In the coastal lowlands of Ecuador, oceanic

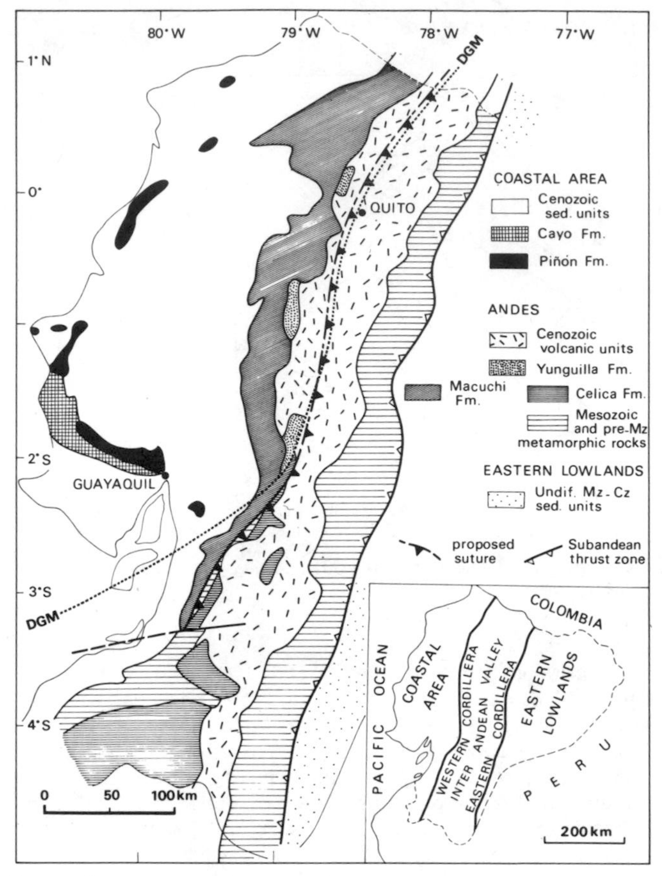

Fig. 6. Schematic geological map of Ecuador after Lebrat et al. (1986). Based on the geological maps of Ecuador, scale 1:1,000,000; 1st and 2nd editions (Servicio Nac. Geol. Min., 1969; Dirección Nac. Geol. Min., 1982). DGM=Dolores-Guayaquil megashear zone.

basement crops out at various places and is called the Piñón Formation (Figs. 6, 7). It consists of massive diabase and pillow-lava which have typical N-type MORB composition according to major element and REE analyses (Goossens and Rose, 1973; Lebrat, 1985), cumulate gabbros associated with the former rocks, and harzburgites in isolated outcrops. K-Ar dates on plagioclase separates from the basalts are 113 ± 10 and 107 ± 15 Ma (Goossens and Rose, 1973; Kennerley, 1980). This ophiolitic complex is overlain by the turbiditic Cayo Formation that is rich in mass-flow deposits containing blocks of volcanic rocks ranging from basic andesitic lavas to rhyolitic welded tuffs. The basal unit of the Cayo Formation contains early Senonian microfossils and its top is chert with Maastrichtian to Danian fossils (Faucher and Savoyat, 1973; Bristow and Hoffstetter, 1977). Volcanic centers that have acted as sources for the gravel to block-size material that is common in the Cayo Formation have not been observed in the coastal area. The composition of the clasts roughly matches that of the partly coeval Macuchi volcanic rocks of the Western Cordillera, suggesting the Macuchi arc as a possible source area. The Piñón-Cayo assemblage may thus represent fore-arc deposits on basement. Calcareous turbidites of Middle Eocene age overlie the Cayo Formation with slight disconformity. According to Feininger and Bristow (1980), they are derived from the Eocene limestones of the upper part of the Macuchi arc. The Santa Elena olistostrome complex (Colman, 1970) was emplaced on top of the calcareous turbidites at the end of Eocene time. Some olistoliths contain conglomerates with pebbles of vein-quartz and schists of SOAM origin. It is uncertain whether the source area of the clasts is to be found in northwestern Peru, as proposed by Feininger and Bristow (1980) or in the already uplifted Eastern Cordillera of Ecuador.

Rocks of oceanic affinity, but probably not part of the Macuchi arc, form small, fault-bounded masses a few hundred metres to a few kilometres in dimensions, that are located east of the arc terrane. One slice, east of Quito, comprises peridotites, gabbros and diabase dykes of a dismembered ophiolite complex (Juteau et al., 1977). Near latitude 2°S, another slice consists of oceanic tholeiites (Lebrat et al., 1985). These slices mark the suture between SOAM and the Macuchi volcanic terrane, for just east of them is either SOAM basement or volcanic rocks of the Celica Formation, a continental volcanic arc (Fig. 6).

Integral SOAM in the Andes and Oriente of Ecuador. The geology of eastern Ecuador is better known in the oil-producing and only slightly deformed Oriente (the Subandean hills and Amazonian lowlands) than in the strongly tectonized and metamorphosed Eastern Cordillera (Figs. 2,6).

In the Eastern Cordillera, a pre-Mesozoic basement containing both Precambrian metamorphic rocks (Feininger, 1982b) and Paleozoic sedimentary rocks, is overlain by Mesozoic strata, but superposed metamorphism makes it difficult to separate Mesozoic from older strata. Paleozoic metamorphism was local, but Late Cretaceous to earliest Tertiary metamorphism was widespread. K-Ar dates on micas from these rocks range between 60.5 and 53.6 Ma (Herbert, 1983; Feininger, 1982a,b). The late metamorphism is in the greenschist facies, of Barrovian-type (Trouw, 1976; Feininger, 1982b), and coeval with a compressive pulse that generated tight folds trending approximatively north-south. In the west, the folds are upright or alternatively east and west-vergent, and in the east they become east-vergent (Fig. 7). The cleavage forms a fan that grades eastward into schistosity that parallels bedding and is very flat in some places. As noted above, metamorphism makes it difficult to differentiate Paleozoic and Mesozoic rocks, but schist including metamorphosed pillow-lava grade westward into flysch of Maastrichtian age associated with volcanic rocks near Cuenca, latitude 2°40'S (Bris-

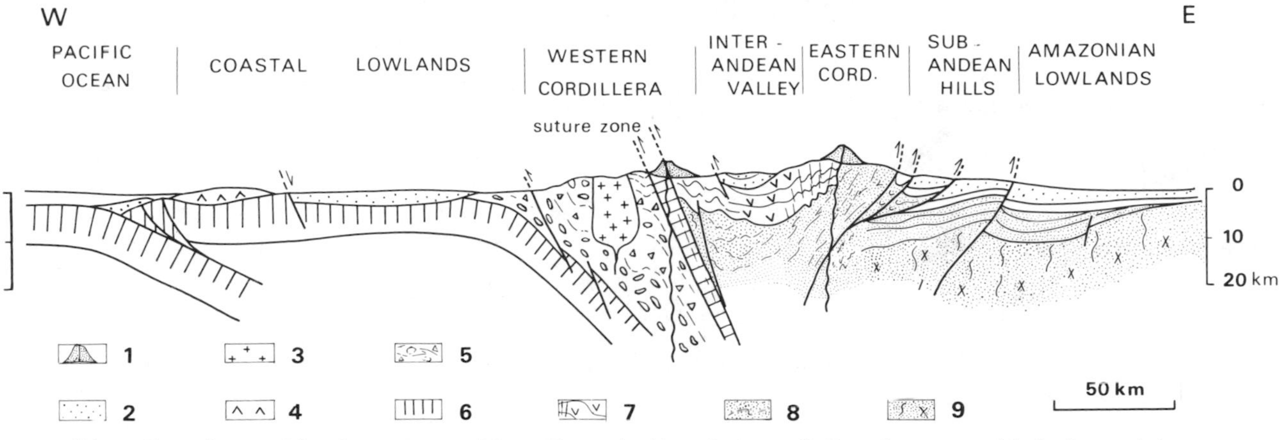

Fig. 7. Generalized cross-section through the Andes of Ecuador, compiled from data collected between latitudes 0° and 2°30'S. All areas: (1) late Cenozoic volcanoes; (2) Tertiary clastic rocks. Pacific Ocean, Coastal lowlands and Western Cordillera: (3) late Tertiary plutons; (4) volcaniclastic Late Cretaceous rocks of the Cayo Fm.; (5) volcanic rocks of the Macuchi island arc and associated sedimentary rocks; (6) oceanic crust. Eastern Cordillera, Subandean hills and Amazonian lowlands: (7) Mesozoic sedimentary rocks, locally cleaved, (V) denotes Celica volcanic rocks; (8) Paleozoic rocks, in part metamorphosed; (9) Precambrian basement.

tow, 1973). Similar rocks, with ages ranging from Cenomanian to Campanian, are widespread in the southern Andes of Ecuador and northwestern Peru (Kennerley, 1980; Baldock, 1982), where their volcanic component is called the Celica Formation (Feininger and Bristow, 1980). Major element and REE analyses (Lebrat, 1985) show that the Celica rocks are calc-alkalic suite typical of intracontinental volcanic arcs. Some large low-potash granitoid stocks were emplaced contemporaneously, within the Celica volcanic cover. The Celica episode of arc activity started in the Aptian and stopped in the Campanian. Lebrat et al. (1986) contend that the cessation of Celica volcanism is related to accretion of the Macuchi arc and associated(?) Piñón fore-arc to SOAM.

In little-deformed rocks of the Oriente, to the east, the Mesozoic record is more complete, and generally similar to that of eastern Peru. There is a transition from shelf carbonate facies to volcanic facies in Lower Jurassic time, and a discrete upper Misahuallí volcanic member, is developed in red beds of Middle Jurassic to earliest Cretaceous age (Tschopp, 1953; Canfield et al., 1982). The Misahuallí volcanic rocks are part of the north northeast-trending, Late Jurassic to earliest Cretaceous Colan-Misahuallí continental volcanic arc (Mourier et al., 1986; see also the section on the Huancabamba Andes). Following deposition of fluvial, braided, Early Cretaceous sandstones, marine sedimentation resumed in Albian time. It ended during the Maastrichtian, and all overlying units consist of clastic rocks deposited in fresh-water or brackish conditions. Conglomerates interbedded with the Cenozoic formations reflect the major pulses of uplift in the Andes. The oldest conglomerate is located at the base of the Eocene Tiyuyacu formation (Tschopp, 1948; Bristow and Hoffstetter, 1977) and postdates slightly the major pulse of deformation and metamorphism recorded in the Eastern Cordillera.

Geometry of the suture and timing of accretion. It is generally considered that the boundary, or suture, between accreted Mesozoic oceanic terranes forming the western parts of Colombia and Ecuador and continental SOAM is the vertical Dolores-Guayaquil megashear (DGM) (Case et al., 1971, 1973; Campbell, 1974b; Feininger and Seguin, 1983). The DGM trends roughly north-south in northern Ecuador and follows the western boundary of the inter-Andean valley of Ecuador (Fig. 2). To the south it bends westward near latitude 2°S and trends southwestward toward the Gulf of Guayaquil (Fig. 2 and 6), which can be interpreted as a late Tertiary to Quaternary pull-apart basin (e.g. Shepherd and Moberly, 1981) related to dextral slip along the DGM.

A few observations in a key area of the western Cordillera of Ecuador south of latitude 2°S, and, south of the westward bend of the DGM, show that the nearly north-south trending suture is distinct from the southwest-trending DGM (Fig. 6). Around latitude 2°S, the suture is a reverse fault-zone (Fig. 7) with, firstly, a discrete western branch dipping 60° east, which separates an eastern upthrust MORB sliver from a western downdropped Macuchi arc assemblage, and, secondly, a loosely-defined eastern branch between the MORB and Celica continental arc volcanics to the east (Lebrat et al., 1985). Around latitude 2°30'S, the suture is a steep east-dipping fault-zone between the Macuchi arc to the west, and slightly metamorphosed basement overlain by

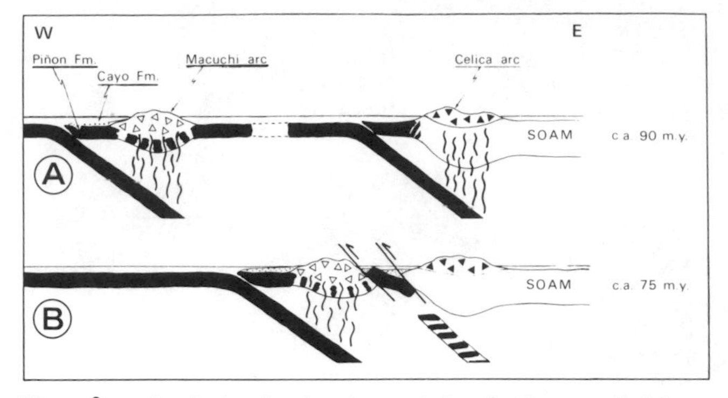

Fig. 8. A plate-tectonic model of the evolution of Ecuador in Cretaceous times, taken from Lebrat et al., 1986.

Celica volcanic rocks to the east. Farther north, the suture is concealed beneath the eastern part of the Cenozoic volcanic cover (Fig. 6). The origin of the ophiolitic and MORB slivers is a matter of discussion, but most authors (Juteau et al., 1977; Feininger and Bristow, 1980; Lebrat et al., 1986) consider that they are slices of oceanic crust which originally lay between the Macuchi arc and sialic SOAM. Figure 8, which is taken from Lebrat et al. (1986) illustrates one possible origin for emplacement of these rocks. These authors agree that SOAM was thrust westward over the accreted Macuchi arc and that accretion occurred during subduction, and not through obduction as seems to be the case for most of the Western Cordillera of Colombia.

Northern Andes: Concluding Remarks

In Ecuador, arguments drawn from time-space relationships of the various volcanic and sedimentary units lead to the proposition of a plausible scenario for the evolution of the Macuchi island arc and its final accretion to SOAM (Fig. 8).

The arc probably was already emergent in Senonian times, when volcaniclastic turbidites from the arc accumulated on top of the Piñón oceanic crust west of the arc, in an area considered to be a fore-arc by Lebrat et al. (1986, see Fig. 8). The area between the Macuchi trench and the western SOAM trench was a marginal oceanic area of unknown extent, distinct from the Farallon plate, that was consumed in the western SOAM trench, and gave rise to the continental Celica arc. Possibly the rate of spreading of the marginal area was lower than the rate at which it was consumed, so that the Macuchi arc collided with SOAM causing the Celica arc to shut-off. The collision progressed from south to north and ended in Eocene time. The Macuchi subduction zone remained active after the collision.

Other scenarios are possible. Feininger and Bristow (1980), suggest that the Macuchi arc was associated with west-dipping subduction in the early stages of its evolution. Another interpretation is that of Kennerley (1980) who suggested that the Macuchi and Celica arcs were parts of a single autochthonous arc related to east-dipping subduction and built partly upon sialic and partly upon intermediate or oceanic crust.

In Colombia, many two-dimensional plate tectonic scenarios leading to the present structure of the Occidente have been proposed, among them those of Toussaint and Restrepo (1976), Barrero (1979), Mooney (1980), Flüh (1983) and Bourgois et al. (1985). However, their soundness is seriously affected by the lack of adequate geochemical information concerning the nature of pre-orogenic magmatism in the accreted terranes. Moreover, parts of the accreted terranes have probably been truncated along north-south wrench faults and have migrated toward the Caribbean on dextral displacements associated with oblique subduction (e.g. Duncan and Hargraves, 1984). In order to understand the evolution of this region, it appears to be necessary to integrate scenarios concerning northwestern South America with those proposed for the Caribbean (e.g. Burke et al., 1984; Duncan and Hargraves, 1984; Mattson, 1984; Leclere-Vanhoeve and Stephan, 1985), which is beyond the scope of this paper.

The Huancabamba Andes

The Huancabamba Andes lie between 3° and 8°S and are characterized by relatively low elevations and by drastic changes in structural trends and morphostructural pattern (Gansser, 1973; Cobbing et al., 1981) with respect to both to Northern and Central Andes (Fig. 9). These differences reflect a distinct pre-Oligocene evolution. Most features that pass gradually from Northern to Central Andes through the Huancabamba segment, such as the Cenozoic volcanic continental arc, the offshore and onshore marine basins near the present continental margin, and the late Miocene-Pliocene Subandean thrust and fold belt, developed after Eocene time.

One distictive feature of the Huancabamba segment is the dramatic relocation of the Mesozoic arc in late Early Cretaceous time. Late Jurassic-Early Cretaceous rocks of Colan-Misahuallí arc strike N10°E and extend over the Subandean area of Ecuador. In Peru, they follow the eastern boundary of the Olmos arch (Fig. 9) and join the coeval arc of the northern Central Andes near Trujillo. The position of the arc jumped in late Early Cretaceous time to a location 150 km northwestward and became the Celica arc, whose activity started in the Aptian. In Albian and also Late Cretaceous time, the Celica arc was connected southward to the Casma arc of central Peru. Mourier et al. (1986) suggest that the sudden relocation of the arc was caused by the accretion in northwestern Peru of a terrane mostly made of continental crust called the Amotape-Tahuin terrane.

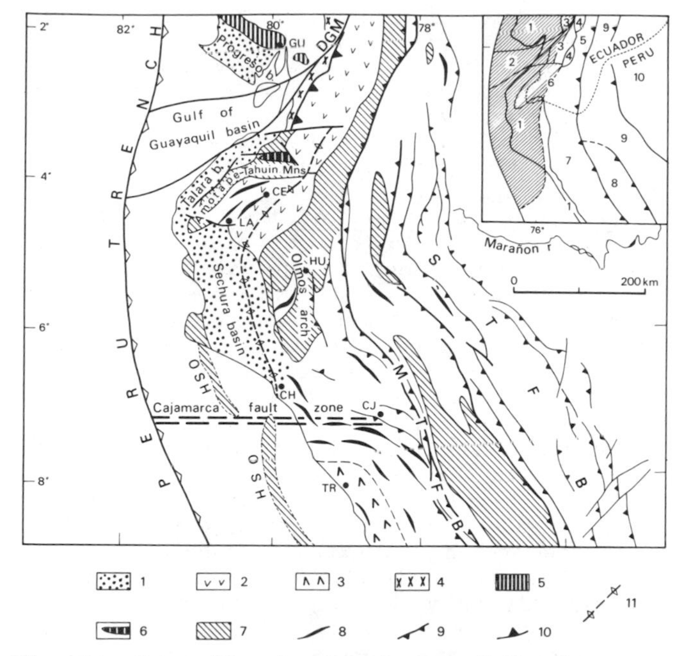

Fig. 9. Schematic structural map of the Huancabamba Andes. CE=Celica, CH=Chiclayo, CJ=Cajamarca, GU=Guayaquil, HU=Huancabamba, LA=Lancones, TR=Trujillo, OSH=outer shelf high, DGM=Dolores-Guayaquil megashear. MTFB and STFB denote respectively the Marañon and Subandean thrust and fold belts. Symbols: (1) Tertiary basins, (2) volcanic rocks of the Celica continental arc, of Aptian to Campanian age, (3) volcanic rocks of the Casma marginal basin, mostly middle Albian in age, (4) volcanic rocks of the Macuchi Island arc, of Late Cretaceous to Eocene age, (5) Piñón ophiolitic assemblage, (6) peridotites and associated schists, (7) pre-Mesozoic rocks, includes Mesozoic metamorphic rocks in the Eastern Cordillera of Ecuador north of 4°S, (8) axes of major folds, (9) overthrusts, (10) Eocene suture zone between SOAM and the Macuchi-Piñón accreted terrane, (11) probable Late Cretaceous suture zone between SOAM and the Amotape-Tahuin terrane. Index map: (1) coastal lowlands and hills; (2) recent basin of the Gulf of Guayaquil; (3) Western Cordillera of Ecuador; (4) inter-Andean valley; (5) Eastern Cordillera of Ecuador; (6) Southern Andes of Ecuador; (7) Western Cordillera of Peru; (8) Eastern Cordillera of Peru; (9) Subandean hills; (10) Amazonian lowlands.

The geology of northwestern Peru and of the contiguous part of southwestern Ecuador is characterized by the following features: (i) an unconformity of Aptian or pre-Aptian age at the base of the Celica volcanics and coeval flysch sequence, (ii) an absence of sedimentary rocks of Triassic to Jurassic age, and (iii) a Devonian to Permian marine series unconformably overlying slates and quartzites of probable early Paleozoic age. This Paleozoic sequence is exposed in the Amotape mountains (Fig. 9). According to Mourier et al. (1986), older Paleozoic rocks of the Amotape mountains grade eastward into rocks in the Tahuin mountains, in which low-pressure metamorphism has been described by Feininger (1978; 1982b), and where a Precambrian basement composed mostly of greenschist and amphibolite facies rocks underlies the Paleozoic metamorphics. Most of these features are specific to northwestern Peru and differ markedly from those of either the eastern part of the Huancabamba Andes or of the Central Andes (next section), and possibly characterize an Amotape-Tahuin terrane.

The assumed suture zone between the possible Amotape-Tahuin accreted terrane and Mesozoic SOAM is concealed almost everywhere by volcanic and sedimentary rocks of the Celica arc and/or by Tertiary strata of the Sechura basin (Fig. 9), but locally may be delineated by mafic and ultramafic rocks. Beaufils (1978) reports a pre-Albian volcano-plutonic complex composed of dominant andesites, pillowed olivine basalts, and gabbros, that is located 25 km southeast of Lancones (Fig. 9) and tentatively interpreted as an island arc remnant trapped along the suture. This interpretation could also apply to the steeply north to north northeast-dipping foliated cumulate gabbros of Morro de Eten, 19 km south of Chiclayo. In the Tahuin mountains and farther north, serpentinite and schists of high pressure metamorphic facies diapirically cut Precambrian amphibolites and Paleozoic rocks (Feininger, 1978, 1980; Dirección Gal. Geol. Min., Map of Ecuador, 1982). They could be slivers of oceanic crust and its sedimentary cover, trapped behind the Amotape-Tahuin terrane, when it collided with SOAM in the Early Cretaceous. Speculatively, the phengite K-Ar age of 132 $\pm$ 5 Ma from schists associated with the serpentinite (Feininger, 1982b) may record the age of high-pressure metamorphism, and hence of collision.

In the western Andes, the southern boundary of the Huancabamba segment is the almost east-west trending possible Cajamarca fault-zone (Fig. 9). This fault zone behaved in a normal sense with respect to the west Peruvian ramp in Late Jurassic time. and as a lateral ramp when the early Tertiary Marañon thrust and fold belt developed. As a consequence, the shortening rate in this Maranon belt decreased markedly north of Cajamarca (Mourier, ms. in prep.).

Central Andes

The Central Andes of Peru and Bolivia and the Southern Andes of Chile and Argentina, excluding the westernmost Magellan belt, seem to be built on sialic Precambrian and/or Paleozoic basement, with no ophiolitic sutures. The Central Andes provide a particularly good example of orogeny related to the long-lived subduction of, successively, Phoenix, Farallon and Nazca plates under continental SOAM. Subduction is evidenced by the Andean magmatic arc that was almost permanently

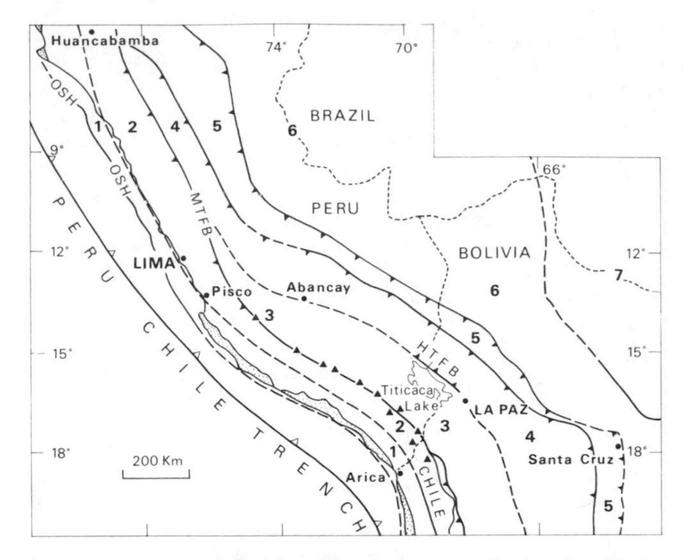

Fig. 10. Morphostructural zonation of the Central Andes between latitudes 7° and 18°S. (1) Coastal area (stippled zone=Coastal Cordillera), (2) Western Cordillera, (3) Altiplano, (4) Eastern Cordillera, (5) Subandean hills, (6) Amazonian lowlands, (7) Brazilian Shield. Triangles along the Western Cordillera-Altiplano boundary denote recent volcanoes. MTFB=Marañon thrust and fold belt, HTFB = Huancané thrust and fold belt, OSH=Outer shelf high.

active at least since Lower Jurassic time and which parallels the grain of the Andean belt and the present Peru trench. It is characterized by a calc-alkaline suite in which andesites, adamellites and tonalites predominate.

The physiographic features and the main morphostructural units shown in Figure 10 developed after the Eocene, much as in the Northern Andes. The roughly east-west trending Pisco deflection, located near latitude 13°30'S, introduces a latitudinal segmentation in the Central Andes. Between latitudes 7°S and 13°30'S the mean trend of the Andes is N35°W. The trend is east-west in the deflection itself, and N50°W to the south of it. It changes again to north-south, south of the Arica-Santa Cruz deflection. Other changes occur across the Pisco deflection. North of it, the Coastal basins develop mostly offshore and the Coastal Cordillera may correspond to the submarine outer shelf high (Thornburg and Kulm, 1981). South of it, the Coastal Cordillera is a horst, that lies west of Coastal basins which are mostly filled with continental sedimentary rocks. Today, Nazca Ridge-Peru Trench intersection lies at about the latitude of the Pisco deflection, but this is a mere accident since the intersection is migrating southward.

The geological history of the Central Andes can be divided into pre-Andean and Andean periods, with the transition taking place in Permian to Early Triassic time. The Andean period is further subdivided into a Late Triassic to Early Cretaceous pre-orogenic interval, and a Late Cretaceous to Recent orogenic interval.

Most of the data perceived in the following sections come from Audebaud et al. (1973, 1976), Mégard (1978; 1986), Cobbing et al. (1981) and Pitcher et al. (1985) for Peru, and from Martinez (1980) for Bolivia.

Pre-Andean evolution of the Central Andes. From latitudes 7°S to 18°S, pre-Mesozoic basement outcrops are common in almost all morphostructural zones (Fig. 11) (Mégard et al., 1971; Dalmayrac et al., 1980; Laubacher and Mégard, 1985). The presence of basement, either onshore or offshore, very close to the Peru trench is important. In northern Peru, around latitude 9°S, pre-Mesozoic metamorphic basement discovered in offshore wells extends to within about 20 km of the trench (Von Huene et al., 1985). South of latitude 13°30'S, Precambrian rocks of the Arequipa massif are present onshore about 120 km from the trench and extend offshore as well. These data show, firstly, that the present day accretionary prism is only a few tens of kilometres wide, and secondly, that no significant volume of accreted rocks of Pacific origin is included in the Central Andes, in contrast with the Northern Andes. If a large accreted prism ever existed, it must subsequently have been either swallowed in the subduction zone (e.g. the "erosional" scheme of Coulbourn, 1981) or removed laterally bg oblique subduction.

The oldest Precambrian rocks of the Central Andes are the 2 Ga granulites of the Arequipa massif in the Coastal Cordillera near latitude 17°S (Cobbing et al., 1977; Dalmayrac et al., 1977) and the 1 Ga granulites of the tiny Pichari massif in the Eastern Cordillera around latitude 12°30'S (Dalmayrac et al., 1980). Most remaining Precambrian rocks were metamorphosed in greenschist and amphibolite facies, successively in Barrovian and low-pressure conditions, and form a belt of metamorphic rocks with 600 Ma dates that underlies a large part of the Andes in Peru and Bolivia. Metabasic tuffs and ultramafics are known in the belt, in which at least one phase of recumbent isoclinal folding has been recognized (Dalmayrac et al., 1980). The belt appears to be bounded to the east by the Pichari granulites in Peru and by 1.3 and 1 Ga-old mobile belts in Bolivia (Litherland et al., 1985), and to the west by the Arequipa massif. Collision with ancestral SOAM of an exotic Arequipa terrane of possibly much larger extent than nowadays would provide a reasonable explanation for the development of this belt. In this case, continental SOAM must already have existed to near its present limit or to a somewhat westward expanded limit at the end of the Precambrian, and Paleozoic and earlier Mesozoic orogens of the Andean area could be viewed as some kind of intracontinental mountain belts.

During Ordovician to Middle and locally Late Devonian time there was sedimentation in an elon-

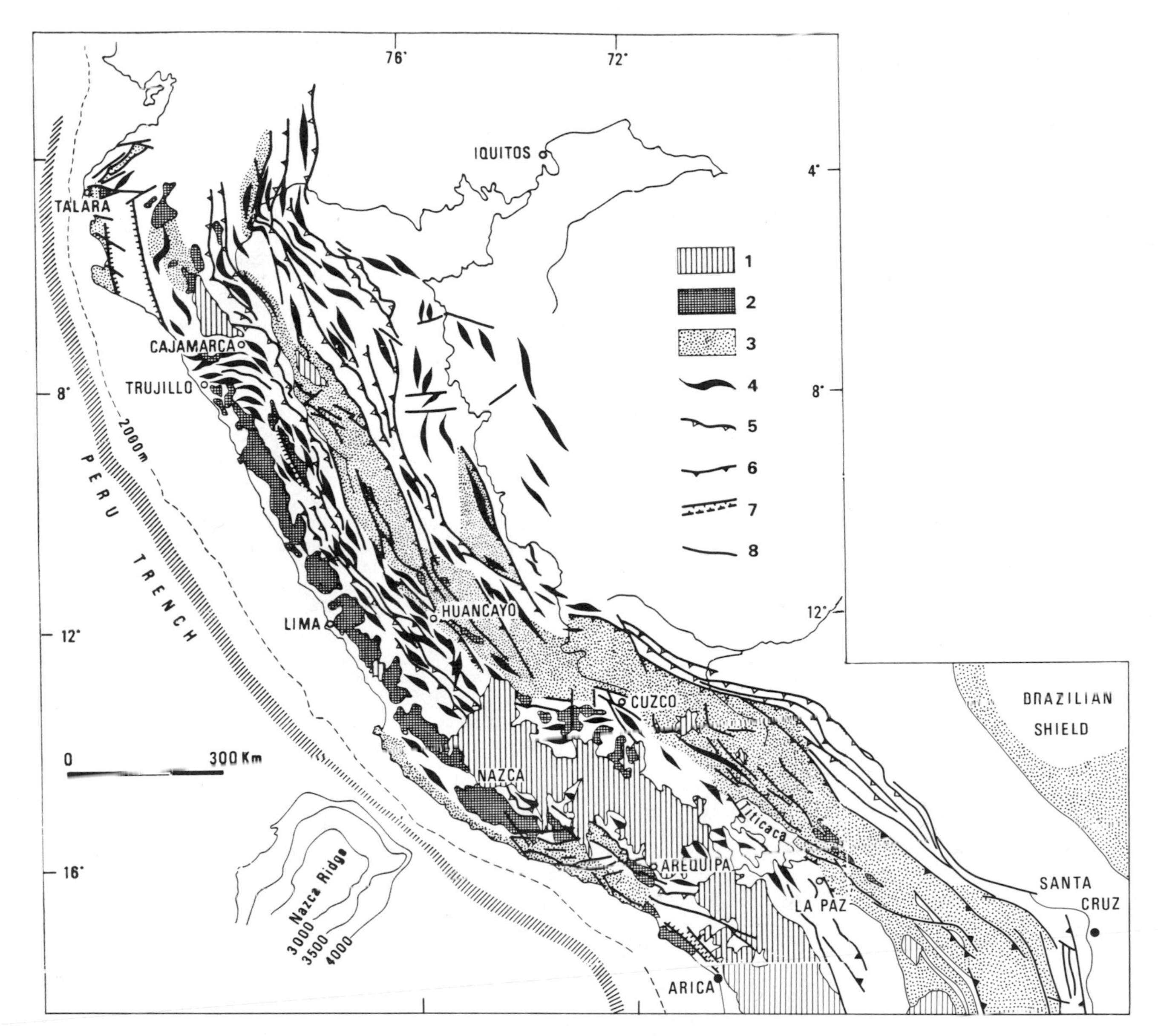

Fig. 11. Structural sketch map of Peru and part of Bolivia, from Mégard (1986), with permission of Princeton Press. (1) Undeformed or slightly deformed late Cenozoic volcanic units; (2) Andean intrusives; (3) pre-Mesozoic rocks; (4) major folds; (5) main overthrusts; (6) main steep thrusts; (7) main normal faults; (8) major wrench faults.

gated northwest-southeast trending tensional basin floored by thinned crust in which thicknesses of up to 10 km of mainly siliciclastic sedimentary units were laid down. Marine volcanic rocks are scarce in this basin but thin pillow basalts and basic tuff layers are present near latitude 12°S along the western Cordillera-High Plateau boundary (Harrison, 1943). The volcanics could be related to the tensional regime indicated by the thickness of the sedimentary pile. A widespread hiatus of the early and middle Ashgillian (latest Ordovician) in southern Peru and a regional disconformity in Bolivia are both followed by deposition of diamictites that in part are of glacial origin of latest Ashgillian to early Llandoverian age (Martinez, 1980; Crowell et al., 1981; Laubacher et al., 1982). The Early Silurian breaks in the succession may be related to diastrophism of the "Taconic" event recorded in Colombia. The major Paleozoic tectonic event in the Central Andes is the early Hercynian phase, during which a 200 km-wide fold and thrust belt formed in latest Devonian or Early Mississippian times. In southern Peru, the belt is southwest-vergent and includes in the southwest, a province of downward-flattening thrusts, probably rooted in a large décollement (Laubacher and Mégard, 1985). Low-grade metamorphism of greenschist facies is common, and grades into amphibolite facies in the vicinity of synkinematic granitic bodies. Such plutons, as well as post-kinematic granitoids, are scarce. The nature of the early Hercynian orogeny is enigmatic. It appears to be ensialic and may have developed in response to stresses originating along the western boundary of the rigid Arequipa massif and transmitted to the thinned sialic crust of the intracratonic Paleozoic basin.

Mississippian continental sandstones and conglomerates, were deposited on eroded early Hercynian structures, and followed by mostly shallow marine terrigenous and carbonate strata of Pennsylvanian and Early Permian ages. The cumulate thickness of upper Paleozoic strata reaches 3 km in a northwest-trending trough centered upon the Eastern Cordillera between latitudes 12° and 14°S. These sedimentary rocks were deformed in late Leonardian time by a late Hercynian pulse. In Bolivia, southern Peru and part of central Peru, this deformation is represented by folds with an axial-plane fracture cleavage and by later, near vertical, faults. Elsewhere only steep faults are observed with large throws and possible large horizontal components.

Subsequent to this late Hercynian deformation, a tensional regime appeared in the Andes and probably lasted until the Early Triassic. Late Permian-Early Triassic red beds intercalated with lava and pyroclastic flows overlap Carboniferous-Early Permian strata. For the first time in the Paleozoic, large quantities of magma were erupted and intruded, mostly in the Eastern Cordillera. Among the volcanic rocks are alkaline, peralkaline and shoshonitic lavas, that commonly are related to continental rifting (Noble et al., 1978; Kontak et al., 1985). Between latitudes 10° and 16°30'S, large biotite-granitoid stocks were emplaced. In Peru they are dated radiometrically as Late Permian and Triassic, but in Bolivia the youngest dates are Early Jurassic. Kontak et al. (1985) relate this magmatism, and part of the Andean magmatism of the Eastern Cordillera, to a persistent discontinuity between the passive craton and the mobile continental margin that is largely independent of plate boundary processes. They suggest that a continental rift developed over the discontinuity in Late Permian time.

Pre-Orogenic Andean evolution. Norian transgressive shelf limestones unconformably overlie rocks of Late Permian and Early Triassic ages in central and northern Peru but are absent between the Pisco deflection and latitude 18°S. Carbonate sedimentation persisted in the north into Lower Jurassic time. At that time, the sea was widespread and even extended over the Coastal area and Western Cordillera of southern Peru. During the Sinemurian, the southern Peru carbonate shelf basin graded westward into a volcanic belt (Vicente, 1981) of dominant andesitic composition, whose geochemistry is typical of a continental volcanic arc (James et al., 1975). Tuff and ash layers in central Peru (Mégard, 1978), near Chiclayo in northern Peru (Prinz, 1985), together with Sinemurian volcanic rocks of the Ecuadorian Oriente (Tschopp, 1953), are evidence of a probably discontinuous Lower Jurassic volcanic arc that was related to east-dipping ocean subduction beneath SOAM.

Marine, Middle and Late Jurassic, sequences are well-developed in southern Peru but are restricted to the Coastal area, the Western Cordillera and the Eastern Altiplano, and are unknown in Bolivia. Evidence for basin evolution is provided by a succession, from base to top, of shelf carbonates, bathyal spongolites, deep sea fan deposits, platformal siliciclastic facies, and uppermost Tithonian reefal limestones (Vicente et al., 1982). Sporadic intercalations of tuff, and the development of thick volcanic and volcaniclastic sequences southwest of the northwest-trending turbiditic basin, show that a volcanic arc was active at least recurrently in the Coastal area.

Jurassic rocks of southern Peru extend northward until about the Pisco deflection, north of which, up to latitude 12°S, only a few outcrops of sandstones and limestones of Bajocian age are present in the Altiplano. Sedimentary and volcanic units similar to those of southern Peru or of the Huancabamba Andes may have been deposited in the Western Cordillera and Coastal area of Peru between latitudes 7° and 13°30'S, but have not been recognized.

The facies distribution of the various sequences of Tithonian and Cretaceous age in Peru, between latitudes 7° and 13°30'S, shows that a new paleotectonic pattern developed between Lower Jurassic and Tithonian times (Benavides, 1956; Wilson, 1963). Two elongate basins, the East Peruvian and West Peruvian basins, were separated by the Marañon geanticline (Fig. 12). The West Peruvian basin was probably bounded to the southwest by a threshold in which pre-Norian rocks outcropped (Mégard, 1978; Cobbing et al., 1981), and can be subdivided into a southwestern trough in which 5 to 12 km of dominantly volcanic and volcaniclastic strata were laid down in late Tithonian to Albian times, and a northeastern platform on which was deposited an eastward-tapering wedge of Neocomian sandstone, overlain by Albian to Santonian carbonates.

According to Atherton et al. (1983, 1985), the late Tithonian basalts and basaltic andesites of the southwestern trough have both within-plate and island-arc characteristics. Some may have formed during the initial stages of back-arc rifting or spreading. Volcanic rocks are also interspersed with terrigenous and carbonate rocks

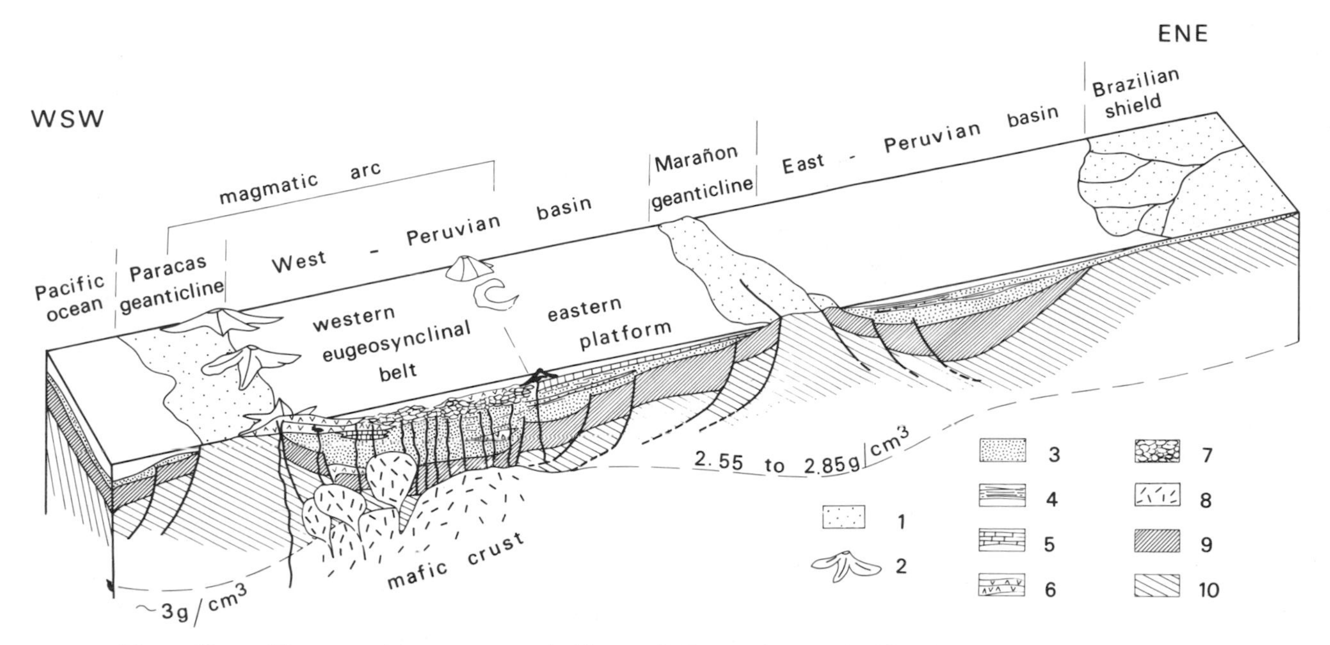

Fig. 12. Diagrammatic representation of Neocomian and Albian facies relations in northern and central Peru, modified from Mégard (1986) with permission of Princeton Press. Physiographic features in middle Albian time: (1) emergent land; (2) volcanoes. Rocks of Albian and Neocomian age: (3) sandstones; (4) shales; (5) limestones; (6) volcanic flows and sills; (7) pillow lavas; (8) gabbros. Older rocks: (9) of Liassic and Triassic ages; (10) of Paleozoic and Precambrian age.

of Early Cretaceous age and are succeeded by up to 10 km of middle Albian, largely pillowed, basic volcanic rocks associated with gabbros (Regan, 1985). The hydrothermal-burial metamorphism in the Albian volcanics correspond to a high geothermal gradient, which may be related to extreme thinning of the continental crust (Aquirre and Offler, 1985). Atherton et al. (1983, 1985) interpret the volcanic rocks to be mantle-derived and extruded in an ensialic marginal basin, which was floored by thinned sialic crust (Fig. 12). This interpretation is backed by modelling of the gravity anomalies (Atherton et al., 1983; Wilson, 1985), which reveals an arch-like structure of 3.0 g/cm^3 material under the southwestern trough (Fig. 12). The Albian marginal basin closed progressively southward near latitude 13°S.

Late Albian mild folding of the volcano-sedimentary succession deposited in the western, sinking, part of the West Peruvian basin north of Lima (Myers, 1974), was followed in Late Cretaceous time by deposition above an unconformity, of tuffs, and by emplacement of high-level, plutons of the Coastal Batholith. During the same period, conformable shelf sedimentation was going on across the adjacent platform to the east.

A thin disconformable marine, brackish and continental sequence was deposited directly upon Lower Jurassic or older rocks of the Marañon geanticline during the Cretaceous.

In the East Peruvian basin, an eastward-tapering wedge of Triassic to Quaternary sedimentary rocks accumulated, whose total thickness reaches 10 km close to the Andes in the Marañon sub-basin of northern Peru, where the Amazonian trough (Fig. 1) and the East Peruvian basin meet. The basin contains continental red clastics of Middle(?) Jurassic to earliest Cretaceous ages, overlain disconformably by Cretaceous sandstones, shales and limestones. In Neocomian times, the influx of quartzose sand from the shield areas to the east was so large that the basin was totally filled and excess sand was distributed across the Maranon geanticline into the West Peruvian basin. During Aptian to Senonian times, the sea invaded the East Peruvian basin, and sedimentation to the west and continental to brackish conditions to the east, with a weaker sand supply from the shield (Soto, 1979).

The Cretaceous evolution the Central Andes south of the Pisco deflection is different from that to the north. The west Peruvian basin was completely filled in Late Jurassic times and unconformable platform sediments accumulated upon it. After a discrete Neocomian terrigenous depositional episode, shelf carbonates were laid down in Albian to Coniacian times, followed by regressive red beds, gypsum and minor carbonates in Santonian times (Vicente, 1981). Volcanism was active west of the platform in Aptian times, and its presence later is indicated by sporadic tuff intercalations in limestones of Albian to Coniacian age. Volcanism resumed in latest

Cretaceous-Paleocene times, after the first widespread pulse of Andean deformation. In the area of the southern Peru and Bolivian altiplanos, and part of the Eastern Cordillera, a continental platform developed in Cretaceous times, with local actively subsiding areas in which up to 2 km of sandstones and red beds accumulated. Two limestone marker beds indicate transgression from the west in Cenomanian and Maastrichtian time (review in Martinez, 1980). A topographic high similar to the Marañon geanticline separated this western continental platform from an eastern basin connected with that of northern Peru. This basin did not undergo much subsidence, with the cumulate thickness of the Cretaceous strata of about 800 m in northern Bolivia. The only fossiliferous strata in the basin are related to a short-lived marine transgression of Maastrichtian age.

In summary, major features of the Central Andes in Norian to Senonian times are, firstly, a magmatic arc located in the vicinity of the present coast and represented by a partly submarine volcanic belt and early plutons of the large calc-alkaline Coastal Batholith, and, secondly, actively sinking troughs parallel with the arc and to the east of it. Differences in evolution recorded north and south of the Pisco deflection may reflect differences in the nature and thickness of the crust that were related to its pre-Andean evolution, and/or to segmentation of the subduction zone.

Late Albian deformation seems to be relatively minor, but its extent and the amount of shortening it caused, are difficult to estimate, since the structures of the Mesozoic volcaniclastic and volcanic series are not well known. One of the major problems is the unknown nature of the contact between the rocks and those of the platform to the east. It could well be a large overthrust that is concealed by intrusions and Tertiary volcanics, and obscured by later steep faults.

Orogenic Andean evolution. The Andean orogenic period is generally thought to begin with a Santonian "Peruvian" phase of compression, followed by an "Incaic" phase of Middle to Late Eocene age, and finally by three Miocene "Quechua" phases, according to the classical phase chronology of Steinmann (1929), as modified by McKee and Noble (1982) and Mégard et al. (1984). These orogenic pulses affect discrete structural zones and die out very rapidly when crossing the boundaries of these zones. Successive pulses also migrate progressively with time toward the foreland, as in most mountain belts. Finally, the major compressive stress associated with most of these pulses is nearly horizontal and strikes northeast-southwest, so that most folds and thrusts trend northwest-southeast. In this section, I will only mention strikes of structures when they deviate from this overall direction.

North of the Pisco deflection, evidence for a Peruvian phase of Santonian age at most places is provided by the rapid transition from marine shelf carbonate or terrigenous strata, to terrestrial red beds, in both West and East Peruvian basins. A slight unconformity is recognized at the base of the red beds in the Cajamarca area of northern Peru, and cobbles of Neocomian quartzites in the red beds of the Altiplano suggest that uplift and possibly also folding occurred in the Western Cordillera. However, structural features directly related to this phase are difficult to document, except in part of the Eastern Cordillera and the eastern Altiplano of central Peru, where upright folds and related axial-plane cleavage developed in late Paleozoic and Mesozoic strata (Mégard, 1978).

South of the Pisco deflection, the Peruvian phase is well-developed in the Western Cordillera, and created folds and at least one large flat overthrust affecting the rigid Precambrian basement and its Mesozoic cover (Vicente et al., 1979). Volcanic rocks of late Late Cretaceous and Paleocene ages overlie Peruvian phase structures unconformably. East of the area of Peruvian deformation, where a widespread marine transgression is recorded in Campanian to Maastrichtian sedimentary rocks of the Altiplano and Subandean basins.

The Incaic phase is commonly considered to be responsible for most of the shortening recorded in the Central Andes. During it, strata as young as Senonian and their undated cover were folded in the Western Cordillera. Incaic structures were then eroded and unconformably overlain by conglomerates and volcanic rocks dated at about 40 Ma (Noble et al., 1974, 1979). The age of the Incaic phase is therefore considered to be Early to Middle Eocene. However, an older pulse of Paleogene deformation may exist near to the coast in southern Peru (Noble et al., 1985) and in central Peru (e.g. Cobbing et al., 1981).

North of the Pisco deflection, the major Incaic structures are folds and overthrusts in the eastern part of the Western Cordillera, that comprise the Marañon thrust and fold belt (MTFB; Figs. 10,13) which was studied by Wilson et al. (1967), Coney (1971), Romani (1982) and Mégard (1984b). The width of the MTFB ranges up to 50 km. It consists of an eastern, eastward-directed imbricate thrust belt in which gently southwest-dipping thrust sheets have been stacked upon the autochthonous basement and its thin Mesozoic cover, and a western province in which the thrusts are scarce and steeper, the vergence of the folds is alternately to the northeast and the southwest, and a steep slaty cleavage is commonly found. My interpretation is that the large décollement flooring the imbricate province is rooted in a steeper fault under the western province (Mégard, 1984b).

South of the Pisco deflection, Incaic deformation caused the oldest continental clastic layers in the coastal basins to be locally deformed in response to reactivation of old faults (Sébrier et al., 1979). Some Peruvian structures

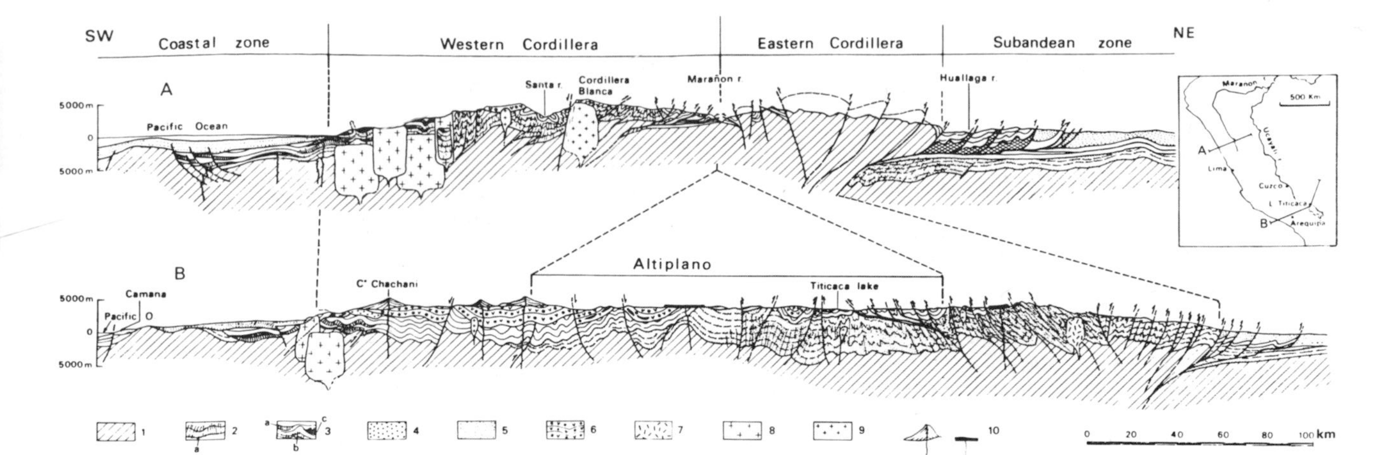

Fig. 13. Cross-sections through the Central Andes in northern Peru (A) and southern Peru (B), modified from Mégard, 1986, (in press), with permission of Princeton Press. Section B after Audebaud et al. (1973, 1976) and Vicente et al. (1979), modified. Symbols: (1) Precambrian basement, may include some Paleozoic rocks in the Western Cordillera; (2) early and middle Paleozoic siliclastic series; (3) late Paleozoic and Mesozoic rock units; symbol (a) denotes volcanic sills and flows, (b) indicates slaty cleavage, (c) denotes evaporites; (4) Red beds of Late Cretaceous to Eocene age in the Western Cordillera and the Altiplano; (5) Red beds of Late Cretáceous to Pliocene age in the Subandean area and Tertiary marine and continental sedimentary rocks of the Coastal zone; (6) Volcanic and continental units of Oligo-Miocene age in the Western Cordillera and the Altiplano; (7) Hercynian plutonic rocks; (8) Coastal batholith, 100 to 32 Ma; (9) Late Tertiary plutonic rocks; (10) Plio-Quaternary stratovolcanoes and tuff plateaux.

were also reactivated in the Western Cordillera (Audebaud et al., 1976), but generally there was little effect in the Altiplano. Along the Altiplano-Eastern Cordillera boundary, the southwesterly-directed Huancané thrust and fold belt developed (Fig. 10,13) (Newell, 1949; Audebaud, 1973; Laubacher, 1978). It is possibly related to reactivation of Hercynian flat thrusts. It extends along strike for 300 km, from latitudes 14°S to 15°30'S, is up to 15 km wide, and can be interpreted as a southwest-vergent imbricate fan (Mégard, 1986). Red molassic deposits dated as Oligocene at the base unconformably overlie the Huancané belt (Chanove et al., 1969). In Bolivia, according to Martinez (1980), the few remnants of Mesozoic cover of the Eastern Cordillera have been thrown into parallel folds that are upright or asymmetric and can then be either southwest or northeast-vergent, and parallel Hercynian structures, many of which have been reactivated. All of these structures predate sedimentation of unconformably overlying Oligocene molassic rocks.

The Quechua phases 1, 2 and 3 are clearly identified by unconformities in central Peru, particularly in some intermontane basins and along the Marañon thrust and fold belt. The Quechua 1 phase is bracketed between 21.5 and 12.5 Ma, but coarse conglomerates interspersed with tuffs and lavas suggest active deformation between 21.5 and 17 Ma (McKee and Noble, 1982). North of the Pisco deflection, Mesozoic rocks and overlying red beds, including their volcaniclastic upper part, were thrown into open upright folds, and some of the Incaic structures of the Western Cordillera were coaxially reactivated. The Quechua 1 phase may also have affected part of the Eastern Cordillera. The Quechua 2 phase caused mainly dextral slip along the many longitudinal northwest-southeast trending faults (Soulas and Mégard, unpub. data). Tectonic analysis of these structures indicates that shortening was subhorizontal and in a north-south direction (Soulas, 1977). This phase is dated between 9.5 and 8.5 Ma in the intermontane Ayacucho basin where both folding and wrench faulting occur (Mégard et al., 1984). The younger tuffs and volcaniclastic strata of this basin were later folded by the Quechua 3 phase, in part during their deposition, which is dated at about 6 Ma in the basin, but is considered to be Pliocene in age in the Subandean zone where it ave rise to the Subandean thrust and fold belt (STFB, Figs. 10,13). The STFB is part of a late Late Miocene to Pliocene belt of deformation that includes the Mérida Andes, the Eastern Cordillera of Colombia, and the Subandean areas of Peru, Bolivia and northern Argentina. Thin-skinned tectonics characterize this belt in northern Peru (Pardo, 1982; Mégard, 1984b), northern Argentina (Mingram et al., 1979; Jordan et al., 1983), and probably also in most of Bolivia and southern

Peru (Roeder, 1982). In northern Peru (Fig. 11 and 13) the flattening at depth of steep reverse faults recognized at surface, was shown by seismic profiling. The STFB varies in width between 100 to 150 km between latitudes 3°S and 8°S, but it decreases in width to only 50 km at latitude 11°S. At about latitude 5°S, it consists of a major flat sole-thrust that lies at a depth of about 9 km at the top of the Paleozoic series and to which all the faults are connected (Fig. 13). Back-thrusts are common near the foreland, giving rise to a triangle zone similar to that of Alberta (e.g. Price, 1981; Jones, 1982).

South of the Pisco deflection, the chronology of the various post-Eocene deformation phases is different. A "30 Ma" phase is loosely bracketed between volcanic formations dated at 34 and 26 Ma in the Bolivian Altiplano (Evernden et al., 1977; Martinez, 1980). A second phase occurred at precisely 16 Ma, but is restricted to the Western Cordillera (Vatin-Pérignon et al., 1982). The Quechua 3 phase corresponds to two distinct compressive pulses. The first occurred between 6.8 and 6.2 Ma in the Western Cordillera of southern Peru, and between 7.25 and 6.4 Ma on the Bolivian Altiplano (Evernden et al. 1977; Martinez, 1980). The second compressive pulse is bracketed at 5.4 and 2.5 Ma in the Bolivian Altiplano, and between 3.9 and 1.8 Ma in southern Peru. These post-Eocene phases mostly reactivated earlier structures, and also gave rise to large reverse faults and to complex flat overthrusts of local extent. Elsewhere, structures related to these pulses are parallel, open, upright folds that involve terrigenous volcaniclastic and volcanic layers whose aggregate thickness reaches 10 km in the Tertiary basins of the Altiplano. As in northern Peru, the Subandean thrust and fold belt is mainly related to the late Miocene Quechua 3 phase of deformation (Martinez, 1980).

The Coastal Batholith of the Central Andes. The Coastal Batholith of Peru has been comprehensively studied during the last 20 years by a team of British geologists led by W.S. Pitcher and E. J. Cobbing. The book "Magmatism at a plate edge: the Peruvian Andes" (Pitcher et al., 1985) provides contributions on the different features of the batholith. I shall thus only give a few general data, most of which are extracted from this book or from earlier articles quoted in it.

The Coastal Batholith of Peru is a multiple and composite intrusion composed dominantly of tonalite and granodiorite. It is 1600 km long and up to 65 km across. It parallels the present-day Peru trench and is not influenced by the upper crust heterogeneities reflected in the Pisco and Huancabamba deflections. Between latitudes 7°S and 13°30'S the plutons occupy the axis of the western marginal basin. However, north of latitude 7°S, the batholith crosses the anomalous east-west trending structures of the Cajamarca fault zone and further north is intruded both into Paleozoic or Mesozoic rocks (Fig. 11). South of latitude 13°30'S, the Coastal Batholith is emplaced in basement of the Arequipa massif. The independence shown by the Coastal Batholith of both surface geology and crustal heterogeneities (Atherton et al. 1983), suggests that its emplacement can be related to a line of infracrustal cells that produced magma at regular intervals from about 101 to 32 Ma. A homogeneous deep-seated source is indicated by uniform geochemical features, including low $Sr^{87}:Sr^{86}$ initial ratios, low δO values, and Nd values near zero.

The granitoids of the batholith are grouped into units according to their relative age, modal variation, texture and xenolith composition. Closely related units form superunits, which are compositionally consanguineous and display distinctive geochemical signatures. Each superunit involves a basic-to-acid sequence and the batholith is made of a number of such sequences, that correspond to emplacement of distinct magma batches. Specific assemblages of units characterize the different segments forming the batholith. This plutonic segmentation does not relate to the structural segmentation of the Andes, since the Arequipa segment clearly crosses the Pisco deflection.

The magma cooled at high, subvolcanic, levels in the crust, perhaps within 3 or 4 km of the surface. Intrusion was controlled by growth-fractures, and individual plutons were emplaced by permissive cauldron subsidence and associated piecemeal stoping. Tonalites and granodiorites form great lenticular-shaped complexes, whereas subordinate "granites" commonly occur as nested assemblages of plutons and ring dykes. Some of the later "granites" may well have vented to the surface, but it seems that most of the volcanic cover of the batholith could have been fed by the numerous synplutonic basic andesite dykes.

In the Lima segment of the batholith, west-to-east younging of the time of intrusion is observed.

The Cenozoic Volcanic Arc of the Central Andes. Along with the late plutons in the eastern part of the Coastal Batholith, and younger plutons located farther east, the Cenozoic plateau volcanics of central and southern Peru represent the main magmatic activity of the orogenic period. Magmatic activity is no longer centered on the Coast but instead in the Western Cordillera or the Altiplano, and it is exclusively terrestrial (e.g. Mégard, 1986).

The volcanic rocks range from calc-alkaline basic andesites with olivine to rhyolites, with andesities and dacites being most widespread. Northeast of the calc-alkaline rocks, flows of shoshonitic composition are found (Lefèvre, 1979; Noble et al., 1975). Farther to the northeast a belt of highly peraluminous S-type volcanic and plutonic rocks of middle to late Cenozoic age is present (Noble et al., 1984; Clark et al., 1983). Volcanism remained active during the whole Tertiary, but was of widely variable intensity. It started in latest Cretaceous time in southern

Peru and may have begun as early as Paleocene in central Peru (Webb, in Cobbing et al., 1981). It was very intense during the Late Eocene and Early Oligocene (40 to 35 Ma) and again during the Miocene and Pliocene, with peaks of volcanic activity around 22 and 10 Ma (Noble et al., 1974; 1979). Quaternary volcanism is known only south of latitude 13°S. From 35 Ma to 7 Ma plutonic activity also migrated eastward, and gave rise to plutons of gabbroic to granodioritic composition. All of these plutons are stock-size, excepting the 150 km-long and 15 km-wide Cordillera Blanca batholith. Some smaller plutons probably consist of hypabyssal material that fed coeval volcanics. On a larger scale, a strong case can be made for contemporaneous arc volcanism and batholithic emplacement in Middle and Late Miocene time (Noble et al., 1975; McKee and Noble, 1982).

The Central Andes: Concluding Remarks

One of the major features of the Central Andes is the magmatic arc, which probably started in Lower Jurassic time and has been intermittently active since. The arc was always ensialic, even if extreme thinning of the crust occurred at some stages of its development in northern Peru. It occupied the same position from Lower Jurassic to Eocene time and then moved inland farther from the present trench by approximately the same distance everywhere. In my opinion, these observations provide the best argument against a collisional hypothesis for the formation of the Andean orogen in the Central Andes. Collision, if present, should have caused the sudden extinction and subsequent oceanward migration of the magmatic activity related to ocean-continent subduction, as it did several times in the Northern Andes. Even if terranes had been accreted and later shifted along the Pacific margin during the Andean orogenic period, collision should have deeply modified the pattern of subduction, relative to arc magmatism. This was not the case. The Central Andes, therefore, seem to be the best illustration of a long-lived simple marginal orogen, exclusively related to subduction of an oceanic plate beneath a continental margin (e.g. James, 1971; Audebaud et al., 1973; Mégard, 1978).

How did a single process, namely subduction, produce different effects at different times and different places, both with respect to volcanism and deformation? The present patterns of subduction at the trenches along western SOAM and of the related volcanism and deformation in the lip of the continent provide an actualistic solution to this question. As shown by the distribution of seismicity along the Benioff zone bordering western SOAM, the subducting Nazca plate is divided into large segments (e.g. Stauder, 1975; Barazangi and Isacks, 1976; Jordan et al., 1983). From about latitudes 5°S to 15°S and latitudes 27°S to 33°S, the subducting slab is dipping at at 5° to 10° east and lies close to the continental slab, whereas from latitudes 15°S to 24°S, it dips at about 30° east, and seems to be separated from the upper plate by an asthenospheric wedge (Baranzangi and Isacks, 1976). There is a striking correlation between the locations of shallowly-dipping setments and recent volcanic nulls and between 30° dipping slabs and active recent volcanism (Barazangi and Isacks, 1976; Mégard and Philip, 1976). The relationship between subduction and recent strain distribution is much more complicated, partly because of insufficient knowledge of the focal mechanisms of crustal earthquakes. However, the conclusion reached by Mégard and Philip (1976), namely that young compression was dominant north of latitude 15°S and extension was dominant to the south, is supported by recent data on the neotectonics of Peru north and south of this latitude. It appears that south of latitude 15°S (Sébrier et al., 1985), the Coastal zone, the Western Cordillera, the Altiplano and possibly the Eastern Cordillera are subjected to north-south extension at the surface with a low stretching ratio of about 1%. These regions are bounded by two zones of nearly east-west compression, located in the Subandean thrust belt and in the Peru trench inner wall. North of latitude 15°S, compression dominates and extension affects only the Coastal area and high-elevation ranges like the Cordillera Blanca (Sébrier et al., 1982; Dalmayrac and Molnar, 1981). The present-day relationship between flat subduction, magmatism nulls and dominant compression, and "normal" subduction which dips 30° or more, active magmatism and dominant extension, implies that better mechanical coupling of subducting and upper plates is obtained with flat subduction.

These inferences can be used for interpreting past events. For instance, shallowing of the angle of subduction along much of western South America might explain deformation pulses affecting the Andes for several thousands of kilometres along strike. This appears to be the case for the Senonian Peruvian phase, the Middle to Late Eocene Incaic phase and the latest Miocene to Early Pliocene Quechua 3 phase. Of these three pulses, the youngest is common to the Northern and Central Andes and to part of the Southern Andes, as it developed during the late Cenozoic evolution of the whole Andean orogen. By contrast, the Peruvian and Incaic pulses are restricted to the marginal Andes, south of latitude 3°S, because farther north, subduction probably was still occurring beneath offshore island arcs not yet welded to SOAM. Normal subduction with dips equal to or greater than 30° would explain both the pre-orogenic period of the Andes and the periods of tectonic quietness or of extension bracketed by the compressive pulses. The assumed large variations in the dip of the subduction zone under western SOAM must evidently be caused by plate tectonic processes of at least the same order of magnitude. Shallowing of a Benioff zone in a discrete segment of relatively modest size can be explained by the subduction of a buoyant volcanic ridge like the Nazca ridge (Pilger, 1981), or by subduction of a piece of crust younger than that in contiguous segments (Wortel, 1984). However, such processes cannot account for flattening of the Benioff zone along the whole SOAM active margin. Many authors, such as Cross and Pilger (1982) among others, have studied the factors controlling subduction geometry. However, a critical examination made by Yokokura (1981) indicates that the main factors controlling the dip angle are the relative velocity of the converging plate (and not the velocities of each plate separately) and the negative buoyancy force of the descending slab. From the foregoing study, changes in relative velocity appear as a first-order factor and changes in buoyancy as a second-order one, at least when the convergence vector makes a high angle with the plate boundary. By computing rates of spreading and of convergence along western SOAM from magnetic anomalies both in the east Pacific and the South Atlantic oceans, Frutos (1981) suggests that the beginning of the periods of higher rates are related to the compressive events. As proposed earlier by Mégard (1978) on the basis of observations by Luyendyck (1970), this would mean that the periods of the strongest coupling of the slabs and thus of major stress release correspond with the times transition from "normal" to flat subduction.

In contrast, stretches of time like the Oligocene to Late Miocene would be very similar to the present one, with compressional or extensional events restricted to a single segment. As already noted by Jordan et al. (1983), it is puzzling to see that the segmentation in past times seems to have preferentially happened near inherited boundaries like the Pisco deflection. This could mean that some features of the upper plate like changes in curvature of the trench, or differences in thickness of the upper slab near the trench, may induce differences in dip of the downgoing slab.

Acknowledgments. This work was supported by the Centre National de la Recherche Scientifique, Paris, and by the Institut Francais d'Etudes Andines and the INGEMMET, Lima, Peru. I am particularly indebted to my South American, French and North American colleagues who participated in common field work during the last 25 years. I wish to thank J.W.H. Monger, and T. Feininger for their review of the early drafts of the manuscript. This paper is a contribution to PICG Project No. 202.

References

Aguirre, L., and R. Offler, Burial metamorphism in the Western Peruvian trough: its relation to Andean magmatism and tectonics, in: Pitcher, W. S., et al., eds., Magmatism at a plate edge: the Peruvian Andes, Blackie, Glasgow and Halsted Press, New York, 59-71, 1985.

Arango, J. L., T. Kassem, and H. Yduque, Mapa Geologico de Colombia, escale 1:1,500,000, Inst. Nat. Inv. Geol. Min., Colombia, 1976.

Atherton, M. P., W. S. Pitcher, and A. Warden, The Mesozoic marginal basin of Central Peru, Nature, 305, 303-306, 1983.

Atherton, M. P., A. Warden, and L. M. Sanderson, The Mesozoic marginal basin of Central Peru: a geochemical study of within-plate-edge volcanism, in: Pitcher, W. S., et al., eds., "Magmatism at a plate edge: the Peruvian Andes", Blackie, Glasgow and Halsted Press, New York, 47-58, 1985.

Aubouin, J., Chaînes liminaires (andines) et chaînes geosynclinales (alpines), 24th Int. Geol. Cong., Montreal, sec. 3, 438-461, 1972.

Audebaud, E., Geología de los cuadrángulos de Ocongate y Sicuani, Bol. Serv. Geol. Min. Lima, 25, 1973.

Audebaud, E., R. Capdevila, B. Dalmayrac, J. Debelmas, G. Laubacher, C. Lefevre, R. Marocco, C. Martinez, M. Mattauer, F. Mégard, J. Paredes and P. Tomasi, Les traits geologiques essentiels des Andes centrales (Pérou-Bolivie), Rev. Geog. Phys. Geol. Dyn., Paris, 15, 73-114, 1973.

Audebaud, E., G. Laubacher, and R. Marocco, Coupe géologique des Andes du Sud du Pérou de l'Océan Pacifique au bouclier brésilien, Geol. Rdsch., 65, 223-264, 1976.

Baldock, J., Geología del Ecuador, Dir. Gral. Geol. Min., Quito, 1-70, 1982.

Barazangi, M., and B. L., Isacks, Spatial distribution of earthquakes and subduction of the Nazca plate beneath South America, Geology, 4, 686-692, 1976.

Barrero, D., Geology of the Central western Cordillera, west of Buga and Roldanillo, Colombia, Ingeominas, Publ. Geol. Espec., 4, 75 p., 1979.

Beaufils, G., Recherches de concentrations métalliques sulfurées et cartographie géologique dans la région de Tambo Grande, BRGM-Perou, unpub. rept. and map, 1978.

Benavides, V., Cretaceous system in Northern Peru, Am. Mus. Nat. Hist. Bull., 108, 252-494, 1956.

Bourgois, J., J. Azema, J. Tournon, H. Bellon, B. Calle, E. Parra, J. F. Toussaint, G. Glacon, H. Feinberg, P. De Wever, and I. Origlia, Ages et structures des complexes basiques et ultrabasiques de la facade pacifique entre 3°N et 12°N (Colombie, Panama et Costa-Rica), Bull. Soc. Geol. France, 24, 545-554, 1982.

Bourgois, J., J. F. Toussaint, H. Gonzales, A. Orrego, J. Azema, B. Calle, A. Desmet, A. Murcia, A. Pablo, E. Parra, and J. Tournon, Les ophiolites des Andes de Colombie: évolution structurale et signification géodynamique, in: Mascle, A., ed., Géodynamique des Caraîbes, 475-493, Technip., Paris, 1985.

Bristow, C. R., Guide to the geology of the Cuenca basin, Southern Ecuador, Ecuadorian Geol. and Geophys. Soc., 54 p., Quito, 1973.

Bristow, C. R., and R. Hoffstetter, Ecuador, Lexique stratigraphique int., V, 5-a2, Centre Nat. Rech. Sci., Paris, 1977.

Burke, K., C. Cooper, J. F. Dewey, P. Mann, and J. L. Pindell, Carribean plate tectonics and relative plate motions, in: Bonini, W. E., et al., eds., "The Caribbean-South American plate boundary and regional tectonics", Geol. Soc. Am. Mem., 162, 31-64, 1984.

Campbell, C. J., Colombian Andes, in: Spencer, A. M., ed., "Mesozoic-Cenozoic orogenic belts", Spec. Pub. Geol. Soc., London, 4, 705-724, 1974a.

Campbell, C. J., Ecuadorian Andes, in: Spencer, A. M., ed., "Mesozoic-Cenozoic orogenic belts", Spec. Pub. Geol. Soc., London, 4, 725-732, 1974b.

Canfield, R. W., G. Bonilla, and R. K. Robbins, Sacha oil field of Ecuadorian oriente, Am. Ass. Pet. Geol. Bull., 66, 1076-1090, 1982.

Case, J. E., J. Barnes, G. París, H. González, and A. Viña, Trans-Andean geophysical profile, southern Colombia, Geol. Soc. Am. Bull., 84, 2895-2904, 1973.

Case, J. E., L. G. Durán, A. López, and W. R. Moore, Tectonic investigations in western Colombia and eastern Panama, Geol. Soc. Am. Bull., 82, 2685-2712, 1971.

Chanove, G., M. Mattauer, and F. Mégard, Précisions sur la tectonique tangentielle des terrains secondaires du massif de Pirin (nord-ouest du lac Titicaca, Pérou), C. R. Acad. Sci., 270, 1088-1091, 1969.

Clark, A. H., V. V. Palma, D. A. Archibald, E. Farrar, M. J. Arenas, and R. C. R. Robertson, Occurrence and age of tin mineralization in the Cordillera Oriental, Southern Peru, Econ. Geol. 78, 514-520, 1983.

Cobbing, E. J., J. M. Ozard, and N. J. Snelling, Reconnaissance geochronology of the cristalline basement rocks of the Coastal Cordillera of Southern Peru, Geol. Soc. Am. Bull., 88, 241-246, 1977.

Cobbing, E. J., W. S. Pitcher, J. J. Wilson, J. W. Baldock, W. P. Taylor, W. McCount, and N. J. Snelling, The geology of the Western Cordillera of northern Peru, Inst. Geol. Sci., Overseas Mem., 5, London, 1981.

Colman, J. A. R. Guidebook to the geology of the Santa Elena Peninsula, Ecuadorian Geol. and Geophys. Soc., Quito, 1970.

Coney, P. J., Structural evolution of the Cordillera Huayhuash, Andes of Peru, Geol. Soc. Am. Bull., 82, 1863-1884, 1971.

Coulbourn, W. I., Tectonics of the Nazca plate and the continental margin of western South America 18° to 23°S, in: Kulm, L. D., et al., eds., "Nazca plate: crustal formation and Andean convergence, Geol. Soc. Am. Mem., 154, 587-618, 1981.

Cross, T. A., and R. H. Pilger, Controls of subduction geometry, location of magmatic arcs, and tectonics of arc and back-arc regions, Geol. Soc. Am. Bull., 93, 545-562, 1982.

Crowell, J. C., A. Suárez-Soruco and A. C. Rocha-Campos, The Silurian Cancaniri (Zapla) Formation of Bolivia, Argentina and Peru, in: Hambrey, M. J., and W. V. Harland, eds., "Earth's pre-Pleistocene glacial record", Cambridge U. Press, 902-907, 1981.

Dalmayrac, B., J. R. Lancelot, and A. Leyreloup, Two billion-year granulites in the late Precambrian metamorphic basement along the Southern Peruvian coast, Science, 198, 49-50, 1977.

Dalmayrac, B., G. Laubacher, and R. Marocco, Caractères généraux de l'évolution géologique des Andes péruviennes, Trav. Doc. ORSTOM, 122, Paris, 1980.

Dalmayrac, B., and P. Molnar, Parallel thrust and normal faulting in Peru and constraints on the state of stress, Earth Planet. Sci. Lett., 55, 473-481, 1981.

De Souza, H. A., A. Espinosa, and M. Delaloye, K-Ar ages of basic rocks in the Patía valley, southwest Colombia, Tectonophysics, 107, 135-145, 1984.

Dirección General de Geología y Minería, Mapa geológico nacional de la república del Ecuador, escala 1/1,000,000, Quito, 1982.

Duncan, R. A., and R. B. Hargraves, Plate tectonic evolution of the Caribbean region in the mantle reference frame, in: Bonini, E. E., et al., eds., "The Caribbean-South American plate boundary and regional tectonics", Geol. Soc. Am. Mem., 162, 81-94, 1984.

Espinosa, A., Sur les roches basiques et ultrabasiques du bassin du Patía", unpub. Ph.D. thesis, U. Genève, 1980.

Evernden, J. F., S. J. Kriz, and C. M. Cherroni, Potassium-Argon ages of some Bolivian rocks, Econ. Geol., 72, 1042-1061, 1977.

Faucher, B., and E. Savoyat, Esquisse geologique des Andes de l'Equateur, Rev. Geogr. Phys. et Geol. Dyn., 15, 115-142, 1973.

Feininger, T., Mapa gravimétrico de anomalías Bouguer simples del Ecuador: 1/1,000,000, Instituto Geográfico Militar, Quito, 1977.

Feininger, T., Mapa geológico de la parte occidental de la provincia de El Oro: escala 1/-50,000, 1978.

Feininger, T., Eclogite and high-pressure regional metamorphic rocks from the Andes of Ecuador, J. Petrology, 21, 107-140, 1980.

Feininger, T., Glaucophane schist in the Andes at Jambaló, Colombia, Can. Mineralogist, 20, 41-47, 1982a.

Feininger, T., The metamorphic "basement" of Ecuador, Geol. Soc. Am. Bull,, 93, 87-92, 1982b.

Feininger, T., D. Barrero and N. Castro, Geología de parte de los departamentos de Antioquia y Caldas (sub-zona IIB), Bol. Geol., XX, 2, 1-173, Bogota, 1972.

Feininger, T., and C. R. Bristow, Cretaceous and

Paleogene geologic history of coastal Ecuador, Geol. Rdsch., 69, 849-874, 1980.

Feininger, T., and M. K. Seguin, Simple Bouguer gravity anomaly field and the inferred crustal structure of continental Ecuador, Geology, 11, 40-44, 1983.

Flüh, E. R., Geodynamische Entwicklung der noerdlichen Anden, Unpub. Ph.D. thesis, U. of Kiel, W. Germany, 1982.

Flüh, E.R., The basic igneous complex, trace of an ancient Galapagos hotspot aseismic ridge?, Zbl. Geol. Palaont., 291-303, 1983.

Frutos, J., Andean tectonics as a consequence of sea-floor spreading, Tectonophysics, 72, T21-T32, 1981.

Gansser, A., Facts and theories on the Andes, J. Geol. Soc. Lon., 129, 93-131, 1973.

Geyer, O. F., Die mesozoische Magnafazies-Abfolge in den nördlichen Anden (Peru, Ekuador, Kolumbien), Geol. Rdsch., 69, 875-891, 1980.

Göbel, V. W., and Stibane, F. R., K/Ar hornblende ages of tonalite plutons, Cordillera occidental, Colombia, Pub. Esp., Geol., U. Nac. Colombia, Medellin, 19, 1-2, 1979.

Goossens, P. J., La geología de la costa ecuatoriana entre Manta y Guayaquil, Bol. estud. Geol. Serv. Nac. Geol. Min., 1, 5-17, Quito, 1968.

Goossens, P. J., and Rose, W. I., Chemical composition and age determination of tholeiitic rocks in the basic igneous complex, Ecuador, Geol. Soc. Am. Bull., 84, 1043-1052, 1973.

Harrison, J. V., The geology of the Central Andes in part of the province of Junín, Quart. J. Geol. Soc. London, 99, 1-136, 1943.

Henderson, W. G., Cretaceous to Eocene volcanic arc activity in the Andes of northern Ecuador, J. Geol. Soc. London, 136, 367-378, 1979.

Herbert, H. J., Die kristallinen Gesteine aus der nördlichen Halfte der E-Kordillera Ecuadors, Geoteckt. Forsch., 65, 1-77, 1983.

Irving, E. M., Structural evolution of the northernmost Andes, Columbia, U.S. Geol. Surv. Prof. Pap., 846, 1975.

James. D. E., Plate tectonic model for the evolution of the central Andes, Geol. Soc. Am. Bull. 82, 3325-3346, 1971.

James, D., C. Brooks, and A. Cuyubamba, Early evolution of the Central Andean volcanic arc, Carnegie Inst. Yearbook, 74, 247-250, 1975.

Jones, P. B., Oil and gas beneath east-dipping underthrust faults in the Alberta foothills, in: Powers, R. B., ed., "Geologic studies of the Cordilleran thrust belt", Rocky Mount. Assoc. Geol., 1, 61-74, 1982.

Jordan, T. E., B. L. Isacks, R. W. Allmendinger, J. A. Brewer, V. A. Ramos, and C. J. Ando, Andean tectonics related to geometry of subducted Nazca plate, Geol. Soc. Am. Bull., 94, 341-361, 1983.

Julivert, M., Colombie (lere partie), Lexique stratigraphique int., V, 4a, Centre Nat. Rech. Sci., Paris, 1968.

Julivert, M., Cover and basement tectonics in the Cordillera Oriental of Colombia, South America, and a comparison with some other folded chains, Geol. Soc. Am. Bull., 81, 3623-3646, 1970.

Julivert, M., Les traits structuraux et l'evolution des Andes colombiennes, Rev. Geog. Phys. Geol. Dyn., Paris, 15, 143-156, 1973.

Juteau, T., F. Mégard, L. Raharison and H. Whitechurch, Les assemblages ophiolitiques de l'Occident équatorien: nature pétrographique et position structurale, Bull. Soc. Geol. France, 19, 1127-1132, 1977.

Kennerley, J. B., Outline of the Geology of Ecuador, Overseas Geol. and Min. Resour., 55, London, 1980.

Kontak, D. J., A. H. Clark, E. Farrar, and D. F. Strong, The rift-associated Permo-Triassic magmatism of the eastern Cordillera: a precursor to the Andean orogeny, in: Pitcher, W. S., et al., eds., Magmatism at a plate edge: the Peruvian Andes, Blakie, Glasgow and Halsted Press, New York, 36-44, 1985.

Kroonenberg, S., A. Grenvillian granulite belt in the Colombian Andes and its relation to the Guiana shield, Geol. in Mijnb., 82, 325-333, 1982.

Laubacher, G., Géologie de la Cordillère orientale et de l'altiplano au nord et au nord-ouest du lac Titicaca (Pérou), Trav. et Doc. ORSTOM, 95, Paris, 1978.

Laubacher, G., A. J. Boucot, and J. Gray, Additions to Silurian stratigraphy, lithofacies, biogeography and paleontology of Bolivia and southern Peru, J. Paleont., 56, 1138-1170, 1982.

Laubacher, G., and F. Mégard, The Hercynian basement: a review, in: Pitcher, W. S., et al., eds., Magmatism at a plate edge, the Peruvian Andes, Blakie, Glasgow and Halsted Press, New York, 29-35, 1985.

Lebrat, M., Caracterisation géochimique du volcanisme anté-orogénique de l'Occident équatorien: implications géodynamiques, Doc. Trav. Centre Géol. Geophys., Montpellier, 6, 118 p., 1985.

Lebrat, M., F. Mégard, C. Dupuy, and J. Dostal, Geochemistry of the Cretaceous volcanic rocks of Ecuador: geodynamic implications, subm. to Geol. Soc. Am. Bull., 1986.

Lebrat, M., F. Mégard, T. Juteau, and J. Calle, Pre-Orogenic volcanic assemblages and structure in the western Cordillera of Ecuador between 1°40'S and 2°20'S, Geol. Rdsch., 74, 343-351, 1985.

Leclere-Vanhoeve, A., and J. F. Stephan, Evolution géodynamique des Caraîbes dans le système points chauds, in: Mascle, A., et al., eds., Geodynamique des Caraîbes, 1, 321-34, Ed. Technip., Paris, 1985.

Lefevre, C., Un exemple de marge active daus les Andes du Perou Sud du Miocene a l'actuel, unpub. Sc.D. thesis, U. Sci. Tech. Languedoc, Montpellier, 1979.

Litherland, M., B. A. Klinck, E. A. O'Connor, and P. E. J. Pitfield, Andean-trending mobile belts in the Brazilian shields, Nature, 314, 345-348, 1985.

Luyendyck, B. P., Dips of downgoing lithospheric plates beneath island arcs, Geol. Soc. Am. Bull., 81, 3411-3416, 1970.

McCourt, W. J., J. A. Apsden, and M. Brook, New geological and geochronological data from the Colombian Andes: continental growth by multiple accretion, J. Geol. Soc. London, 141, 831-845, 1984.

McKee, E. H., and D. C. Noble, Miocene volcanism and deformation in the Western Cordillera and high plateaux of south-central Peru, Geol. Soc. Am. Bull., 93, 657-662, 1982.

Marriner, G. F., and D. Millward, The petrology and geochemistry of Cretaceous to Recent volcanism in Colombia, J. Geol. Soc. London, 141, 473-486, 1984.

Martinez, C., Structure et evolution de lachaîne hercynienne et de la chaîne andine dans le nord de la cordillere des Andes de Bolivie, Trav. et Doc., ORSTOM, 119, Paris, 1980.

Mattson, P., Caribbean structural breaks and plate movements, in: Bonini, W. E., et al., eds., The Caribbean-South American plate boundary and regional tectonics, Geol. Soc. Am. Mem. 162, 131-152, 1984.

Mégard, F., Etude géologique des Andes du Pérou central, Mem. ORSTOM, 86, Paris, 1978.

Mégard, F., Andine (chaîne) Encyclopaedia Universalis, 2d ed., 80-84, Paris, 1984a.

Mégard, F., The Andean orogenic period and its major structures in central and northern Peru, J. Geol. Soc. London, 141, 893-900, 1984b.

Mégard, F., Structure and evolution of the Peruvian Andes, in: Schaer, J. P., and J. Rodgers, eds., "Anatomy of mountain ranges", Princeton University Press, in press, 1986.

Mégard, F., B. Dalmayrac, G. Laubacher, C. Martinez, J. Paredes, and P. Tomasi, La chaîne hercynienne au Pérou et en Bolivie: premiers résultats, Cah. ORSTOM, Sér. Géol., III, 5-44, 1971.

Mégard, F., D. C. Noble, E. H. McKee, and H. Bellon, Multiple pulses of Neogene compressive deformation in the Ayacucho intermontane basin, Andes of Central Peru, Geol. Soc. Am. Bull., 95, 1108-1117, 1984.

Mégard, F., and H. Philip, Plio-Quaternary tectono-magmatic zonation and plate tectonics in the Central Andes, Earth Planet. Sci. Lett., 33, 231-238, 1976.

Meissner, R., E. R. Flüh, F. Stibane and E. Berg, Dynamics of the active plate boundary in southwest Colombia according to recent geophysical measurements, Tectonophysics, 35, 115-136, 1976

Meyer, R. P., W. D. Mooney, A. L. Hales, C. E. Helsley, C. P. Woolard, D. M. Hussong, and J. E. Ramírez, Refraction observations across a leading edge, Malpelo Island to the Colombian Cordillera occidental, in: Sutton, G. H., et al., eds., "The geophysics of the Pacific ocean basin and its margins", Am. Geophys. Union, Monograph 19, 105-132, 1976.

Millward, D., G. F. Marriner, and A. D. Saunders, Cretaceous tholeiitic volcanic rocks from the western Cordillera of Colombia, J. Geol. Soc. London, 141, 847-860, 1984.

Mingramm, A., A. Russo, A. Pozzo, and L. Cazau, Sierras subandinas, in: "Segundo simposio de Geologia regional argentina": Cordoba, Acad. Nac. Ciencas, 1, 95-130, 1979.

Mojica, J., An outline on the Jurassic in Colombia, Geol. Colombiana, 13, 129-136, 1984.

Mooney, W. D., An east-Pacific Caribbean ridge during the Jurassic and Cretaceous and the evolution of western Colombia, in: Pilger, R. J., ed., "The origin of the Gulf of Mexico and the early opening of the Central North Atlantic Ocean", Louisiana State Univ., Raton Rouge, 55-73, 1980.

Mooney, W. D., R. P. Meyer, J. P. Laurence, H. Meyer, and E. Ramírez, Seismic refraction studies of the Western Cordillera, Colombia, Bull. Seismol. Soc. Am., 69, 1745-1761, 1979.

Mourier, T., F. Mégard, O. A. Pardo and L. Reyes, L'évolution mésozoïques des Andes de Huancabamba (3°-8°S) et l'hypothèse de l'accretion du microcontinent Amotape-Tahin, subm. to Bull. Soc. Géol. Fr., 1986.

Myers, J. S., Cretaceous stratigraphy and structure, Western Andes of Peru between latitudes 10°-10°30'S, Am. Assoc. Petrol. Geol. Bull., 58, 474-487, 1974.

Nelson, H. W., Contribution to the geology of the central and western Cordillera of Colombia in the sector between Ibagué and Cali, Leidse Geol. Mededellngen, 22, 1-76, 1957.

Newell, N. D., Geology of the lake Titicaca region, Peru and Bolivia, Geol. Soc. Am. Mem., 36, 111 p., 1949.

Noble, D. C., H. R. Bowman, A. J. Hebert, M. L. Silbermann, C. E. Heropoulos, B. P. Fabbi and C. E. Hedge, Chemical and isotopic constraints on the origin of low-silica latite and andesite from the Andes of central Peru, Geology, 3, 501-504, 1975.

Noble, D. C., E. H. McKee, E. Farrar, and U. Petersen, Episodic Cenozoic volcanism and tectonism in the Andes of Peru, Earth Planet. Sci. Lett., 21, 213-220, 1974.

Noble, D. C., E. H. McKee, and F. Mégard, Early Tertiary "Incaic" tectonism, uplift and volcanic activity, Andes of central Peru, Geol. Soc. Am. Bull., 90, 903-937, 1979.

Noble, D. C., M. Sébrier, F. Mégard, and E. H. McKee, Demonstration of two pulses of Paleogene deformation in the Andes of Peru, Earth Planet. Sci. Lett., 73, 345-349, 1985.

Noble, D. C., M. L. Silbermann, F. Mégard, and H. R. Bowman, Comendite (peralkaline rhyolite) in the Mitu Group Central Peru: evidence of Permian-Triassic crustal extension in the Central Andes, J. Res. U.S. Geol. Surv., 6, 453-457, 1978.

Noble, D. C., T. A. Vogel, P. S. Peterson, G. A. Landis, N. K. Grant, P. A. Jezek, and E. H. McKee, Rare-element enriched, S-type ash-flow tuffs containing phenocrysts of muscovite, andalusite and sillimanite, southeastern Peru, Geology, 12, 25-39, 1984.

Orrego, L. A., H. V. Cepeda, and S. G. I. Rodriguez, Esquistos glaucofánicos en el área de Jambaló, Cauca (Colombia), Geol. Norandina, 1, 5-10, 1980.

Pardo, A., Caracteristicas estructurales de la faja subandina del Norte del Peru, in: Simp. "Exploracion petrolera en las cuencas subandinas de Venezuela, Colombia, Ecuador y Peru", Asoc. Colomb. Geol. Geofis. del Petroleo, Bogota, 1982.

Pérez Téllez, G., Evolución geológica de la cuenca pacífica (Geosynclinal de Bolivar), sector noroccidental de Suramérica, Bol. Geol., 14 (28), 25-44, Bucaramanga, Colombia, 1980.

Pichler, H., F. R. Stibane and R. Weyl, Basischer Magmatismus, in: Sudlichen Mittelamerika, Kolumbien und Ecuador, Neues Jahrb. Geol. Palaont. 2, 102-105, 1974.

Pilger, R. J., Plate reconstructions, aseismic ridges and low-angle subduction beneath the Andes, Geol. Soc. Am. Bull., 86, 1639-1653, 1981.

Pitcher, W. S., M. P. Atherton, E. J. Cobbing and R. B. Beckinsale, (eds.), "Magmatism at a plate edge: the Peruvian Andes", Blackie, Glasgow and Halsted Press, New York, 1985.

Price, R. A., The Cordilleran foreland thrust and fold belt in the southern Canadian Rocky Mountains, in: McClay, K. N. and N. J. Price, eds., "Thrust and Nappe tectonics", Spec. Publ. Geol. Soc. London, 9, 427-448, 1981.

Prinz, P., Stratigraphie und Ammonitenfauna der Pucara-Gruppe (Obertrias-Unterjura) von Nord-Peru, Paleontographica, A188, 153-197, 1985.

Regan, P. F., The early basic intrusions, in: Pitcher, W. S., et al., eds., "Magmatism at a plate edge: the Peruvian Andes", Blackie, Glasgow and Halsted Press, New York, 72-89, 1985.

Restrepo, J. J., and J. F. Toussaint, Edades radiométricas de algunas rocas de Antioquia, Colombia, Pub. Espec. Geol., U. Nac. Colombia, Medellín, 6, 1-24, 1975.

Roeder, D., Geodynamic model of the Subandean zone in Alto Beni area, Bolivia, in: Simp. "Exploración petrolera en las cuencas subandinas de Venezuela, Colombia, Ecuador y Perú", Asoc. Colomb. Geol. Geofis. del Petroleo, Bogotá, 1982.

Romani, M., Géologie de la région minière Uchucchacua-Hacienda Otuto, Pérou, unpub. Thesis, 3rd cycle D, Univ. Grenoble, 1982.

Sébrier, M., D. Huaman, J. L. Blanc, J. Macharé, D. Bonnot, and J. Cabrera, Observaciones acerca de la neotectónica del Perú, unpub. report, Inst. Geofis. Perú, Lima, 1982.

Sébrier, M., R. Marocco, J. J. Gross, S. Macedo, and M. Montoya, Evolución néogena del piedemonte Pacîfico de los Andes del Sur del Perú, 2 do Congr. Geol. Chile, Arica, Actas 3, 171-188, 1979.

Sébrier, M., J. L. Mercier, F. Mégard, G. Laubacher, and E. Carey-Gailhardis, Quaternary normal and reverse faulting and the state of stress in the central Andes of southern Peru, Tectonics, 4, 739-780, 1985.

Servicio Nacional de Geologia y Mineria, Mapa geologico de la republica del Ecuador, escala 1/1,000,000, Quito, 1969.

Shepherd, G. L., and R. Moberly, Coastal structure of the continental margin, northwest Peru and southwest Ecuador, in: Kulm, L. D., et al., eds., "Nazca plate: crustal formation and Andean convergence", Geol. Soc. Am., Mem., 154, 351-392, 1981.

Soto, F., Facies y ambientes deposicionales Cretácicos, Area centro-sur de la cuenca Marañon Bol. Soc. Geol. Perú, 60, 233-250, 1979.

Soulas, J. P., Les phases tectoniques andines du Tertiaire supérieur, résultat d'une transversale Pisco-Ayacucho (Pérou central), C. R. Acad. Sci., Paris, 284D, 2207-2210, 1977.

Stauder, W., Subduction of the Nazca plate under Peru as evidenced by focal mechanisms and by seismicity, J. Geophys. Res., 80, 1053-1064, 1975.

Steinmann, G., Geologie von Peru, Karl Winter, Heidelberg, 1929.

Stibane, F. R., Paläogeographie und Tektogenese der kolumbianischen Anden, Geol. Rdsch., 56, 629-642, 1967.

Thery, J. M., Constitution du Nord-Ouest du continent sud-americain avant les tectoniques andines, unpub. Thesis, U. Bordeaux III, 1982.

Thornburg, T., and L. D. Kulm, Sedimentary basins of the Peru continental margin. Structure, stratigraphy and Cenozoic tectonics from 6°S to 16°S latitude, in: Kulm, L. D., ed., "Nazca plate: crustal formation and Andean convergence", Geol. Soc. Am. Mem., 154, 393-422, 1981.

Toussaint, J. F., G. Botero, and J. L. Restrepo, Datación K-Ar del batolito de Buga, Publ. Esp., Geol. U. Nac. Colombia, Medellín, 13, 1-3, 1978.

Toussaint, J. F., and J. J. Restrepo, Modelos orogénicos de tectónica de placas en los Andes colombianos, Bol. Cienc. Tierra, U. Nac., Medellîn, 1, 1-47, 1976.

Toussaint, J. F., and J. L. Restrepo, Magmatic evolution of the northwestern Andes of Colombia, Earth Sci. Rev., 18, 205-213, 1982.

Trouw, R., Cuatro cortes por la faja metamórfica de la Cordillera Real, Ecuador, Bol. Cient. Tecno., 1, 40 p., Esc. Sup. Politec. Litoral, Guayaquil, 1976.

Tschopp, H. J., Geologische Skizze von Ecuador, Bull. Ver. Schweiz. Petroleumgeol. und Ing., 15(48), 14-45, 1948.

Tschopp, H. J., Oil explorations in the Oriente of Ecuador 1938-1950, Bull. Am. Ass. Petrol. Geol., 37, 2303-2347, 1953.

Van der Hammen, T., Late Cretaceous and Tertiary stratigraphy and tectogenesis of the Colombian Andes, Geol. in Mijn., 40, 181-188, 1961.

Vatin-Pérignon, N., G. Vivier, M. Sébrier, and M.

Fornari, Les derniers évènements andins marqués par le volcanisme cénozoique de la Cordillère Occidentale sud-péruvienne et de son piémont pacifique entre 15°45' et 18°S, Bull. Soc. Geol. France, 24, 649-650, 1982.

Vicente, J. C., Elementos de la estratigrafia mesozoica sur-peruana, in: Volkheimer, W., and E. A. Musacchio, "Cuencas sedimentarias del Jurásico y Cretácico de América del Sur", Comité sudamer. Jura Cretac, 1, 319-351, 1981.

Vicente, J. C., B. Beaudoin, A. Chavez, and I. Léon, La cuenca de Arequipa (Sur Peru) durante el Jurásico-Cretácico inferior, in: 5th Cong. Latinoam., Geol., Buenos Aires, Actas 1, 121-153, 1982.

Vicente, J. G., F. Sequeiros, M. A. Valdivia, and J. Zavala, El sobre-escurrimiento de Cincha-Lluta: elemento del accidente mayor andino al noroeste de Arequipa, Bol. Soc. Geol. Peru, 61, 67-100, 1979.

Von Huene, R., L. K. Kulm, and J. Miller, Structure of the frontal part of the Andean convergent margin, J. Geophys. Res., 90, 5429-5442, 1985.

Wilson, D. V., The deeper structure of the Central Andes and some geophysical constraints, in: Pitcher, W. S., et al., eds., "Magmatism at a plate edge: the Peruvian Andes", Blackie, Glasgow and Halsted Press, New York, 13-18, 1985.

Wilson, J. J., Cretaceous stratigraphy of Central Andes of Peru, Am. Ass. Petrol. Geol. Bull., 47, 1-34, 1963.

Wilson, J. J., L. Reyes, and J. Garayar, Geología de los cuadrángulos de Mollebamba, Tayabamba, Pomabamba y Huari, Bol. Serv. Geol. Min. Lima, 16, 1967.

Wortel, M. J. R., Spatial and temporal variations in the Andean subduction zone, J. Geol. Soc. London, 141, 783-791, 1984.

Yokokura, T., On subduction dip angles, Tectonophysics, 77, 63-77, 1981.

Zeil, W., The Andes: a geological review, Geb. Borntraeger, Berlin, 1979.

A GENERAL VIEW ON THE CHILEAN-ARGENTINE ANDES, WITH EMPHASIS ON THEIR EARLY HISTORY

Francisco Hervé,[1] Estanislao Godoy,[1] Miguel A. Parada,[1] Víctor Ramos,[2]
Carlos Rapela,[3] Constantino Mpodozis,[4] and John Davidson[4]

Abstract. The Andes are generally thought of as a young mountain belt associated with eastward subduction of the Nazca, or pre-existing oceanic plates, beneath the South American plate. However, the Meso-Cenozoic Andes are built over a highly complex basement which spans Precambrian to Triassic time. Collision of continental blocks seems to have occurred in the early Paleozoic in northern and central Argentina and Chile. A well defined late Paleozoic subduction complex is present along the Coastal Range south of latitude 30°S. This is associated with a Carboniferous to Permian magmatic belt and related marine basins constructed inboard of the southwest Gondwana continental margin. Paired metamorphic belts with local blueschists were developed during the accretion processes. The Meso-Cenozoic history is mainly one of plutovolcanic "accretion" via episodic buildups of narrow, linear, north-south trending, calc-alkaline magmatic belts. Those of Jurassic to Early Cretaceous age are related to back-arc ensialic basins which, in the southernmost Andes, evolved into short lived, marginal basins floored by oceanic-type crust. Longitudinal changes in stratigraphy and tectonism along the belt appear to reflect complex interactions of the converging palaeo-plates which have present day expression in the volcanic and tectonic segmentation of the Andes. The huge late Tertiary to Recent volcanic deposits testify to the widespread partial melting of subcontinental mantle associated with subduction of the Nazca plate. The latter may have tectonically eroded parts of the Paleozoic complexes and Meso-Cenozoic fore-arc assemblages. The existence of major faults roughly parallel with the continental margin, one of which is associated with the subducting Chile Rise, suggests that other rises or ridges have been consumed earlier in the geological history of the Andean chain.

[1]Departamento de Geología, Universidad de Chile, Casilla 13518, Correo 21, Santiago, Chile
[2]Servicio Geológico Nacional, Avda. Sta. Fe 1548, 1060 Buenos Aires, Argentina
[3]División Geología, Universidad Nacional de La Plata, Paseo del Bosque 1000, Argentina
[4]Servicio Nacional de Geología y Minería, Casilla 10465, Santiago, Chile

Introduction

The Andes are located along the western margin of the South American plate, beneath which Pacific Ocean lithosphere is being subducted. They constitute a favourable place for the study of processes taking place today, such as magmatic and seismic activity, and oceanic trench development. However, the Andes result from complex processes that extend back in time to the Paleozoic, and contain a changing geological record of the interaction between continental and oceanic plates. These interactions can be interpreted in terms of the main geotectonic elements, related to magmatic arcs, fore-arc and back-arc sedimentary basins, metamorphism and deformation. It is the purpose of this paper to present synthetically the succession of events and tectonic settings, using wherever suitable the conceptual model of an active continental margin, that has been recorded in this region since early Paleozoic time.

So fast has been the accumulation of geologic data in the last few decades that it is difficult for one person to fully understand or record them. The authors are acquainted with modern research throughout this region, but a full recounting of all of it is beyond the scope of the paper. The less known pre-Mesozoic development of the Andes is emphasized herein and the Meso-Cenozoic evolution is presented only as a concise synthesis, as more detailed information can be obtained in other recent reviews.

Synthetic information on the geotectonic organization will be given for each of the generally accepted orogenic periods of the Andes, and will point out the distribution and nature of magmatic, tectonic and metamorphic processes. No detailed account of stratigraphy is presented, and local names are avoided as much as possible. Reference will be made mainly to modern syntheses in which the interested reader may pursue more detailed accounts of regional or particular geologic aspects.

Unpublished information in Chile and Argentina produced by research and mapping projects of the

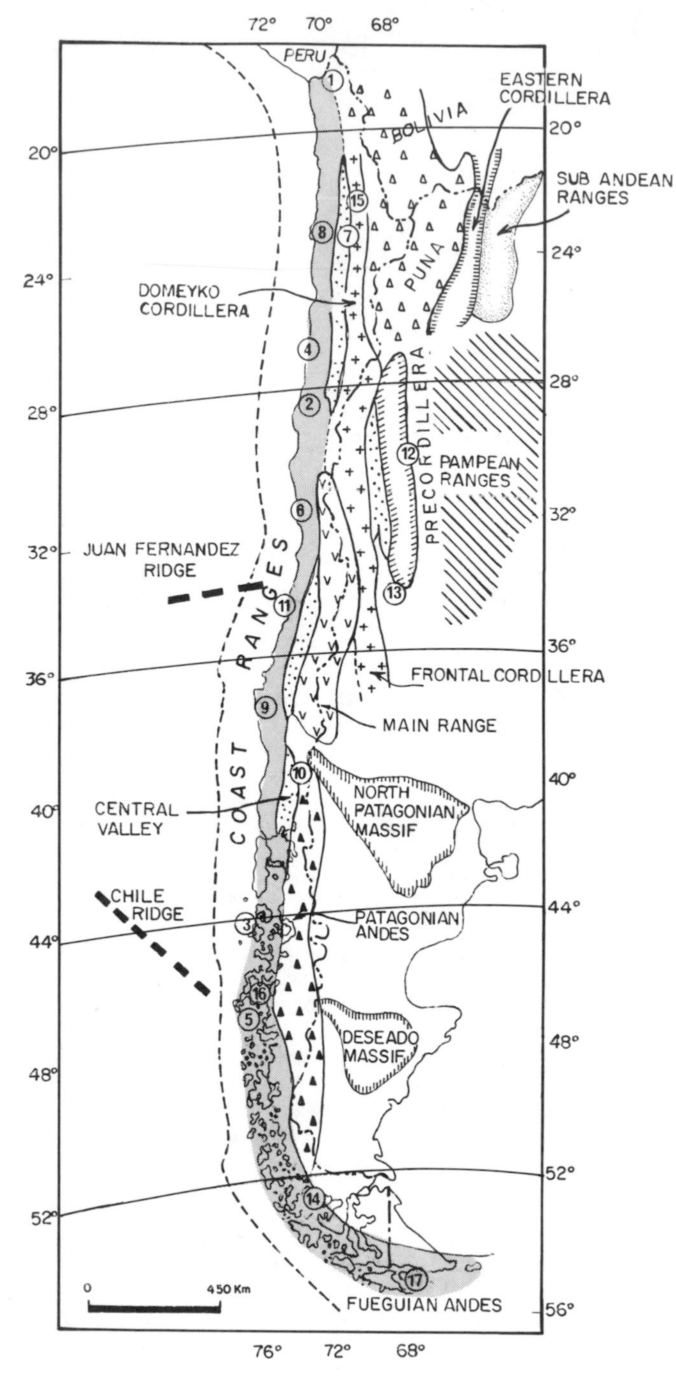

Fig. 1. Morphostructural elements of the Chilean-Argentinian Andes used in this paper indicated by different patterns. Patterns indicate different elements. Circled numbers indicate locality names as follows: (1) Belén; (2) Chañaral; (3) Chonos Archipelago; (4) El Toco; (5) Golfo de Penas; (6) Huentelauquén; (7) Limón Verde; (8) Mejillones; (9) Nahuelbuta Mountains; (10) Panguipulli; (11) Pichilemu; (12) San Juan; (13) San Rafael; (14) Sarmiento; (15) Sierra Moreno; (16) Taitao; (17) Tortuga.

National Geological Surveys, Universities and international cooperative projects are also included.

Morphostructural Characteristics

The southern part of the Andes has the morphologic and structural elements shown in Figure 1.

The coastal Cordillera is continuous along the continental margin, with a discontinuous tectonic depression east of it (Central Valley of Fig. 1), which separates it from higher portions of the range. East of this depression, from latitudes 28° to 36°S, three ranges are distinguished from west to east: the Main Range, the Frontal Cordillera, and the Precordillera which exists only between these latitudes. The Frontal Cordillera, a morphostructural north northwest-trending unit mainly composed of upper Paleozoic rocks, is oblique to the continental drainage divide, and is entirely within Chilean territory north of latitude 28°, where it is named Cordillera de Domeyko, and flanks the Puna plateau on the west. The Puna is a high altitude graben, with the Eastern Cordillera and Subandean ranges east of it.

South of latitude 36°S, the Andes consists of a single mountain belt, termed Patagonian Andes between latitudes 40° and about 52°S, and Fueguian Andes in the more east-west trending southern extremities.

The Pampean Ranges are composed of fault blocks with a crystalline basement of Proterozoic age, and are exposed east of the Andes proper at latitudes 28°-33°S, and share with them Paleozoic plutonic activity and late Cenozoic tectonism.

Precambrian Terranes of the Foreland and the Andean Chain

The basement of the Chilean-Argentine Andes is mainly composed of Paleozoic rocks. Outcrops of Precambrian rocks are known in the foreland east of the mountain belt and Proterozoic basement occurs in the northern part of the part of the Andes considered herein.

Precambrian units of the foreland can be traced from the Sierras Pampeanas to Tierra del Fuego. The rocks in Sierras Pampeanas (Fig. 2), have yielded geochronological information indicating metamorphism at about 1000 Ma (Caminos et al., 1982), and early Paleozoic intrusive activity. The Deseado and North Patagonian massifs (Fig. 1) have poorly exposed crystalline basement of suspected Precambrian age. The subsurface gneisses of Tierra del Fuego, have yielded isotopic information which indicates their latest Proterozoic age.

The schists of Belén (Fig. 1), dated 1000 Ma (Pacci et al., 1981), crop out in a tectonically uplifted block in the western border of the Altiplano near latitude 20°S. They consist of mica schist, biotite gneiss and amphibolite which are intruded by ultramafic, mafic and intermediate

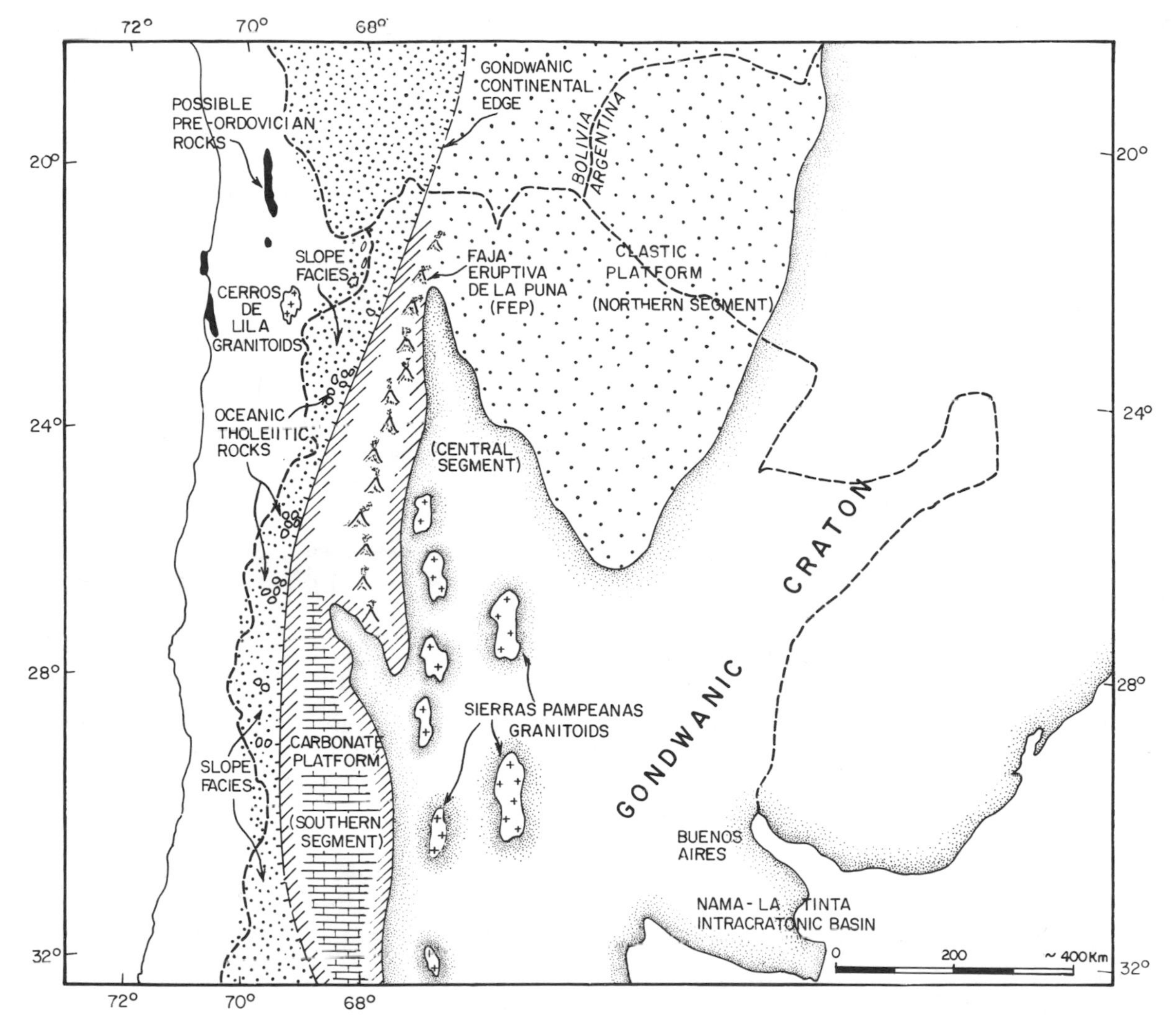

Fig. 2. Paleogeographic setting during the Ordovician. Note that the continental (granitic) areas west of the Gondwana continental edge could be allochthonous (based on Aceñolaza et al., 1984; Baldis et al., 1982; Ramos, 1982).

plutons, which at least in part are dated by K-Ar and yield early Paleozoic ages (Mpodozis et al., 1983). These are the only well documented Precambrian rocks in the Andes proper. Outcrops of metamorphic rocks along the western slope of the Andes, e.g. at Mejillones, Limón Verde and Sierra Moreno (Fig. 1), have failed so far to produce well defined Proterozoic radiometric ages, though U/Pb 1260 ± 30 Ma ages are mentioned for the latter locality by Breitkreutz and Zeil (1984).

The western limit of Precambrian basement underlying the Andes, is as yet undefined, but it is obvious that it did not reach the present continental border at all the considered latitudes.

Early Paleozoic

The distribution and nature of early Paleozoic metamorphic, plutonic, volcanic and sedimentary rocks allow a broad characterization of the geotectonic setting of the continental margin of Gondwana to be made. Figure 2 is a sketch of the paleogeography during the Ordovician, in which three segments are distinguished from north to south.

The northern segment crops out in the Eastern Cordillera and Puna (Figs. 1,2), and is characterized by the development of a Cambrian to Ordovician miogeoclinal prism consisting mainly of mature clastic sediments of platform facies. These sediments overlie Precambrian rocks with marked angular unconformity. Their eastern limit is probably the easternmost shoreline of the time, for sedimentary thickness increases towards the west, and higher, Ordovician, stratigraphic levels consist of alternating clastic and carbonate deposits hundred of metres thick. The western facies consist of volcaniclastic flysch, with turbidite sequences thousands of metres thick. These grade in turn westwards into calc-alkaline volcanics, tuffs and breccias of the "Faja Eruptiva de la Puna" (FEP), which have isotopic ages of around 470 Ma (Omarini et al., 1984) and fairly high Sr^{87}/Sr^{86} ratios (0.710 to 0.718). The FEP has been interpreted as a volcanic arc related to subduction of oceanic crust (Coira et al., 1982; Allmendinger et al., 1983). To the west of the volcanic arc, the Ordovician sequence is associated with mafic and ultramafic rocks, deformed and slightly metamorphosed, which have been interpreted as an ophiolitic association by Allmendinger et al. (1982; 1983).

The southern segment corresponds to the Precordillera (Figs. 1,2), where an extensive carbonate platform developed contemporaneously with the miogeoclinal wedge of the northern segment. This carbonate platform, several hundred metres thick, interfingers to the west with offshelf, talus deposits (Baldis et al., 1982) represented by limestones, pelites and turbiditic sandstones. In places, as in western parts of the Precordillera of San Juan, pelagic deposits of chert and finely laminated siltstone are present. The thickness of the sequence in the western portions attains several thousand metres (Cuerda et al., 1983).

The central segment (Fig. 2) is less well known, more complex and includes the Famatima System (or Sierras Transpampeanas). In this segment it is not possible to reconstruct a platform regime of sedimentation, because outcrops are disconnected from each other. They consist of lithofacies of more unstable environments, comprising sandstone and pelite, with an important component of pyroclastic deposits with rhyolite and andesite clasts, and a Lower Ordovician fauna. Tuffs become more important towards the top of the sequence, culminating with thick dacite sheets of Late Ordovician age. Dacite porphyries and Paleozoic granitoids intrude the sequence, which in places bears the imprint of greenschist facies metamorphism.

Probable Oceanic Rocks

Steiglitz (1914), Keidel (1925) and Borello (1969) recognized an ophiolitic sequence in the western margin of the Precordillera. More recently, petrographic (Quartino et al., 1971), and geochemical studies (Kay et al., 1984) show their oceanic affinities of these rocks. The sequence comprises basaltic pillow lavas, turbiditic and pelagic sediments, gabbros, and hypabyssal mafic to ultramafic bodies, whose lowermost parts are foliated ultramafic tectonites (Haller and Ramos, 1984). The basaltic rocks are TiO_2 rich, highly evolved rocks, with chemical affinities to those of back-arc basins or transitional mid-ocean ridges of the Reykjanes type (Kay et al., 1984).

In addition, a volcano-sedimentary complex containing pillow lavas has been identified on the Chilean side of the Andes, at latitude 24°S, south of Salar de Atacama (Niemeyer, in prep.) for which an early Paleozoic, pre-Late Ordovician age is inferred.

Granitoid Plutons

Early Paleozoic granitoid plutons are widespread in Sierras Pampeanas (Fig. 2) where they have been studied in some detail. After an initial, poorly developed, gabbroic and ultramafic episode (Toselli et al., 1978; Reissinger, 1983), three main (Cambrian, Ordovician, Early Carboniferous) granitoid emplacement episodes lacking coeval volcanics are recognized (Rapela, 1976a,b; Reissinger, op. cit.). Most Pampean granitoid bodies are peraluminous regardless of their age (Rapela and Shaw, 1979; Reissinger, op. cit.) with the exception of some high-K mildly metaluminous stocks.

The most striking geochemical characteristic within the Pampean granitoids is the progressive enrichment of incompatible elements with decreasing age (Rapela, 1981, 1982; Reissinger, op. cit.). The evolution culminates in post-orogenic units that show extreme enrichment in the LIL elements, Nb and K_2O, and are highly fractionated in terms of K/Rb and Rb/Sr. In addition such younger units have relatively high initial Sr^{87}/Sr^{86} ratios (>0.710) compared with the older ones (Knüver and Reissinger, 1981; Rapela et al., 1982), implying that the former units were derived by partial melting of crustal rocks or somehow acquired a component of crustal strontium during their evolution in the crust. This indicates an increasing component of crustal rocks with time, so that the lower Paleozoic granites closely resemble the I-type (Caledonian) granites (Pitcher, 1982) and the later S-type encratonic granites.

In Chile outcrops of comparable rocks are scarce but seem to define a belt which lies 100 km west of the Sierras Pampeanas (Fig. 2). At Cerros de Lila the plutons are overlain unconformably by Devonian clastic sediments (Mpodozis et al., 1983). There, the Tucucaro leucocratic monzogranites give a date of 441 ± 8 Ma (Rb-Sr, whole rock isochron), and the Tilopozo biotite and muscovite monzogranite, 452 ± 11 Ma. Both have fairly high initial Sr^{87}/Sr^{86} ratios, 0.7102 and 0.7173 respectively.

Early Paleozoic Orogenesis

Early Paleozoic sediments and volcanic rocks were deformed, metamorphosed and intruded by granitoids before the Devonian, as documented in northwestern Argentina (e.g. Miller and Willner, 1981) and northern Chile.

Late Ordovician tectonism has been referred to as the "Ocloyic phase" by Turner and Mendez (1975) and was more intense in the northern segment, which includes the Puna and Eastern Cordillera, than elsewhere. It is responsible for the

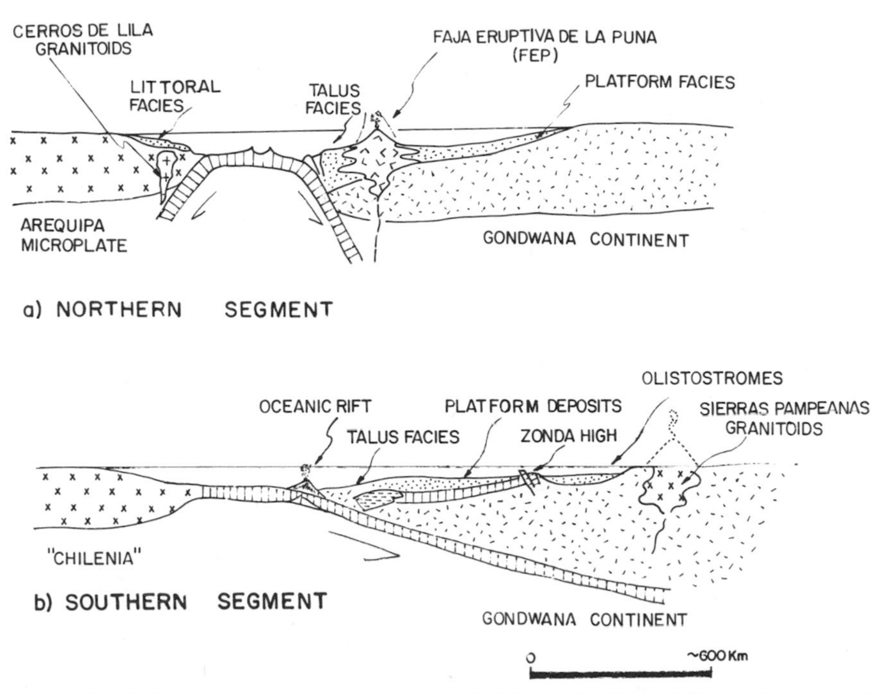

Fig. 3. Tectonic interpretation for the evolution of the different segments of the Ordovician continental margin as presented in Figure 2. Modified from Coira et al. (1982), and Ramos et al. (1984).

first uplift of the Puna, the configuration of the Eastern Cordillera, and was interpreted by Coira et al. (1982), Ramos (1982) and Allmendinger et al. (1982) as the product of collision between an Arequipa microplate to the west and the "Pampean craton" (Fig. 3).

Evidences for this interpretation (Fig. 3) are the presence of an active continental margin with a calc-alkaline magmatic arc (FEP) in the Puna, platformal sedimentary facies to the east of it, and oceanic rocks to the west of the arc, all of which suggest an east-dipping subduction zone. The ocean basin to the west is bounded by the incoming Arequipa microplate in the models of Aceñolaza and Toselli (1984) and Coira et al. (1982). The existence of this old cratonic block is well documented in the southern Peruvian coastal region, and it is here extended to latitude 24° or 25°S on the basis of the existence of isolated outcrops of metamorphic rocks of possible Precambrian age. The predominantly I-type granitoids in Chile 200 km west of the contemporary arc (FEP) may represent either a western extension of the same arc, or else magmatic activity on the eastern margin of the drifting Arequipa microplate. The S-type characteristics of some of these granitoids are compatible with the collision hypothesis. An extensive regional thermic event, which has reset K-Ar ages in the Arequipa massif proper (Shackleton et al., 1979) and in the Belén schists, could also be an expression of continental collision (Mpodozis et al., 1983).

In the southern segment (Precordillera), "Ocloyic phase" movements consist of two separate compressive episodes, which can be bracketed in time very precisely because of good biostratigraphic control. The "Guandacolica" phase (Furque, 1963) is of Llanvirnian age (460 Ma), and the "Villicunica" phase (Baldis et al., 1982) lies between Caradoc and Ashgill (417-433 Ma, in the time scale of Odin, 1982).

A second orogenic event, the "Chanic" phase took place towards the end of the Devonian and marks the end of the Famatinian orogenic cycle of Aceñolaza and Toselli (1973-1976).

In the northern segment this phase is represented by deformation and reactivation of older structures at 374 ± 7 Ma (Omarini et al., 1979) in the arc region (FEP), and by angular unconformity in the back-arc where upper Paleozoic rocks overlie Devonian deposits (Mingrann et al., 1979).

In the southern segment the "Chanic" movements were more intense and produced extreme deformation in the western Precordillera as well as thermal and dynamic metamorphism dated at 357 to 373 Ma by Cucchi (1971).

A tectonic interpretation of the early Paleozoic events has been recently suggested by Ramos et al. (1984), who proposed that deformation is controlled by the collision of a continental mass, "Chilenia", against the Gondwanic craton (Fig. 3). The relationships between the Arequipa microplate and the hypothetical Chilenia are unknown, but at least two possibilities can be raised: they could be parts of the same continental plate accreted in different periods, or two completely independent elements. The recognition of rocks definitely belonging to Chilenia is wanting.

It is worthy to note that after the "Chanic" movements, an important paleogeographic change occurred in the Gondwanic foreland, due to the uplift of the proto-Precordillera (Rolleri and Baldis, 1967), which allowed the further development of extensive intracratonic continental basins. Also, at about this time, Late Devonian (Hervé et al., 1984) metamorphic processes probably related to subduction were active in the coastal areas immediately south of the Precordillera. Their characteristics will be, arbitrarily, dealt with in the late Paleozoic chapter. In the model proposed by Ramos et al. (1984) this subduction episode occurred west of the hypothetical Chilenia microplate that had by this time been welded to the continental margin of Gondwana.

Late Paleozoic

Late Paleozoic rocks reveal the existence of an active continental margin which can be traced along most of the southern part of the Andes. Petrotectonic associations are indicative of the existence of a coastal subduction complex, of a broadly contemporaneous plutovolcanic belt or belts, and of several rather short-lived intra-arc or back-arc basins (Fig. 4).

Late Paleozoic Subduction Complex

A belt of metamorphic rocks which crops out sporadically north of latitude 34°S in the Coast Ranges and continuously south of this latitude, has been interpreted as a subduction complex accreted at the active western margin of Gondwana, mainly during late Paleozoic time. In Patagonia, rocks of this complex also outcrop east of the Main Range of the Andes.

Sparsely distributed fossils in the flyschoid greywacke-pelite element of the complex, indicate Silurian and mainly Devonian ages of sedimentation. Such paleontological evidence exists at El Toco (Maksaev and Marinovic, 1980), Chañaral (Miller, 1973; Bell, 1983), Lumaco (Tavera, 1982), Chonos Archipelago (Miller and Sprechmann, 1978) and San Martin Lake (Nullo et al., 1978; Sepúlveda and Cucchi, 1978). South of Golfo de Penas, however, well documented Late Carboniferous to Early Permian fusulinids (Mpodozis and Forsythe, 1983) occur. The rocks of this complex were deposited either on the ocean floor (Bell, 1983) or on the continental platform and/or rise, in what was possibly a passive-type margin until Devonian time.

Polyphase metamorphism and deformation of the complex, which generally is more recrystallized to the west, spanned from the Devonian to Carboniferous and Permian, and in central Chile is of glaucophanitic greenschist grade. Lawsonite has been identified only in Chiloé island but blue amphiboles are widespread. The eastern limit of the complex is, where it exists, the late Paleozoic batholith. In close association with the batholith, biotite, andalusite, staurolite and sillimanite zones have been mapped especially in central Chile, in the low-pressure half of a paired metamorphic belt.

The structure of the complex has an overall pattern. The eastern flyschoid portions are folded in tight upright folds. A gently east-dipping crenulation foliation, develops gradually to the west, where it obliterates pre-existing primary or tectonic structures. The western schists usually include substantial amounts of metabasites of ocean-floor basalt chemical affinities, metacherts and serpentinites. Mélanges have been described near Chañaral (Bell, 1983), at Pichilemu (Hervé et al., 1984) and in Chonos Archipelago (Hervé et al., 1981).

The radiometric ages on the metamorphics south of latitude 34°S are sparse but clearly indicate that there, the metamorphic basement was remobilized during Gondwanic and later orogenic events. In northern and central Chile, there is stratigraphic evidence that the metamorphics were not penetratively deformed after Late Triassic time. The dates indicate Devonian (380-344 Ma) metamorphism in the easternmost portions of the belt, with an Early Carboniferous event (311± Ma) predominating in the glaucophanitic western portions of it at latitude 34°S (Hervé et al., 1984). In the Chonos Archipelago, metamorphism of the subduction complex may have spanned from Carboniferous to Jurassic time (Godoy et al., in press). Permian and Cretaceous events at Cordillera Darwin, Tierra del Fuego (Hervé et al., 1981) indicate that younger accretion ages and Mesozoic remobilization of the subduction complex were effective in the southernmost parts of the continent.

Late Paleozoic Magmatic Arcs

The products of upper Paleozoic igneous activity in the Southern Andes are distributed in the following belts (Fig. 5): 1. The Marginal "Andean" Belts comprise igneous associations that evolved parallel to and within 200 km of the present continental margin. They comprise (i) the Coastal Plutonic Belt (CPB) which consists of the Northern Coastal Batholith (NCB) and the Southern Coastal Batholith (SCB), and (ii) the Frontal Range Magmatic Belt (FRMB). 2. The Inner "Cratonic" Belts (ICB) consist of (i) Pampean Ranges Plutonic Belt (PRPB), the (ii) Northern Patagonian Magmatic Domain (NPMD) and (iii) the San Rafael-La Pampa Magmatic Belt.

The CPB is interrupted between latitudes 26° and 33°S, dividing it into NCB and SCB. The FRMB parallels the CPB 200 km eastward of it, and outcrops continuously from latitudes 27° to 35°S, but southwards only sporadic outcrops of upper Paleozoic igneous rocks are recognized. The

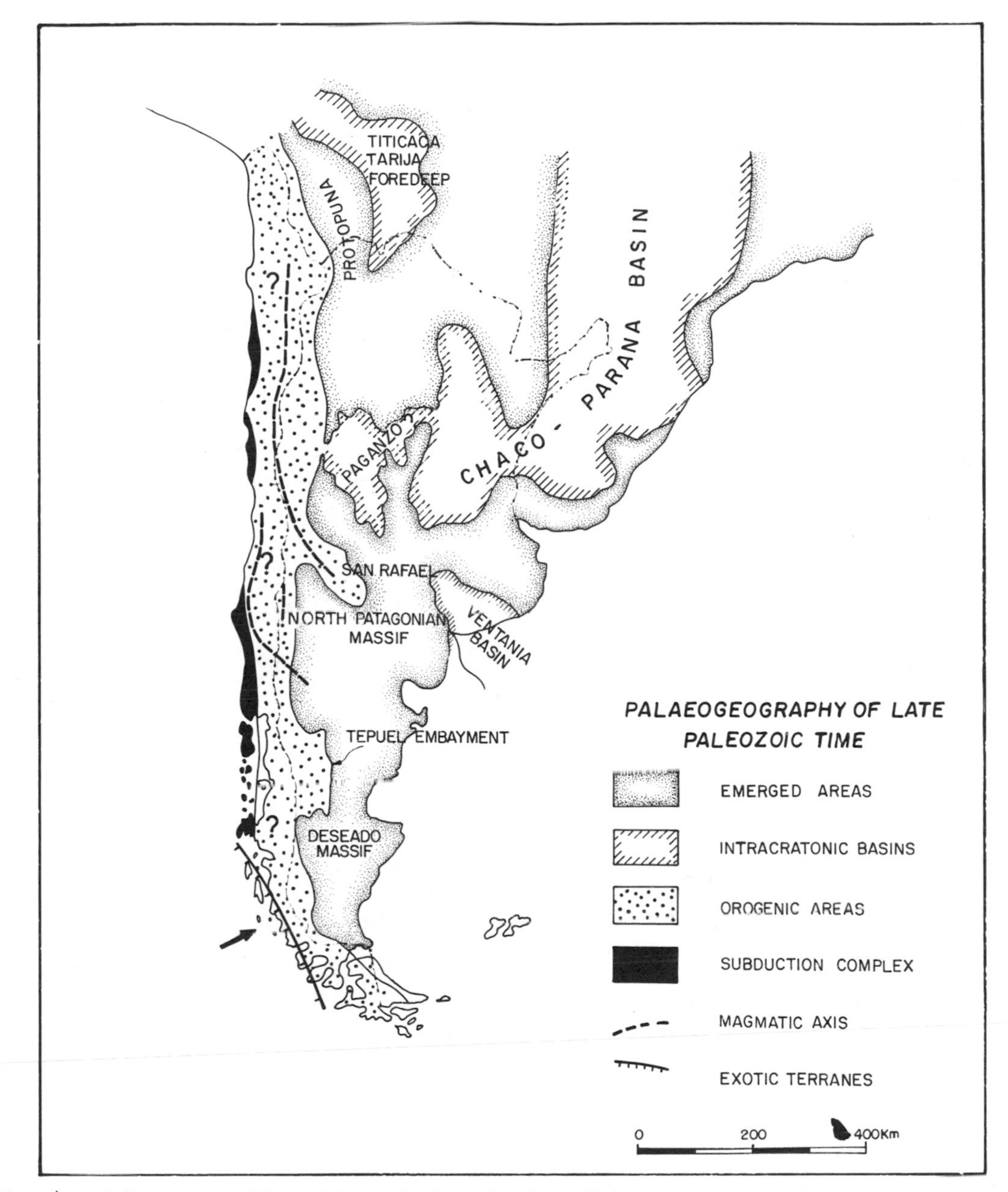

Fig. 4. Paleogeographic setting during the late Paleozoic, indicating the main petro-tectonic associations of the active continental margin.

limits of the interruption in the CPB roughly coincide with the limits in the FRMB in which the igneous activity was most extensively developed. The typical north-south, Andean, igneous alignments show a remarkable change at latitudes 35°S and 39°S where the SLMB and NPMD respectively follow a general southeast direction across the continent towards the Atlantic margin. Upper Paleozoic granitic activity is also recorded as a later event along the lower and middle Paleozoic Pampean Ranges Plutonic Belt (PRPB).

Marginal Andean Belts. The SCB is a north northeast-trending batholith which extends southwards continuously from latitude 32°30'S for about 600 km. It comprises calc-alkaline plutonic rocks that intrude fore-arc sediments which are coevally metamorphosed up to upper amphibolite or lower granulite facies (Hervé, 1977; Hervé et al., 1984). Much of the batholith appears to be of Early Carboniferous age (Hervé et al., 1976; 1984), although Silurian to Jurassic ages have been recorded (Corvalán and Munizaga, 1972; Drake et al., 1982). Lithological variations from the margin to the core of the SCB

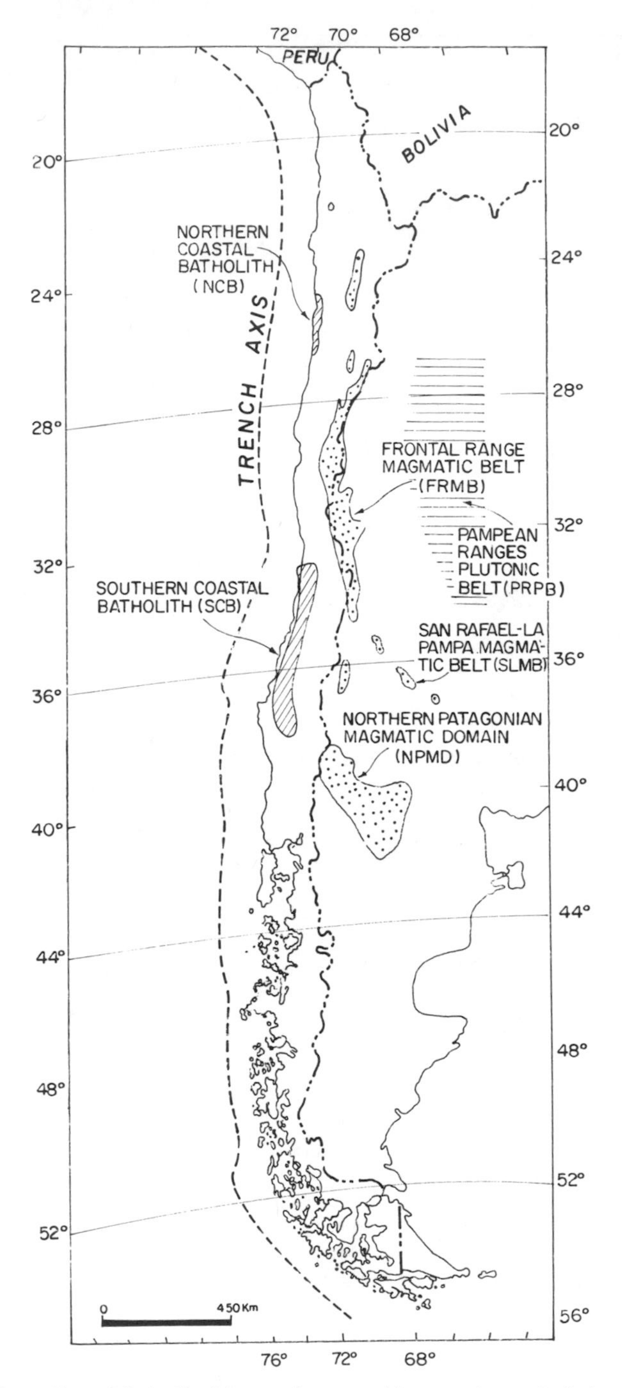

Fig. 5. Distribution of magmatic rocks of late Paleozoic age.

have been observed at the Nahuelbuta Mountains. Marginal facies locally consist of biotite, muscovite (± garnet ± sillimanite) granites and migmatites with sillimanite bearing gneissic bands parallel with the regional structure. The central part of the batholith is more homogenous in composition, and consists mainly of hornblende and biotite tonalites and granodiorites. The northern part of the SCB is characterized by abundant amphibolitic inclusions within tonalites with variably-developed, oriented fabric. Based on limited sampling in the northern extremity of the SCB, López-Escobar et al. (1979) suggest that it had a mantle origin comparable to that of the Mesozoic and Cenozoic plutonic rocks of central Chile.

The NCB can be envisaged as being composed of two composite batholiths spanning in time from late Paleozoic (270 Ma) to Cretaceous (124 Ma) (Berg et al., 1983). Upper Paleozoic granitoids (Albayay superunit) are restricted to the western flank of the batholith, where they intrude Paleozoic fore-arc rocks. The Albayay superunit includes monzogranites, syenogranites and minor alkali-feldspar granites, and is intruded by a mafic dyke swarm. The minimum melt compositions of these rocks, as well as the high Sr^{87}/Sr^{86} initial ratios (>0.710), suggest a crustal origin for them.

The igneous rocks of the FRMB extend continuously for 650 km from the Chilean Cordillera de Domeyko near latitude 27°S, to the Argentinian Frontal Range (Fig. 1). The plutonic rocks constitute a composite batholith of mainly upper Paleozoic rocks. Early plutonic activity of Devonian to Lower Carboniferous age (Caminos et al., 1979; Parada et al., 1981) is represented by a calc-alkaline plutonic association of hornblende + biotite tonalites and granodiorites. Younger plutonism shifts towards more felsic compositions, represented by biotite and biotite-muscovite granites and granodiorites, and finally to biotite, hornblende-bearing leucogranites and porphyries. The one- and two-mica granitoids of the FRMB are peraluminous and exhibit small amounts of metapelitic inclusions as well as cordierite-bearing nodules. These are typical characteristics of S-type granites (Chappell and White, 1974). In contrast, the leucogranitic plutons include high-silica metaluminous, to somewhat alkaline rocks (Parada, 1981), and are too iron-rich relative to magnesium to be considered calc-alkaline. These features, together with their near-eutectic composition, are comparable to A-type granites (Anderson et al., 1980; Collins et al., 1982) derived from quartzo-feldspathic sources. This origin is compatible with the higher Sr^{87}/Sr^{86} initial ratio of the leucogranites (0.7078) compared to the "mantle-type" ratio of the early tonalites (0.7048) (Parada et al., 1981).

The available radiometric values for the late felsic plutons of the FRMB indicate a range between Lower Permian and Triassic (Caminos et al., 1979; Parada et al., 1981; Cornejo et al., 1984). A mafic dyke swarm of regional scale cuts all the plutons, and seems to be partly coeval with the late granite bodies, and although possibly some dykes are as young as Jurassic. Locally, the mafic dyke intrusions represent crustal extension of about 7%. Volcanic rocks mainly related to the late Felsic plutonism are rhyolites (ignimbrites, lava flows, tuffs) that are chemically similar to the leucogranites (Parada, 1983; 1984).

The shift in the magmatic composition of the FRMB may reflect a change from a mantle-dominated source to a crustal one. Factors that probably contributed most to late-stage crustal fusion were preheating and decompression of a water-deficient crust (Parada, in prep.).

Inner Cratonic Belts. The NPMD is mostly composed of upper Paleozoic igneous assemblages that do not display the typical linear pattern of Andean belts but outcrop almost continuously over an area of 500 by 400 km (Fig. 5). Detailed studies recently carried out in two separate areas show their similar petrological evolution (Caminos, 1983; Llambías et al., 1984; Rapela and Llambías, 1985).

A protracted period of igneous activity that started during the Early Carboniferous and lasted until Jurassic times is characteristic for the NPMD. The lithostratigraphic sequence is composed of Carboniferous and Permian granitic complexes overlain by Permian volcanic rocks. These upper Paleozoic complexes are covered in turn by silicic volcanic associations that formed ignimbritic (rhyolitic) shields of Triassic-Jurassic age.

Hornblende-biotite granodiorites and tonalites, biotite granites (often with K-feldspar megacrysts) and leucogranites constitute the main facies of the plutonic complexes. Although most of the previous K-Ar data indicate a Permian age (Stipanicic, 1967; Stipanicic et al., 1968; Stipanicic and Methol, 1972), recent Rb-Sr isochrons obtained on different plutons suggest an older age (317-332 Ma) for the main plutonic episode of the NPMD (Llambías et al., 1984). The granitoid units are cut by a swarm of dikes of dioritic composition (52-54% SiO_2). Rb-Sr and K-Ar data on the overlying volcanic complexes indicate an age of 265-277 Ma for the older units while some ignimbrites and leucogranites probably are as young as Late Permian (Llambías et al., op. cit.).

The main geochemical characteristics of the Paleozoic NPMD are summarized below:

a. The older units of individual complexes are composed by high-K calc-alkalic metaluminous associations while the younger are mostly high-K peraluminous sequences. Some young leucogranites and associated dikes also show transitions to mildly peralkaline acmite-normative compositions.

b. There is a lack of basic rocks. With the exception of the dioritic dike swarm noted above all analyzed units have silica content higher than 64%.

c. Both plutonic and volcanic complexes show a generally increasing acidity with decreasing age.

d. Clear consanguinous relationships relate rocks of a given petrographic unit, but the trends of the different units of a complex do not follow a common evolutionary pattern. Each complex appears to be composed of several progressively more acidic magmatic pulses emplaced at epizonal levels (plutonic complexes), subvolcanic and surface levels (volcanic complexes).

e. Chemical trends of plutonic and volcanic complexes show remarkable similarities suggesting that both types were controlled by the same evolutionary mechanism.

f. Initial Sr^{87}/Sr^{86} ratios range from 0.7050 (hornblende-biotite granodiorites) to 0.7080 (leucogranites).

The geological and chemical evolution of the plutonic and volcanic complexes of the NPMD resembles closely those described for the FRMB, suggesting that both belts were part of a single extensive magmatic province (Rapela and Llambías, op. cit.). Nevertheless while the Frontal Range underwent several tectonic episodes during the Mesozoic and Tertiary Andean orogenies, the North Patagonian Massif has remained anorogenic since the Permian. Cenozoic alkalic members of the Patagonian Plateau basalts overlie most upper Paleozoic and lower Mesozoic magmatic units (Corbella, 1984).

The petrological characteristics of volcanic-plutonic rocks (mostly rhyolitic complexes and ignimbritic shields) of the SLMB are very similar to those described above for the volcanic complexes of the NPMD (Llambías and Leveratto, 1975; Rapela and Llambías, op. cit.). Rb-Sr and K-Ar data of several units of the SLB yielded ages of 210 to 270 Ma and an initial Sr^{87}/Sr^{86} ratio of 0.7053, for a 238 ± 5 Ma isochron (Linares et al., 1980).

The back-arc and foreland basins. Several independent basins developed east of the magmatic arcs, on the cratonic foreland. These basins have an independent sedimentary evolution but have an generally synchronous diastrophic history (Azcui, 1983). They are referred below to from north to south.

The northern back-arc basins extend into the Subandean areas of Bolivia and northern Argentina (Tarija-Titicaca basin; Fig. 4). The sedimentary sequences start with clastic marine deposits that grade upwards into continental molasse facies with glacial horizons (Mingrann et al., 1979; Amos and Lopez-Gamundi, 1981), which formed during much of Late Carboniferous-Early Permian time. This basin was initiated by the Late Devonian "Chanic" orogenic movements (Turner and Mendez, 1975). Further south, and west of the Precordillera, there are marine facies (Rolleri and Baldis, 1967) which grade west into prodeltaic deposits with thick turbiditic intervals (Ramos et al., 1984). These deposits constitute the substratum of part of the upper Paleozoic acid lavas and tuffs of the Frontal Cordillera and Domeyko Cordillera (Fig. 1) at its southern extension, in the San Rafael Block, the facies are similar to the ones in the central part. Here, the strong unconformity which separates the underlying earliest Paleozoic deposits from the Lower Carboniferous platform deposits reflects the "Chanic" orogenic movements (Polanski, 1970).

The foreland basins develop over the Pampean foreland, mainly in the depocenters of Paganzo and Chaco-Paraná (Fig. 4). Initiation of deposition in the latter is Carboniferous and is synchronous with that of the Paganzo basin, but its deposits continue until Late Permian time. The Paganzo basin was in part connected with the back-arc basins of the Precordillera. It contains fluvial continental deposits, proximal and distal alluvial fans with sparse alkali basalts (Spalletti, 1979). Paleoflora indicates a mid-Carboniferous to Early Permian range of deposition (Azcui, 1983). The deformation of the basin is moderate, increasing to the west.

The fore-arc basins. Clastic and limy sequences, 200-500 m thick, with brachiopods and foraminifera of Late Carboniferous to Early Permian age, outcrop sparsely between latitudes 18° and 32°S. While at Augusta Victoria and Huentelauquén-Mincha they were deposited across the recently accreted subduction complex, to the east (e.g. Sierra de Fraga) they lie directly on Paleozoic granitic basement. At Huentelauquén their base is formed by talus breccias that formerly were described as tillites.

This carbonate platform facies is well known in the Peru-Bolivian Andes and has been related to a subtropical position of the western margin of Gondwana during the late Carboniferous (Valencio, 1975; Davidson et al., 1981).

The Panguipulli basin (Fig. 1), of limited areal extent, developed over the subduction complex at latitude 40°S, and was infilled by terrigenous sediments containing a Permian flora. The magmatic arc and the already metamorphosed subduction complex provided detritus to this basin, which was strongly deformed before the deposition of Late Triassic rocks.

Similar lithologies constitute the Tepuel-Genoa basins (Fig. 4; latitudes 42°-46°S) developed east of the present day Patagonian Andes (Ugarte, 1966). Fossiliferous marine turbiditic sequences, with Carboniferous and Early Permian fossils, grade into continental sequence which include glacial deposits. The presence of large late Paleozoic gabbroic bodies associated with the basins (Franchi and Page, 1980), suggest crustal thinning, possibly similar to that occurring in modern marginal basins (Ramos, 1983).

The Mesozoic-Paleogene Geologic Development

Mesozoic to Recent evolution is summarized in Figure 6. Middle to Late Triassic time is marked by a marine incursion that progressed from the west over the late Paleozoic subduction complex, the Permian carbonate platforms, and, north of latitude 29°S, the late Paleozoic plutonic complexes. Marine facies are restricted to the present day fore-arc in Chile, and only continental facies are known in Argentina.

Thick deltaic, proximal marine turbidites and alluvial deposits accumulated together with volcanoclastic units and andesitic lavas, in a series of fault-bounded basins in north and central Chile and in western central Argentina. During earliest Jurassic time, this volcanism migrated westwards giving way to a longitudinal magmatic arc that in Middle Jurassic time acted as the western border to a series of shallow, to pelagic north-south elongate ensialic back-arc basins and carbonate platforms. These developed as a continuous series of depocenters north of latitude 40°S (Coira et al., 1982). Uplift of these basins in Late Jurassic time, preceded the development of a new series of volcanic arc back-arc systems in the Early Cretaceous, which partly overlapped the Jurassic systems.

The different patterns of geologic evolution of the back-arc areas is documented for different segments of the Andes during the Lower Cretaceous. The Fueguian Andes (Dalziel, 1981) provide a well established model of the tectonic setting during this period. The Cretaceous calc-alkaline magmatic arc, represented mainly by granitoid plutons of the Patagonian Batholith, migrated away from the continent with the subsequent generation of a marginal basin floored by oceanic-type crust. At least two en-echelon ophiolitic bodies, the Sarmiento and the Tortuga complexes, represent the floor of this marginal basin.

A marine basin east of the magmatic arc existed also in the Aysen region of the Patagonian Andes (Baker et al., 1981), where back-arc extension allowed intrusion of numerous dykes into the pelitic sediments of the basin. Further north at Lago Fontana, an active intra-arc basin was developed during Neocomian times (Ramos and Palma, 1982). In the Patagonian Andes and Magallanes Cordillera a trend of increasing southward extension in the back-arc area is well defined, and controlled the transition between back-arc basins built over attenuated continental crust, to that of a true marginal basin floored with quasioceanic to oceanic crust at the southern extremity of the Andes. In contrast to the central and northern Chilean-Argentine Andes, in the southern area, Mesozoic-Cenozoic evolution leading to the modern Andes began in the Middle to Upper Jurassic with extrusion of voluminous acid tuffs and lavas (Tobifera Formation). The origin of these seems to be related to crustal extension and anatexis predating the opening of the Atlantic and the Magellan marginal basin (Bruhn et al., 1978).

The "Main Andean", "Subhercynian" or "Peruvian" tectonic phase of mid-Cretaceous age (Vicente et al., 1973; Dalziel, 1981; Coira et al., 1982), is contemporaneous with the final opening of the Atlantic Ocean and produces a major break in the geological evolution of the whole Andean domain. In northern Chile the "Proto-Cordillera de Domeyko" was uplifted and the products of its erosion shed to the east where a wide foreland basin was formed that began to be infilled with terrestrial red beds and marine calcareous deposits (Purilactis formation,

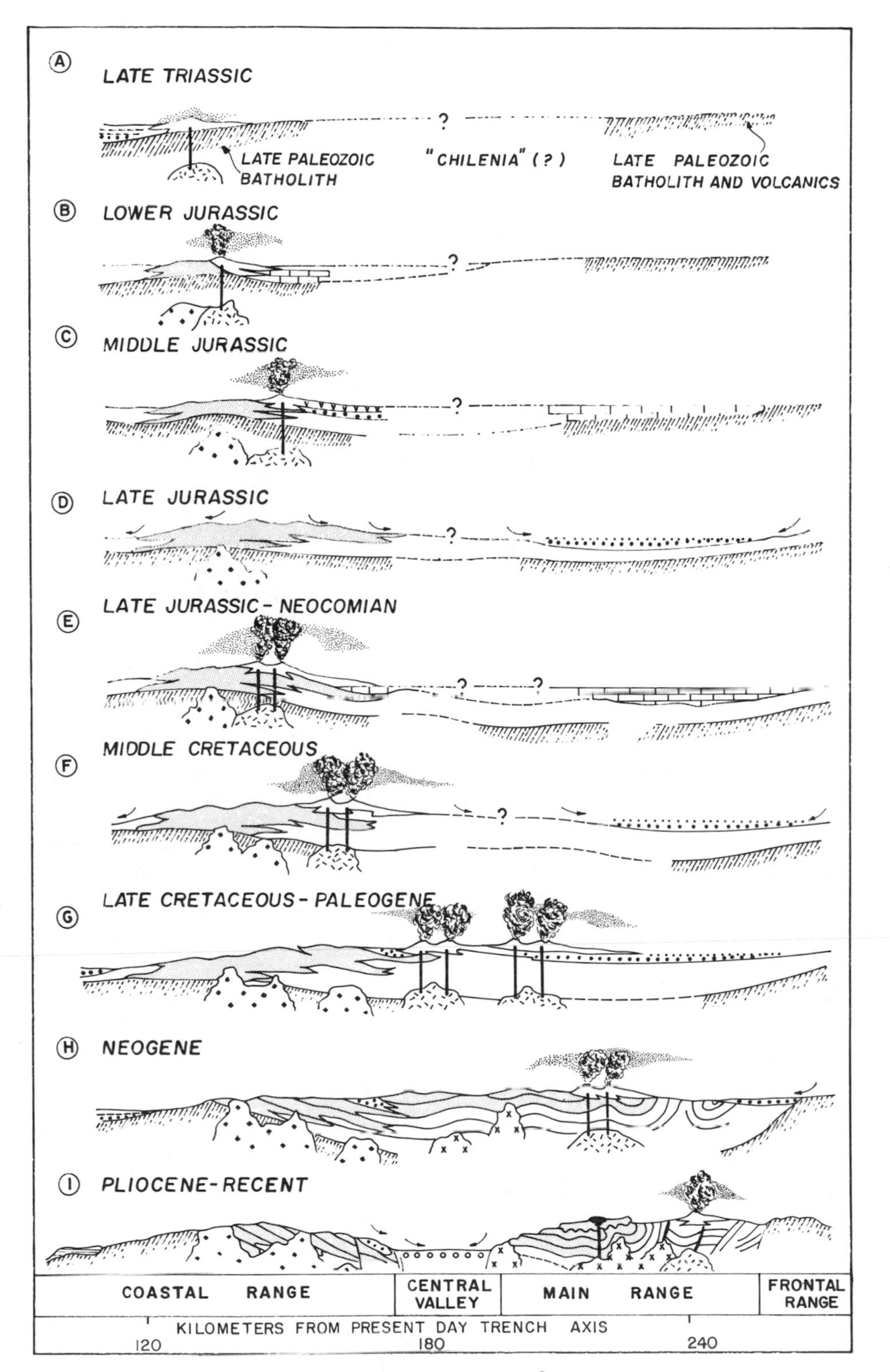

Fig. 6. Migration of magmatic foci at latitude 33°S during the Meso-Cenozoic. Modified after Thiele and Nasi (1982b).

Salta Group; Salfity, 1982; Marinovic and Lahsen, in press). In Central Argentina and Chile the back-arc basins disappeared (Davidson and Vicente, 1973; Nasi and Thiele, 1982). In the Magellan Cordillera, where the marginal basin was closed by collision of the magmatic arc with the stable South American platform, penetrative regional deformation and metamorphism overprinted, at Cordillera Darwin, the Paleozoic structures of the fore-arc assemblages (Nelson et al., 1980; Hervé et al., 1981).

Following the mid-Cretaceous orogeny the tectonic regime changed, back-arc basins disappeared, and from then until the present the calc-alkaline magmatic arc has remained as the major tectonic element. Along different segments of the Andean Chain the magmatic arcs have: a) migrated continuously to the east, as is the case in northern and central Chile (Fig. 6) (Farrar et al., 1970; Thiele and Nasi, 1982b; Munizaga and Vicente, 1982); b) remained stationary as is the case in the northern Patagonian Andes (Munizaga et al., 1984); c) moved east and then west as in the Aysen Region of the southern Patagonian Andes; d) Migrated south (oceanwards) as in the Magellan Cordillera (Hervé et al., 1984). The migration of the magmatic foci proceeded as discrete "jumps" punctuated by short-lived compressive deformational events, some of which are clearly time-related to major changes in the plate tectonic regime of the Pacific Ocean (e.g. Charrier, 1973; Rona and Richardson, 1978; Mammerickx et al., 1980; Cande et al., 1982).

These differences in the behaviour of the magmatic foci show the "segmented" nature of the Andean margin during the Late Cretaceous and Tertiary. The segmented model emerges not only from the above mentioned phenomena but also from the discontinuous nature of the magmatic activity. Upper Cretaceous volcanism is scarce or nonexistent north of latitude 27°S while very important activity is registered at that time south of latitude 27°S. Oligocene volcanism occurred along two discrete zones of the Andes in northern Chile separated by a "magmatic" quiet area between latitudes 20 and 28°S (Coira et al., 1982; Jordan, 1984). This phenomenen could have been related to and controlled by the dip of the Benioff zone in a way similar to that occurring in the modern Andes (Isacks and Baranzangi, 1977; Jordan et al., 1983; Boyd et al., 1983). "Flat slab" segments and nonvolcanic zones in the upper plate could have existed at different times in different locations during, at least, the Mesozoic and Cenozoic history of the Andean chain.

"Subduction-erosion" (Scholl et al., 1980) is another possible phenomenon emerging from the geological history of the Chilean-Argentine Andes. No fore-arc assemblages or subduction complexes of Meso-Cenozoic age have been reported from the region. Moreover the distance between Jurassic volcanic arcs exposed in the Coast Ranges of Northern Chile, and the modern Peru-Chile Trench is too short for the latter to have been the site of subduction at that time. Jurassic trenches need to have been located, at least 100 km oceanward of the modern trench. Removal of the missing arc-trench gap, by subduction/erosion or by strike-slip faulting has been proposed (Rutland, 1971; Coira et al., 1982; Mégard, 1984). Subduction/erosion could explain the observed Meso- and Cenozoic migration toward the east of the magmatic foci, the thick crustal "keel" of the Andes of northern Chile and Argentina as well as the "crustal contamination" imprint observed in most modern lavas of the central Andes.

The Neogene Orogenesis

Recent seismological studies have established that the geometry of the subducting Nazca plate is complex in that it does not have a uniform inclination, and that segments of differing angles of dip can be established along the subduction zone (Swift and Carr, 1974; Isacks and Barazangi, 1977). These segments are characterized by different tectonic and magmatic features of the South American plate (Jordan et al., 1983, 1984).

The segmentation is evidenced in the oceanic trench (Schweller et al., 1981) by the longitudinal alternation of sectors with or without incipient subduction complexes, the existence of the latter being favoured by a shallower dip of the Benioff zone (Isacks et al., 1982). Also, a magmatic arc is controlled by a minimum dip angle of subduction, below which the arc will tend to disappear (Isacks and Barazangi, 1977).

However, the effects of longitudinal segmentation of the chain are more evident in the back-arc areas close to the foreland. Between latitudes 15° and 24°S, with a 30° dip of the Benioff zone, the magmatic arc is well developed, and a thin-skinned thrust belt is present, with frontal thrusts in the foreland 700 km east of the oceanic trench.

Between latitudes 28 and 33°S, there is no late Cenozoic magmatic arc, and the compressive regime generated by the subhorizontal Benioff zone has produced strong deformation in the foreland, affecting the crystalline basement and giving rise to the Pampean Ranges as a compressive block structure related to the Andean orogen.

South of latitude 33°S, the Sierras Pampeanas disappear as a major morphostructural unit, and extensional basins develop like those of Beazley and Mercedes (Criado Roque et al., 1981). Here, the orogenic front is localized only 350 km east of the oceanic trench. The magmatic arc is accompanied, east of the orogenic front, by the important back-arc volcanism of the Patagonian basalts, that are characterized by bimodal volcanic sequences, with low Sr^{87}/Sr^{86} initial ratios, and controlled by an extensional regime.

Diverse possible explanations have been brought up for the mechanism responsible for this segmentation. Agreement seems to exist on the influence of subduction of aseismic ridges below the continental margin (Nur and Ben-Avraham 1981; Pilger, 1981; Jordan et al., 1983). The collision of these structures would produce a progressive shallowing of the dip of the Benioff zone and thus could lead to the end of magmatism. However, the mechanism is very complex, because segments with similar dip angles (latitudes 15-24°S, south of latitude 34°S) exhibit different tectonic behaviour.

The distances from the trench to the orogenic front is apparently controlled by the age of the subducting lithosphere. The colder and older oceanic crust in the northern part of the Chilean-Argentina Andes allows the deformation to take place far to the east, in the foreland, while the younger, hotter crust in the south does not advance the orogenic front but facilitates the development of the back-arc volcanism.

Differing behaviour of the Patagonian Andes is evident north and south of the Taitao triple point (latitude 47°S). To the north, volcanism is prominent, together with Neogene to Recent high angle or transcurrent faults, as in the Liquine-Ofqui fault zone. South of the triple point, the Cenozoic sequences are affected by low-angle faulting (Ramos; 1979, 1981).

Final Remarks

The continental margin of Gondwana, and then South America, displays a continuous Phanerozoic history of magmatic and tectonic activity. However, the nature of these processes has not been homogenous in time and space. Some major heterogeneities are briefly pointed out below.

During most of the Paleozoic, the Andes of Argentina and Chile showed an apparent history of continental growth by accretion of "exotic terrains" and construction of accretionary prisms along the subduction zone of the Pacific margin of Gondwana. A dramatic change occurred at the end of the Paleozoic and during the Mesozoic and Cenozoic, when the newly built continental border began slowly to be removed or eroded by subduction, a process that is still operating. Accretion during Meso-Cenozoic times appears to have operated through mainly volcanic and magmatic processes transferring material from mantle to crust. The change from continental margin growth to continental margin tectonic erosion was also accompanied by a major change in the magnitude of the magmatic activity. The Paleozoic magmatic "arcs" were broad zones hundreds of kilometres wide. The Mesozoic and Cenozoic "arcs", are linear magmatic features only of a few tens of kilometres wide, an order of magnitude less than the Paleozoic ones. Bulk chemistry comparison among southern Andes Phanerozoic granites of different belts and ages shows a remarkable overall variation from mostly peraluminous (Pampean Ranges granitoids) through transitional metaluminous-peraluminous (Frontal Cordillera and north Patagonian upper Paleozoic batholiths), to mostly metaluminous (circum-Pacific Andean Cretaceous batholiths) (Rapela and Llambías, 1985).

The difference in behaviour between the Paleozoic and the Mesozoic and Cenozoic Andean lithosphere appears to relate to "hotter" and "softer" conditions during the Paleozoic, when the lithosphere was possibly thinner.

Tectonic segments and "flat slabs" with no volcanism in the overriding plate, such as occur today, seem to have existed in different areas of the Andes since at least the Cretaceous, and possibly since the late Paleozoic, as indicated by the gaps in the magmatic belts.

The mid-Cretaceous orogeny marks a turning point from Jurassic-Lower Cretaceous geologic development, when back-arc extension with the generation of marine basins was a rule, to Upper Cretaceous-Tertiary, when there is no evidence of back-arc extension. During the first period, which predates the final opening of the Atlantic Ocean, subduction along the Pacific margin of South America possibly took place in a "Mariana-type" mode after the classification of Uyeda and Kanamori (1979) and Uyeda (1982). Contrarily during the later period, active advance of the South America plate over the Pacific plates precluded any type of extension in the back-arc and gave rise to the "Chilean-type" of subduction of Uyeda (1982).

Acknowledgments. The ILP provided the opportunity of bringing together the facts and ideas presented in this paper. Grant E-1300 of DIB, Universidad de Chile, allowed the Chilean authors to attend the IX Argentinian Geological Congress in Bariloche where the authors had the only opportunity to discuss, personally, initial versions of the paper. T. Jordan gave advice on the manuscript. Cristina Maureira and Bev Vanlier diligently typed different versions of the manuscript.

References

Aceñolaza, G., and A. Toselli, Consideraciones estratigráficas y tectónicas sobre el Paleozoico inferior del noroeste argentino, II Congr. Latinoamericano de Geología Mem. II:755-763, Caracas, 1973-1976.

Aceñolaza, G., and A. Toselli, Lower Ordovician volcanism in northwest Argentina, in: Aspects of the Ordovician system, Bruton, D. L., ed., Palaeontological Contributions from the Univ. of Oslo, N°295, 203-209, Universitetsforlaget, 1984.

Allmendinger, R., T. Jordan, M. Palma, and V. Ramos, Perfil estructural de la Puna Catamarquena (25-27°S), Argentina, V Congr. Latinoamericano de Geología, 1:499-518, Buenos Aires, 1982.

Allmendinger, R., V. Ramos, T. Jordan, M. Palma, and B. Isacks, Paleogeography and Andean structural geometry, northwest Argentina, Tectonics, 2, 1-16, 1983.

Amos, A. J. and O. Lopez-Gamundi, Las diamictitas del Paleozoico superior en Argentina: su edad e interpretación, 8° Congr. Geol. Argentino, T3, 11-58, Buenos Aires, 1981.

Anderson, J. L., R. L. Culler, and W. R. Van Schumus, Anorogenic metaluminous and peraluminous granite plutonism in the mid-Proterozoic of Wisconsin, USA, Contr. Mineral. Petrol., 74, 311-328, 1980.

Azcui, C. L., Paleogeography and stratigraphy of Late Carboniferous of Argentina, X Int. Congr. Carbon. Strat. Geol., (pre-print), 1983.

Baker, P., W. J. Rea, J. Skarmeta, R. Caminos and D. Rex, Igneous history of the Andean cordillera and Patagonian plateau around latitude 46°S, Phil. Trans. Roy. Soc. London, 303, 1474, 105-149, 1981.

Baldis, B., M. Beresi, O. Bordonaro, and A. Vaca, Síntesis evolutiva de la Precordillera Argentina, V Congr. Geol. Latin., Actas IV:339-445, Buenos Aires, 1982.

Bell, M., The lower Palaeozoic metasedimentary basement of the Coastal Range of Chile between 25°30' and 27°S, Rev. Geol. Chile, 17, 21-29, 1983.

Berg, K., C. Breitkreuz, K. Damm, S. Pichowiak, and W. Zeil, The north Chilean Coast Range - an example for the development of an active continental margin, Geologische Rundschau, 72, 2, 715-731, 1983.

Borello, A. V., Los Geosinclinales de la Argentina An. Dir. Nac. Geol. Min., 14:1-188, Buenos Aires, 1969.

Boyd, T. M., J. A. Snoke, I. S. Sacks, and A. Rodríguez, High resolution determination of the Benioff zone geometry beneath southern Peru, Ann. Rept. Direct. Dept. Terrestrial Magnetism, Carnegie Inst., 500-505, Washington, 1983.

Breitkreuz, C., and W. Zeil, Geodynamic and magmatic stages on a traverse through the Andes between 20° and 24°S (N Chile, S Bolivia, NW Argentina), J. Geol. Soc. London, 141, 861-868, 1984.

Bruhn, R. L., Ch. Stern and M. De Witt, New field and geochemical data bearing on the development of a Mesozoic volcanotectonic rift zone and back arc basin in Southernmost South America, Earth Plan. Sci. Let., 41, 32-45, 1978.

Caminos, R., Descripción Geologica de las Hojas 39g, Cerro Talinay y. 39h Chipauquil, provincia de Río Negro, Serv. Geol. Nac., Inédito, 1983.

Caminos, R., U. Cordani, and E. Linares, Geología y geocronología de las rocas metamórficas y eruptivas de la precordillera y cordillera frontal de Mendoza, República Argentina, 2° Congr. Geol. Chil., Actas 1, F43-F60, 1979.

Caminos, R., C. Cingolani, F. Hervé, and E. Linares, Geochronology of the pre-Andean metamorphism and magmatism in the Andean Cordillera between latitudes 30° and 36°S, Earth Sci. Rev., 18, 333-352, 1982.

Cande, S., et al., The early Cenozoic tectonic history of the southeast Pacific, Earth Plan. Sci. Let., 57, 63-74, 1982.

Chappell, B. W., and A. J. R. White, Two contrasting granite types, Pacific Geol., 8, 173-174, 1974.

Charrier, R., Interruptions of spreading and the compressive tectonic phases of the Meridional Andes, Earth Plan. Sci. Let., 20, 2, 247-249, 1973.

Coira, B., J. Davidson, C. Mpodozis, and V. Ramos Tectonic and magmatic evolution of the Andes of northern Argentina and Chile, Earth Sci. Rev., 18, 3/4, 303-352, 1982.

Collins, W. J., S. D. Beams, A. J. R. White, and B. W. Chappell, Nature and origin of A-type granites with particular reference to southeastern Australia, Contrib. Mineral. Petrol., 80, 189-200, 1982.

Corbella, H., El vulcanismo de la altiplanicie del Somuncurá, IX Congr. Geol. Arg. (S. C. de Bariloche) Relatoria I 10:267-300, 1984.

Cornejo, P., C. Nasi, and C. Mpodozis, Geología y alteración hidrotermal de la Alta Cordillera entre Copiapó y Ovalle, in: Seminario de Actualización de la Geología de Chile, Servicio Nacional de Geología y Minería, H1-H45, Santiago, 1984.

Corvalán, J., and F. Munizaga, Edades radiométricas de rocas intrusivas y metamórficas de la Hoja Valparaiso-San Antonio, Bol. 28, IIG, Santiago, 40 p., 1972.

Criado Roque, P., C. Mombrú and V. Ramos, Estructura e interpretación tectonica, VII Congr. Geol. Arg., Relatoria, 155-192, Buenos Aires, 1981.

Cucchi, R., Edades radiométricas y correlación de metamorfitas de la Precordillera, San Juan, Mendoza, Rep. Argentina, Asoc. Geol. Arg. Rev., XXVI(4), 503-515, Buenos Aires, 1971.

Cuerda, A., C. Cingolani, and R. Varela, Las graptofaunas de la Formación Los Sombreros, Ordovícico inferior de la vertiente oriental de la Sierra de Tontal, precordillera de San Juan, Ameghiniana, XX(3-4):239-260, 1983.

Dalziel, I., Back arc extension in the southern Andes: a review and critical reappraisal, Phil. Trans. R. Soc. London, 300, 319-335, 1981.

Davidson, J., and J. C. Vicente, Características paleogeográficas y estructurales del área fronteriza de las nacientes del Teno (Chile) y Santa Elena (Argentina) (Cordillera Principal, 35° a 35°15' de Lat. S), Actas V Congr. Geol. Argentino, T5, 11-55, Buenos Aires, 1973.

Davidson, J., C. Mpodozis, and S. Rivano, Evidencias de tectonogénesis del Devónico Superior-Carbonífero inferior al oeste de Augusta

Victoria, Antofagasta, Chile, Rev. Geol. Chile, 12, 79-86, 1981.
Drake, R., M. Vergara, F. Munizaga, and J. C. Vicente, Geochronology of Mesozoic-Cenozoic magmatism in central Chile, lat. 31°-36°S, Earth Sci. Rev., 18, 3/4, 353-364, 1982.
Farrar, E., A. Clark, S. Haynes, G. Quirt, H. Conn, and M. Zentilli, K-Ar evidence for the post-Palaeozoic migration of granitic intrusion foci in the Andes of northern Chile, Earth Plan. Sci. Lett., 10, 60-66, 1970.
Franchi, M., and R. Page, Los basaltos Cretácicos y la evolución magmatica del Chubut occidental. Rev. Asoc. Geol. Arg., XXXV, 2, 208-229, Buenos Aires, 1980.
Furque, G., Descripción geológica de la Hoja 176 Guandacol, prov. de La Rioja., Div. Nac. Geol. Min., Bol. 92:1-104, Buenos Aires, 1963.
Godoy, E., J. Davidson, F. Hervé, C. Mpodozis, and K. Kawashita, Deformación sobreimpuesta y metamorfismo progresivo en un prisma de acreción Paleozoico: archipiélago de los Chonos, Aysén, Chile, IX Congr. Geol. Arg., Acta IV, Bariloche (en prensa).
Haller, M., and V. Ramos, Las ofiolitas famatinianas (eopaleozoico) de las provincias de San Juan y Mendoza, IX Congr. Geol. Arg., Actas II, 66-83, 1984.
Hervé, F. Petrology of the crystalline basement of the Nahuelbuta Mountains, south central Chile, in: Comparative studies on the Geology of the Circum-Pacific Orogonic Belt in Japan and Chile, 1st Rept., Ishikawa, T., and L. Agquirre, eds., 1-51, Tokyo, 1977.
Hervé, F., F. Munizaga, M. Mantovani and M. Hervé Edades Rb-Sr neopaleozoicas del basamento cristalino de la Cordillera de Nahuelbuta, 1er Congr. Geol. Chileno, Actas, F19-F26, 1976.
Hervé, F., C. Mpodozis, J. Davidson and E. Godoy, Observaciones estructurales y petrográficas en el Basamento metamórfico del Archipiélago de los Chonos, entre el Canal King y el canal Ninualac, Aisén, Rev. Geol. Chile, 13-14, 3-16, 1981.
Hervé, F., E. Nelson, K. Kawashita and M. Suarez, New isotopic ages and the timing of orogenic events in the Cordillera Darwin, southernmost Chilean Andes, Earth Plan. Sci. Lett., 55, 257-265, 1981.
Hervé, F., K. Kawashita, F. Munizaga, and M. Bassei, Rb-Sr isotopic ages from late Paleozoic metamorphic rocks of central Chile, Jour. Geol. Soc. London, 141, 877-884, 1984.
Hervé, M., M. Suarez, and A. Puig, The Patagonian batholith S of Tierra del Fuego, Chile: timing and tectonic implications, J. Geol. Soc. London 141, 909-917, 1984.
Isacks, B., and G. Barazangi, Geometry of Benioff zones: lateral segmentation and downward bending of the subducting lithosphere, in: Island arcs, deep sea trenches and back arc basins, Amer. Geophys. Unions, Maurice Ewing Series, 1, 99-144, 1977.
Isacks, B., R. Jordan, R. Allmendinger and V. Ramos, La segmentación tectónica de los Anges Centrales y su relación con la geometría de la placa de Nazca subductada, V Congr. Latinoam. Geol., Actas III:587-606, Buenos Aires, 1982.
Jordan, T., Cuencas, vulcanismo y acortamientos cenozoicos, Argentina, Bolivia y Chile, 20-28°S, IX Congr. Geol. Argentino, Actas III, 7-24, 1984.
Jordan, T., B. Isacks, V. Ramos, and R. Allmendinger, Mountain building in the Central Andes, Epidoses, 3, 20-26, 1983.
Jordan, T., B. Isacks, R. Allmendinger, J. Brewer V. Ramos, and C. Ando, Andean tectonics related to geometry of subducted plates, Geol. Soc. Amer. Bull., 94(3):341-361, 1983.
Jordan, T., B. Isacks, V. Ramos, and R. Allmendinger, La formacion de los Andes Centrales, Comunicación YPF, 5(227):16-23, Buenos Aires, 1984.
Kay, S., V. Ramos, and R. Kay, Elementos mayoritarios y trazas de las vulcanitas ordovícicas de la precordillera occidental, basaltos de rift oceánico temprano(?) próximos al margen continental, Actas IX Congreso Geológica Arg., II, 48-65, 1984.
Keidel, J., Sobre el desarrollo paleogeográfico de las grandes unidades geológicas de la Argentina, Soc. Arg. Est. Geogr. GAEA, Anales, 4, 251-312, Buenos Aires, 1925.
Knüver, M., and M. Reissinger, The plutonic and metamorphic history of the Sierra de Ancasti (Catamarca Province, Argentina), Zbl. Geol. Palaont., I, 285-297, Stuttgart, 1981.
Linares, E., E. Llambías, and C. Latorre, Geología de la provincia de La Pampa, República Argentina y geocronología de sus rocas metamórficas y eruptivas, Asoc. Geol. Arg. Rev., 35:87-146, 1980.
Llambías, E., and M. A. Leveratto, El "plateau" riolítico de la provincia de La Pampa, República Argentina, II Congr. Iberoamer. Geol. Econom., I:99-114, Buenos Aires, 1975.
Llambías, E., R. Caminos, and C. Rapela, Las plutonitas y volcanitas del ciclo eruptivo gondwanico, IX Congr. Geol. Arg., Relatorio I(4):85-117, Bariloche, 1984.
López-Escobar, L., F. Frey, and J. Oyarzún, Geochemical characteristics of Central Chile (33°-34°S) Granitoids, Contrib. Mineral. Petrol, 70, 439-450, 1979.
Maksaev, V., and N. Marinovic, Cuadrángulos Cerro de la Mica, Quillagua, Cerro Posada y Oficina Prosperidad: Carta Geológica de Chile 1:50,000, Inst. Invest. Geol., 45-48, Chile, 1980.
Mammerickx, J., E. Herron, and L. Dorman, Evidence for two fossil spreading ridges in the southeast Pacific, Geol. Soc. Am. Bull., 91, no. 5, pt. I, 263-271, 1980.
Marinovic, N., and A. Lahsen, Geologìa del sector cordillerano cuadrángulo Putana-Licancabur, cerros de Guayaques, Hoja Calama, Sernageomin Depto. de Geología, (in press).

Mégard, F., The Andean orogenic period and its major structures in central and northern Peru, J. Geol. Soc. London, 141, 893-900, 1984.

Miller, H., Características estructurales del basamento geológico chileno, V Congr. Geol. Argentino, Actas, 4, 101-115, 1973.

Miller, H., and P. Sprechman, Eine devonische Faunula aus dem Chonos-Archipel, region Aisen, Chile, und ihere stratigraphische Bedeutung, Geol. Jahrb., B28, 37-45, 1978.

Miller, H., and A. Willner, Poliphase metamorphism in the Sierra de Ancasti (Pampean Ranges, NW Argentina) and its relation to deformation, V Congr. Latinoamer. Geol., III:441-455, 1981.

Miller, H., and R. Willner, The Sierra de Ancasti (Catamarca Province, Argentina), an example of polyphase deformation of lower Paleozoic age in the Pampean Ranges, Zbl. Geol. Paläont., Tail I 3/4, 272-284, 1981.

Mingrann, A., A. Russo, A. Pozzo, and L. Cazan, Sierras Subandinas, in: Geología Reg. Arg., T. 1, 95-137, Acad. Nac. Ciencias Córdoba Córdoba, 1979.

Mpodozis, C., F. Hervé, J. Davidson and S. Rivano Los granitoides de cerros de Lila, manifestaciones de un episodio intrusivo y termal del Paleozoico inferior en los Andes del Norte de Chile, Rev. Geol. de Chile, 18:3-14, 1983.

Mpodozis, C., and R. Forsythe, Stratigraphy and Geochemistry of accreted fragments of the ancestral Pacific Floor in southern South America, Palaeogeogr., Palaeclim., Palaeocol., 41, 103-124, Amsterdam, 1983.

Munizaga, F., and J. C. Vicente, Acerca de la zonación plutónica y del volcanismo miocénico en los Andes de Aconcagua (Lat. 32-33°S), Datos radiométricos K-Ar, Rev. Geol. Chile, 16:3-21, 1982.

Munizaga, F., F. Hervé, and R. Drake, Geochronología K-Ar del extremo septentrional del Batolito Patagónico en la Región de los Lagos, Chile, Actas IX Congr. Geol. Argentino, Tomo III:133-145 (Bariloche), 1984.

Nasi, C., and R. Thiele, Estratigrafía del Jurásico y Cretácico de la Cordillera de la Costa, al sur del río Maipo, entre Melipilla y Laguna de Aculeo (Chile Central), Rev. Geol. Chile, 16, 81-99, 1982.

Nelson, E., I. Dalziel, and A. Milnes, Structural geology of Cordillera Darwin-collisional style orogenesis in the southernmost Chilean Andes, Eclogae Geol. Helv., 73, 727-751, 1980.

Nullo, F. E., C. A. Proserpio and V. A. Ramos, Estratigrafía y tectonica de la vertiente este del Hielo Patagónico Argentina y Chile, Actas VII Cong. Geol. Arg., I:455-470, Buenos Aires, 1978.

Nur, A., and Z. Ben-Avraham, Volcanic gaps and the consumption of aseismic ridges in South America, Geol. Soc. Amer., Mem. 154, 729-740, 1981.

Odin, G. S., Numerical dating in stratigraphy, John Wiley & Sons, 2 vol., 1040 p., 1982.

Omarini, R., U. Cordani, J. Viramonte, J. Salfity and K. Kawashita, Estudio isotópico Rb-Sr de la "Faja Eruptiva de la Puna" a los 22°35'L.S., Argentina, 2° Congr. Geol. Chil., Actas III, E257-267, Arica, 1979.

Omarini, R., J. Viramonte, U. Cordani, J. Salfity and Kawashita, K., Estudio geocronológico Rb-Sr de la Faja Eruptiva de la Puna en el sector de San Antonio de los Cobres, Provincia de Salta, Actas IV Congr. Geol. Arg., III, 146-158, 1984.

Pacci, D., F. Hervé, F. Munizaga, K. Kawashita, and U. Cordani, Acerca de la edad Rb-Sr precámbrica de las rocas de la Formación Esquistos de Belén, Departamento de Parinacota, Chile, Rev. Geol. Chile, II, 43-50, 1981.

Parada, M. A., Lower Triassic Alkaline Granites of Central Chile (30°S) in the High Andean Cordillera, Geologische Rundschau, 70, n°3, 1043-1053, 1981.

Parada, M. A., Crystallization conditions of Epizonal Leucogranites Plutons in the light of compositional zoning of plagioclase, High Andes (30°S), Chile, Rev. Geol. Chile, N°18, 43-54, 1983.

Parada, M. A. Caracterización geoquímica de Elementos Mayores de las Rocas Igneas Hercínicas de la Cordillera Frontal entre los 30° y 33°S, IX Congr. Geol. Arg., III, 159-170, 1984.

Parada, M. A., F. Munizaga, and K. Kawashita, Edades Rb-Sr roca total del Batolito Compuesto de los ríos Elqui-Limarí a la latitud 30°S, Rev. Geol. Chile, 13-14, 87-93, 1981.

Pilger, R. H. J., Plate constructions, aesismic ridges and low angle subduction beneath the Andes, Geol. Soc. Amer. Bull., 92, 1981.

Pitcher, W. S., Granite type and tectonic environment, in: Hsu, K. J., Mountain Building Processes, Academic Press, 1-3:19-40, London, 1982.

Polanski, J., Carbónico y Pérmico de la Argentina Eudaba Ed., 216 p., Buenos Aires, 1970.

Quartino, B., R. Zardini, and A. Amos, Estudio y exploración geológica de la región Barreal-Calingasta, Asoc. Geol. Arg., Monogr. 1, 1-184, Bs. Aires, 1971.

Ramos, V., Tectónica de la región del río y lago Belgrano, Cordillera Patagónica, Argentina, II Congr. Geol. Chil., Actas I, B1-32, Arica, 1979.

Ramos, V., Descripción geológica de la Hoja 47 a-b Lago Fontana, provincia de Chubut, Servicio Geol. Nac. Bol., 183, 1-130, Buenos Aires, 1981.

Ramos, V. El diastrofismo oclóyico: un ejemplo de tectónica de colisión durante el Eopaleozoico en el Noroeste Argentino, Rev. Inst. Cienc. Geol., 6, (in press), S.S. de Jujuy, 1982.

Ramos, V., Evolución tectónica y metalogénesis de la Cordillera Patagónica, Seg. Congr. Nac. Geol. Econom., Actas I:108-124, San Juan, 1983.

Ramos, V., and M. Palma, Las lutitas pizarreñas fosilíferas del Cerro Dedo y su evolución tectónica, Lago La Plata, Provincia de Chubut, Asoc. Geol. Arg., Revista XXXVIII(2):148-160, 1982.

Ramos, V., R. Jordán, R. Allmendinger, S. Kay, J. Cortes, and M. Palma, Chilenia, un terreno alóctono en la evolución paleozoica de los Andes centrales, IX Congr. Geol. Arg., Actas II, 84-106, Bariloche, 1984.

Rapela, C., Las rocas granitoides de la región de Cafayate, provincia de Salta, Aspectos petrológicos y geoquímicos, Asoc. Geol. Arg. Rev., 31:260-278, 1976a.

Rapela, C., el basamento metamórfico de la regíon de Cafayate, provincia de Salta, Aspectos petrológicos y geoquímicos, Asoc. Geol. Arg. Rev., 31:203-222, 1976b.

Rapela, B., Características geoquímicas y geocronológicas de la evolución ígneo-metamórfica de las Sierras Pampeanas, Primera Reun. Com. Cient. Centro de Investigaciones Geol. (abstr.) p. 14, 1981.

Rapela, C., Aspectos geoquímicos y petrológicos del batolito de Achala, provincia de Córdoba, Asoc. Geol. Arg. Rev., 37:313-330, 1982.

Rapela, C., and D. M. Shaw, Trace and major element models of granitoid genesis in the Pampean Ranges, Argentina, Geochim. Cosmochim., Acta 43, 1117-1129, 1979.

Rapela, C., L. Heaman, and R. MacNutt, Rb-Sr geochronology of granitoid rocks from the Pampean Ranges, Argentina, J. Geol., 90, 574-582, 1982.

Rapela, C., and Llambías, E., Evolución magmática y relaciones regionales de los complejos eruptivos de La Esperanza, provincia de Rio Negro, Asoc. Geol. Arg. Rev., 1985.

Reissinger, M., Geología de la Sierra de Ancasti, Evolución geoquímica de las rocas plutónicas, Münster Forsch. Geol. Paläont., 59:101-112, 1983.

Rolleri, E., and B. Baldis, Paleogeography and distribution of carboniferous deposits in the Argentine Precordillera, I.U.G.A. Coll. on Gondwana Stratigraphy, Ciencias de la Tierra, 2, 1005-1024, Unesco, 1967.

Rona, P., and E. Richardson, Early Cenozoic global plate tectonic reorganization, Earth Plan. Sci. Lett., 40, 1-11, 1978.

Rutland, W. R., Andean orogeny and sea floor spreading, Nature, 233, 252-255, 1971.

Salfity, J., Evolucíon paleogeográfica del Grupo Salta (Cretácico-Eogénico), República Argentina, V Congr. Geol. Lat., Actas I, 11-26, 1982.

Scholl, D. W., R. von Heune, T. L. Vallier, and D. G. Howell, Sedimentary masses and concepts about tectonic processes at underthrust ocean margins, Geology 8, 564-568, 1980.

Schweller, W. I., L. D. Kul, and R. A. Prince, Tectonics, structure and sedimentary framework of the Peru-Chile Trench, Geol. Soc. Am., Mem., 154, 323-350, 1981.

Sepúlveda, E. C., and R. Cucchi, Contribución al conocimiento de las metasedimentitas de la Formación Esquel en los cerros Excursión, Provincia de Chubut, Actas VII Congr. Geol. Arg., 1, 437-443, 1978.

Shackleton, R., A. Ries and M. Coward, Structure, metamorphism and geochronology of the Arequipa Massif of Coastal Peru, J. Geol. Soc. London, 136, 195-214, 1979.

Spalletti, L. A., Paleoambientes de sedimentación de la Formación Patquía (pérmico) en la Sierra de Maz, La Rioja, Bol. Acad. Nac., 53(1-2):167-202, Córdoba, 1979.

Steiglitz, O., Contribución a la petrografía de la Precordillera y Pie de Palo, Div. Gen. Min., Geol., Hidrogeol., Bol. 10, Buenos Aires, 1914.

Stipanicic, P., Consideraciones sobre la edades de algunas fases magmáticas del Neopaleozoico y Mesozoico, Asoc. Geol. Arg. Rev., 22:101-133, 1967.

Stipanicic, P., F. Rodríguez, O. Baulíes, and C. Martínez, Las formaciones presenonianas en el denominado Maciozo Nordpatagónico y regiones adyacentes, Asoc. Geo. Arg. Rev., 23:67-88, 1968.

Stipanicic, P. and E. Methol, Macizo de Somuncurá in: Geología Regional Argentina, Acad. Nac. Ciencias Córdoba, 581-599, 1972.

Swift, S. A. and M. J. Carr, The segmented nature of the Chilean seismic zone, Physics Earth & Planet. Interiors, 9(1974):183-191, Amsterdam, 1974.

Tavera, J., Noticia sobre la presencia de Graptoloideos en rocas del basamento cristalino, Imp. gráfica, 1982.

Thiele, R., C. Nasi, Evolución tectónica de los Andes a la latitud 33° 34°Sur (Chile Central) durante el Mesozoico-Cenozoico, V Congr. Latin. Geol., Actas III, 403-426, Buenos Aires, 1982b.

Toselli, J., A. Toselli, and C. Rapela, El basamento metamórfico de la Sierra de Quilmes, Re. Argentina, Asoc. Geol. Arg. Rev., XXXIII(2): 105-121, Buenos Aires, 1978.

Turner, J., and V. Mendez, Geología del sector oriental de los Departamentos de Santa Victoria e Iruya, Provincia de Salta, Argentina, Bol. Acad. Nac. Cienc., 51(1-2):11-24, Cordoba 1975.

Ugarte, F., La cuenca compuesta carbonífera jurádica de la Patagonia Meridional, An. Univ. Patag. "San Juan Bosco" Geol., I(1):37-68, Com. Rivadavia, 1966.

Uyeda, S., Subduction zones: An introduction to comparative subductology, Tectonophysics, 81, 3-4, 133-159, 1982.

Uyeda, S., and Kanamori, H., Back arc opening and the mode of subduction, Jour. Geophys. Research 84, 1049-1061, 1979.

Valencio, D., Paleogeograffía de Gondwana a partir de datos paleomagnéticos, 2° Congr. Latin. Geol., Actas I, 162-195, 1975.

Vicente, J. C., R. Charrier, J. Davidson, A. Mpodozis, and S. Rivano, La orogénesis subhercínica: Fase mayor de la evolución paleogeográfica y estructural de los Andes argentino-chilenos centrales, Actas V Congr. Geol. Arg., T5, 81-98, 1973.

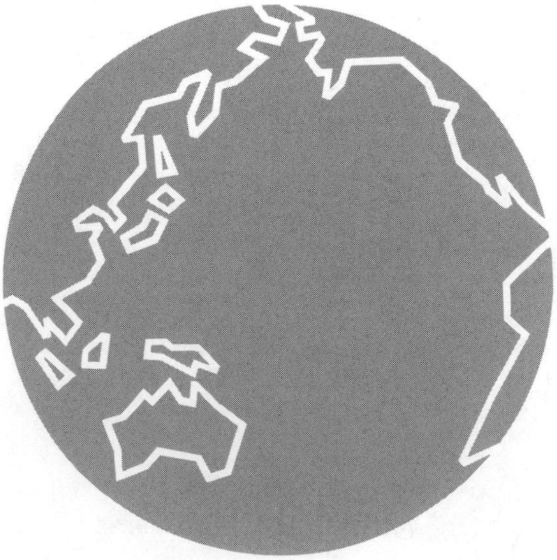

DEVELOPMENT OF THE NEW ZEALAND MICROCONTINENT

Klaus Bernhard Spörli

Department of Geology, University of Auckland, Private Bag, Auckland, New Zealand

Abstract. Three main orogenic or convergent episodes, the Paleozoic "Tuhua event" (400 to 360 Ma), the Mesozoic "Rangitata event" (mainly between 120 and 100 Ma) and the late Cenozoic "Kaikoura event" (25 Ma to Recent), and a major Late Cretaceous to early Tertiary episode of rifting, define the tectonic evolution of New Zealand. Evidence for the Tuhua event is restricted to the western foreland of New Zealand, where early Paleozoic strata have affinities with those of southeastern Australia, although large-scale Paleozoic sinistral faulting is required to make this correlation compatible with the Gondwana position of New Zealand. Plate tectonic processes leading to Tuhua deformation are unknown. The Tuhua event culminated in intrusion of the Karamea batholiths. A complex, Carboniferous through Jurassic greywacke-dominated assemblage lies east of the foreland and represents accreted clastic wedges deposited on oceanic crust, together with probable exotic terranes, one of which, Torlesse terrane, possibly originated in Marie Byrd Land, Antarctica. Collision of the Torlesse with the Gondwana margin may have caused a first phase of the Rangitata event in Late Triassic and Early Jurassic time. The second Rangitata phase, in Neocomian time, is marked by batholiths in the western foreland and a widespread unconformity, and signals cessation of the convergent regime. Up to this time, New Zealand was part of Gondwana, but after protracted rifting in the Cretaceous it separated from the supercontinent in the early Tertiary during opening of the Tasman Sea, when an Eocene-Oligocene rift system joined up newly formed oceanic crust at opposite ends of the continental fragment. With increasing separation of Antarctica and Australia at about 25 Ma, the presently active Pacific/Indian convergent transform plate boundary, including the Alpine Fault, was established, together with formation of the calc-alkaline Taupo Volcanic Zone. Increasingly stronger convergence across the plate boundary led to oroclinal bending of New Zealand and to uplift in excess of 10 mm/y near the Alpine Fault.

Introduction

The geological record of New Zealand is of special interest from the point of view of lithosphere development, for the following reasons. Firstly, it documents separation of a small plate from the Gondwana supercontinent and the incorporation of this plate into the present system of plates. Secondly, together with New Caledonia, it is part of a mainly continental fragment, 90% of which is submerged. Thirdly, it straddles the Pacific/Indian plate boundary, so that the plate boundary must have propagated through the continental fragment, which leads to questions on mechanisms of propagation at all levels of the lithosphere. Fourthly, an orogeny in progress along a transform plate boundary can be studied in an easily accessible region.

Present day tectonism is dominated by this transform plate boundary (Fig. 1; Walcott, 1978a, b). In the South Island it is represented by the dextral strike-slip/collisional Alpine Fault which associated with uplift in excess of 1 cm/y at Mt. Cook. South of New Zealand, the plate boundary is the Puysegur Trench, which marks the site of underthrusting from the west. In the North Island, a full continental arc with an accretion complex, volcanic arc and intra-arc rift system has developed, which is transitional to the Tonga-Kermodec trench system farther north, and is associated with underthrusting from the east. The New Zealand land area is surrounded by basins and ridges resulting from earlier plate activity (Fig. 1).

The part of the continental fragment above sea level can be divided into two zones according to the basement geology, which are separated by the "median tectonic line" (Landis and Coombs, 1967). The western zone acts as a foreland to Mesozoic deformations and contains abundant Paleozoic rocks, major plutonic belts, and features high-pressure, low-temperature metamorphism. The eastern zone contains only scattered Paleozoic rocks, no plutons, and is dominated by greywacke-type sedimentary sequences and derived schists. It includes the narrow Dun Mountain belt of sheared ultramafics (Coombs et al., 1976; Davis

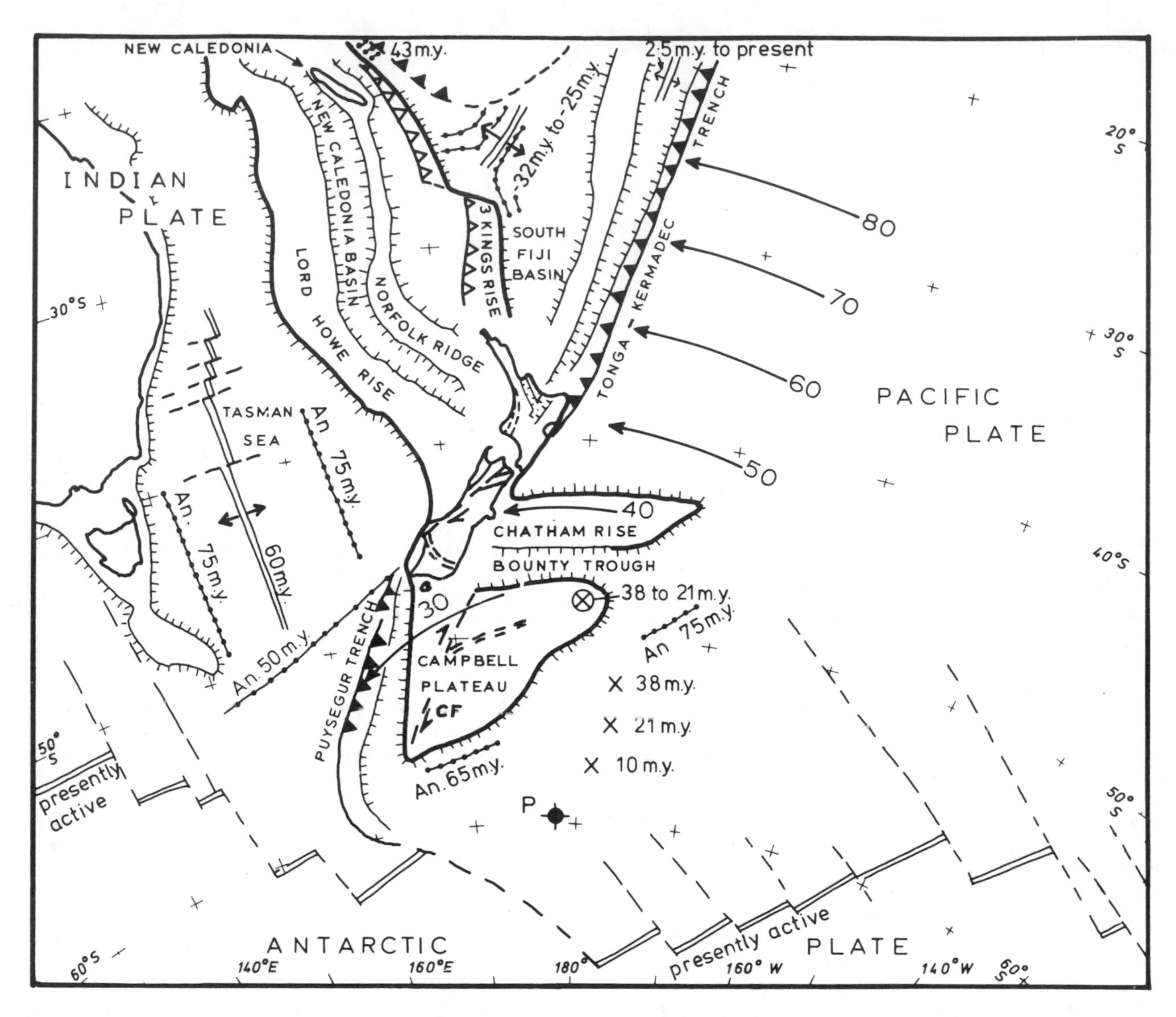

Fig. 1. Present plate tectonic regime of the New Zealand microcontinent (bordered by heavy line with hachure). Lines with filled teeth: Active subduction systems, teeth on upper plate. Open teeth: inactive (Miocene?) subduction system. Lines with hachures: rift systems. Dashed lines: transforms; Double lines: spreading ridges. Dots with lines (+An) important magnetic anomalies with age in millions of years. Broken double line in New Zealand: Stokes magnetic anomaly system, locus of the Dun Mountain ophiolite belt. P: present instantaneous pole; crosses: finite poles (ages indicated); circled cross: stage pole with age span; curved arrows: direction of present convergence (mm/year) after Walcott (1978a). CF=Campbell Fault.

et al., 1980) which can be traced magnetically in the subsurface and offshore (Hatherton, 1975). The zones described above have been distorted into a Z-shape by distributed intracrustal strain during movement on the Alpine Fault transform system.

Traditionally, deformation in the New Zealand region has been described in relation to three major orogenic episodes, the Paleozoic Tuhua orogeny, the Mesozoic Rangitata orogeny and the Cenozoic Kaikoura orogeny. These events can now be interpreted as resulting from processes such as accretion, microplate collision, obduction, and uplift due to isostatic rebound. In addition, a fourth event, Cretaceous-early Tertiary rifting, plays an important part in the tectonic

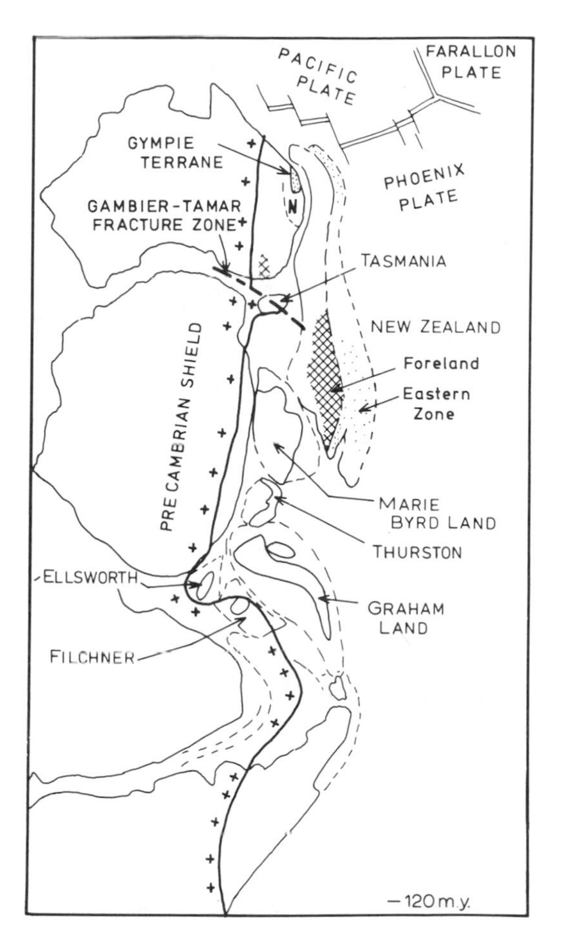

Fig. 2. Reconstruction of the Gondwanan margin approximately 120 million years ago (Late Neocomian). Base combined from De Wit (1977, 1981). Names of west Antarctic microplates after De Wit (1981); N=New England Fold belt.

evolution. In this paper, I subdivide the geological history into four "events" each labelled after its major orogeny or rifting phase.

Gondwana Assembly

Understanding the configuration of the Gondwana continent prior to its break-up is important in deciphering the geological history of New Zealand. Since, with revised dating of seafloor anomalies, Australia initially separated from Antarctica between 85 and 110 million years ago (Stock and Molnar, 1982), a complete reassembly of the New Zealand region must be referred to Neocomian (120-140 Ma) time (Fig. 2). There is considerable evidence (Grindley et al., 1981; Cooper et al., 1982; Grindley and Davey, 1982) that at that time all continental fragments in the region were tightly clustered, with few intervening oceanic areas of any great consequence. Basins such as the Bounty Trough and New Caledonia Basin are all assumed to be closed in the reconstruction. South Island and Campbell Plateau are fitted against the Ross Sea Basin, and the present Marie Byrd Land coast of Antarctica together with Lord Howe Rise and North Island, against the present eastern seaboard of Australia. Geometrically, there are only small differences between various recently-employed reconstructions, and problems arise mainly from the probable existence of a number of microplates in West Antarctica. In New Zealand, only the exact restoration of the Z-bend causes considerable problems. A consensus is emerging that bending is due entirely to late Cenozoic deformation and has no Mesozoic component (Norris, 1979; Walcott and Mumme, 1982) although the greater sharpness of the bend in the South Island may hide an earlier structure (Grindley et al., 1981; Bradshaw et al., 1981).

The major continental fragments were dispersed after Neocomian time, although smaller fragments were separated prior to this time, possibly in the Paleozoic, so that the Paleozoic Tuhua orogeny may indicate an episode of amalgamation of terranes.

While the tracks of fragments are well constrained by paleomagnetism, by matching of seafloor anomalies, and by the fit of fragments for the time interval from Late Cretaceous to Recent, the extent, shape and pattern of movements of pre-late Mesozoic fragments are based on regional geological correlations only.

Paleozoic "Tuhua Event"

The western, Paleozoic foreland is exposed in two major areas, called the Nelson and the Fiordland blocks (Fig. 3). Scattered information is also available from wells drilled off the west coast of the North Island (Wodzicki, 1974; Pilaar and Wakefield, 1978), from Stewart Island, and from dredgings on the Campbell Plateau east of the South Island (Adams et al., 1979). Because of accessibility, the Paleozoic geology of the Nelson block is far better known than that of Fiordland block. In addition, difficulty arises from the fact that Mesozoic and Cenozoic deformations have overprinted Paleozoic structures. The effects of the various deformations have only been separated-out to a limited extent in those places where major Mesozoic and Cenozoic peneplains and well dated Cretaceous plutons and their metamorphic aureoles provide time markers (e.g. Grindley, 1980).

No rocks older than late Precambrian gneisses of the Charleston Group are recognized in New Zealand (Adams, 1975; Hume, 1977). The early Paleozoic sequence of the Nelson Block is divided into three contrasting belts (Fig. 3; Cooper, 1975; 1979). The Eastern Belt represents a non-volcanic trough containing continent-derived

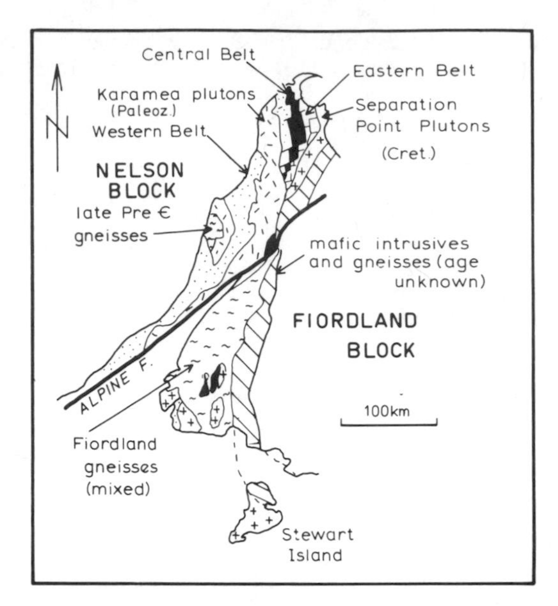

Fig. 3. Elements of the Western foreland, consolidated in the Tuhua event (after Cooper, 1979). Dextral movement on the Alpine Fault is reversed. The two blocks are still internally distorted by post-Paleozoic distributed dextral shear that is parallel to the Alpine Fault.

clastic sedimentary rocks and prominent limestones, which range in age from Middle Ordovician to Late Silurian or Early Devonian. The Central Belt comprises Middle to Late Cambrian basaltic to andesitic assemblages and minor ultramafics, overlain by volcanically-derived sedimentary rocks, including thick conglomerates, and trilobite-bearing limestone, which range up to Early Ordovician age. The Western Belt represents nonvolcanic trough deposits, mainly Lower Ordovician quartz-rich greywacke with no associated limestone or conglomerate, which discordantly overlie complexly-folded late Precambrian Charleston gneisses (Hume, 1977).

A variety of evidence documents Paleozoic tectonism in these rocks. If the formation of the Charleston gneisses occurred in the late Precambrian, this is the oldest record of deformation and metamorphism in New Zealand. A later "Haupiri disturbance" is assumed to account for the influx of coarse Late Cambrian conglomerates (Grindley, 1980). A major unconformity, with onlap of Devonian (Emsian and Siegenian) strata now preserved only in small inliers, marks a strong phase of deformation which probably ended in Late Ordovician or Early Silurian time. On the basis of radiometric dates, it is assumed that the main Tuhua deformation occurred in the Late Devonian or Carboniferous, simultaneously with intrusion of the Middle or Late Devonian and Carboniferous Karamea granitic plutons, which occur mainly in the Western Belt. However, folding and thrusting in Devonian strata can only be constrained as pre-Cenozoic (Bradshaw and Hegan, 1983), and the question remains open whether a pre- or post-Devonian phase represents the main structures seen in the older rocks. Crook and Feary (1982) have interpreted the plutonic rocks as "post-subduction" granitoids.

Controversy exists about the kinematics of structures in early Paleozoic rocks of the Central Belt, with a number of different views held (Bradshaw and Grindley, 1982). In one interpretation, the Central Belt with its east-west trending, reclined, first-phase folds was thrust, as nappes, from south to north, and then was deformed along north-south trending axes (Grindley, 1980). An alternative interpretation is given by Shelley (1984) who relates all folds in the Central and Eastern Belt to one regime of progressive coaxial deformation with a west southwest-east northeast oriented direction of movement. Those folds that are parallel to this direction are interpreted to be sheath folds rotated into the movement direction, in zones of high strain, with a possible derivation of the recumbent nappes from the western or southwestern sector. On the basis of this complex structural model, Shelley questions the interpretation by Crook and Feary (1982) of the Eastern Belt as a "post-arc sedimentary sequence" belonging to a Tuhuan terrain. Shelley (1975) considers the simple deformation in the Western Belt to be due to gravitational sliding of Ordovician Greenland Group sediments off highs created above intruding Tuhuan plutons. It is important to note that, so far, none of the features typical of deformation in subduction zones, such as mélanges, broken formations and ophiolites have been positively identified in Paleozoic rocks of the western foreland in the Nelson Block. The absence of these could indicate a foreland thrust regime or collision of consolidated terranes. Alternatively, there may be hitherto unrecognized melanges such as the "olistostromes" of the Early Cambrian Balloon Formation (Grindley, 1980) and/or in the "soft-sediment slumping" and lensoid limestones of the Middle Cambrian Tasman Formation (Grindley, 1980). The question of whether the Cobb Igneous Complex (Grindley, 1980) is a stratiform ultramafic intrusion (Hunter, 1977) or an allochthonous ophiolite, also must remain open.

In the Nelson Block there is a large gap in the sedimentary record (Cooper, 1975) between the Devonian and a small inlier of Permian conglomerate, sandstone and siltstone, with marine fossils, of the Parapara Group.

It is difficult to obtain a clear picture of the tectonic situation in the Fiordland Block. The extensive areas of high-grade metamorphic and(?) plutonic rock indicate that deeper crustal levels are exposed there, than in the Nelson Block (Oliver and Coggon, 1979). However, graptolite-bearing beds in the southwest of the area (Preservation Formation of Preservation and Chalky inlets) are supracrustal and may be a direct correlative of the early Paleozoic units in the Nelson Block (Cooper, 1979). Regional subdivisions in the rest of the area are different from author to author (Gibson, 1982a,b). Gibson distinguishes a Manapouri Province composed of ortho- and para-gneisses, in part granulitic, together with psammitic and pelitic schists and leucogranites, from a Doubtful Sound Province of mostly paragneisses with some calc-silicate gneisses, anorthosite and gabbros. Middle to upper amphibolite facies metamorphism took place during isoclinal F_1 folding. Metamorphic crystallization continued with large-scale north-south trending F_2 folds and thrusts. F_3 folds are post-metamorphic open folds with steep axial planes. Some granites intrude the metamorphosed cover in the Doubtful Sound Province and have given Rb/Sr isochron ages of 372 Ma (Oliver, 1980). However, large tracts of granulite facies rocks in western Fiordland have recently been found to have been magmatically emplaced only 125-130 million years ago (Mattinson et al., 1984).

On the basis of stratigraphy, faunal association and tectonic evolution, Cooper (1975) considers the three belts of early Paleozoic rocks in the Nelson Block to be direct analogues of the Ballarat and Melbourne troughs and the interposed Heathcote axis of central Victoria, Australia. Because the two areas are not aligned and the Precambrian-cored Tasmanian Block is interposed between them in the Gondwana reconstruction (Fig. 2), Cooper postulates a major sinistral fault at a high-angle to the Gondwana margin, which was active during Silurian or earliest Devonian time.

Correlations with rocks in Antarctica are more tenuous. Lithologically, greywackes of the Western Belt are similar to the Swanson Group of Marie Byrd Land and the Mariner/Leapyear Group in north Victoria Land (Adams, 1981). The migmatites and amphibolites of the Fosdick Mountains in Marie Byrd Land have been compared with the similar rocks of Fiordland (Cooper et al., 1982). Synkinematic and post-kinematic plutons in Marie Byrd Land match the late Devonian to Carboniferous plutons involved in the Tuhua orogeny in New Zealand.

Little is known about the configuration of the Paleo-Pacific off the New Zealand Gondwanan margin. Most of the oceanic crust is presumed to have been lost during Mesozoic subduction along the margin. Crook and Feary (1982) have attempted a reconstruction for the Middle Cambrian in which double subduction zones in Australia (Victorian and Riverina terrains) are facing double subduction zones (northwest Tasmania, Tuhua terrains) across a strongly transform-segmented spreading ridge system trending at a high angle to the Gondwanan margin.

If the correlation of the early Paleozoic rocks of the Nelson Block with coeval rocks of central Victoria is correct, why did the New Zealand assemblage end up at the continental margin of Gondwana, while the Victorian assemblage lay far inland with the Paleozoic rocks of New South Wales interspersed on the Paleo-Pacific side? Cooper (1975) used a Late Devonian dextral fault subparallel to the Gondwanan margin to bring these extra sequences into position. Harrington (1983) on the other hand, postulated a Tamar salient, controlled by the sinistral Gambier-Tamar (or Gambier-Beaconsfield) fracture zone at a high angle to the margin, to explain the different positions. The continuation of this fracture intersects the western foreland of New Zealand somewhere on the now-submerged parts of Lord Howe Rise. Another unsolved problem in relation to Lord Howe Rise is the position of the southern boundary of the late Paleozoic New England fold belt, since there is no evidence of such a zone in New Zealand.

Mesozoic "Rangitata Event"

By the conclusion of the Tuhua orogeny, the western foreland of New Zealand appears to have been completely consolidated. Against this foreland was juxtaposed in the Mesozoic a set of rock units, that are separated into discrete terranes by some authors (Howell, 1980; Bradshaw et al., 1981) or considered to be more continuous by others (e.g. Spörli, 1978). These rock units form seven terranes more-or-less parallel with the Gondwanan margin (Fig. 4), that are listed below, from the foreland eastwards towards the ocean. (1) Brook Street terrane is interpreted to be a Permian volcanic arc. (2) Productus Creek terrane is interpreted to be a Late Permian subsiding-arc margin (and included in Fig. 4 with the Murihiku terrane. (3) Murihiku terrane comprises Triassic and Jurassic volcaniclastic rocks, up to 15 km thick, that are interpreted as forearc sequences. (4) Maitai terrane, of Late Permian to Early Triassic age, consists of carbonate and volcaniclastic rocks that are interpreted as forearc sequences. (5) Dun Mountain terrane, comprises mafic volcanic and ultramafic rocks that are believed to have formed the original floor of the Maitai basin, and now is a major tectonic and geophysical marker throughout New Zealand; where exposed it shows mélange structure (Coombs et al., 1976; Davis et al., 1980). (6) Caples terrane consists of mainly volcaniclastic sediments and volcanics of trench-slope origin that in the South Island yields Permian fossils; its probable equivalent, Waipapa, in North Island, contains Jurassic fossils as well. (7) Torlesse terrane is composed of mainly quartzofeldspathic sandstone and mudstone with minor metavolcanics, chert, and conglomerate, which ranges in age from Late Carboniferous to Late Jurassic and perhaps earliest Cretaceous.

It can be debated whether the first four terranes really should be considered as separate entities, as it is possible to construct a model of a single, persistent arc that contains all of the terranes. However, Williams and Smith (1979) have presented evidence for paired arcs in the

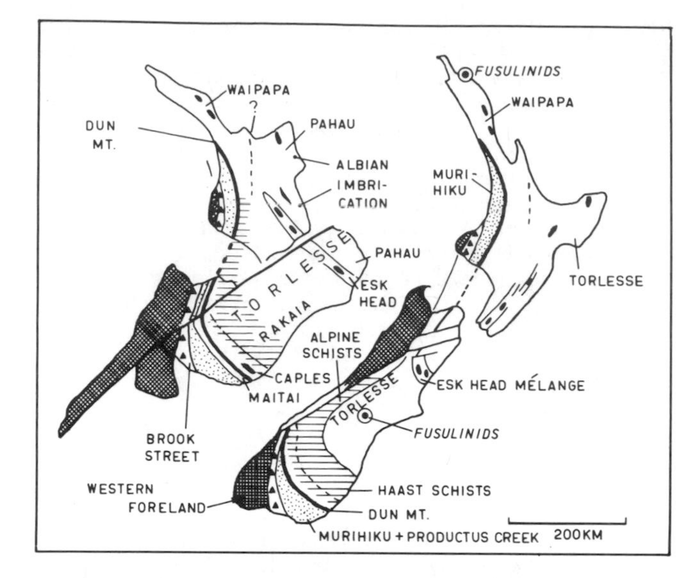

Fig. 4. Elements involved in the Rangitata event after Bradshaw et al., 1981 (slightly modified). The present geography is on the right; the Early Cretaceous on the left. Black lenses denote major zones of mélange and broken formations. Rakaia, Esk Head mélange and Pahau are subdivisions of the Torlesse terrane. "Fusulinids" mark localities with exotic Tethyan (equatorial) Permian ocean floor material.

Permian. All four terranes have a simple structural style, and metamorphism is apparently related to stratigraphic depth. Murihiku terrane with its simple asymmetric Gondwana-ward verging synclinal structure may record continuous accretion during Triassic and Jurassic time.

Caples and Torlesse terranes are complex in structure and show polyphase deformation. Metamorphic isograds cross one or more phases of folding. The terranes contain zones of broken formation and mélange. Torlesse terrane occupies by far the largest area, and has been further subdivided (Fig. 4). Within it beds are steep to vertical over distances of several tens to 100 km across strike, indicating imbrication and rotation. In both Caples and Torlesse terranes, spilitic volcanics of oceanic derivation (either from spreading ridges or intraplate), cherts and rare limestones are commonly associated with the zones of broken formation and mélange (e.g. Turnbull, 1980, Nelson, 1982); this association is interpreted to represent the top of the oceanic crust that was sheared off during subduction, and onto which the terrigenous sandstones and argillites were deposited (Spörli and Bell, 1976; Spörli, 1978).

The age range of the Caples strata is not well known, although the Permian age usually assumed is based on scattered occurrences of *Atomodesma* mostly in limestones associated with volcanics. It is quite possible that the voluminous terrigenous clastics are considerably younger. In the Waipapa terrane (Fig. 4; Spörli, 1978), which is in the analogous position to the Caples but in North Island, Permian Tethyan fusulinids and reef building corals intimately associated with spilites date only the sheared-off seafloor. They are clearly exotic to the New Zealand sector of Gondwana, which was at a polar latitude at that time (Spörli and Gregory, 1981). The clastics in this terrane have only yielded Jurassic fossils (Spörli and Grant-Mackie, 1976).

A persistent problem has been the source of terriginous clastic rocks in the Torlesse terrane. Its quartzo-feldspathic clastic suite is interpreted as originating from continental crust (Coombs et al., 1976; Bradshaw et al., 1981), but Torlesse terrane is situated oceanwards from the volcanic arc suite represented by terranes 1 to 6. Models put forward to explain this relationship include bypassing of continental clastics through the volcanic arc via canyons descending from the Gondwana margin (Landis and Bishop, 1972), an origin in an intracratonic basin (Austin, 1975), and lateral faulting of a more-or-less exotic clastic wedge (and/or accretionary prism) onto the New Zealand sector of Gondwana (Spörli, 1978; Bradshaw et al., 1981). Consensus for the last model seems to be growing, and the source of the exotic clastic wedge is now commonly taken to be Marie Byrd Land (McKinnon, 1983). The Bradshaw et al. (1981) model assumes that the wedge was deposited from a transform-controlled promontory. McKinnon (1983) postulates that the Torlesse clastic sequence was accreted directly onto the Marie Byrd Land sector of Gondwana, but coeval arcs, sedimentary basins and accretionary sequences of the New Zealand sector were separated from the margin by a back-arc basin or marginal sea. The two sets of terranes are assumed to have been eventually superposed by oblique convergence and sinistral strike-slip faulting. The Gulf of Alaska is cited as a modern analogue.

As a generalization, the structural development of the Torlesse and probably Caples and Waipapa terranes proceeded in three stages (Spörli, 1978). Firstly, the top of the oceanic crust was sliced off along shallowly-dipping thrusts which produced mélange and broken formation. Secondly, continued accretion rotated thrusts and bedding to the vertical and refolded pre-existing structures around subhorizontal axes. Thirdly, the vertical beds were affected by strike-slip motions and formed steeply-plunging folds.

A major problem concerns the relationship and development of the schists of South Island to the Caples and Torlesse terranes. Craw (1984) considered that in Otago the schists are a complex collage of different protolithic associations. The junction of the Caples and Torlesse is obscure and within the Haast schists (see Figs. 4,6) of South Island (Wood, 1978; Craw, 1984). The Pounamou ultramafics form a mappable belt within the Haast and (younger) Alpine schists

(Cooper and Reay, 1983; Ireland et al., 1984) but are considered to originate within the Torlesse. Torlesse rocks generally show only sporadic development of cleavage and are metamorphosed in the zeolite to prehnite-pumpellyite facies, whereas the schists grade up to oligoclase-zone amphibolite facies in a Barrovian facies series (Cooper, 1972). Structurally, the schists contrast with the Torlesse in that their major folds are recumbent, which may be due to the existence of a megaculmination within the schist which refolds earlier mainly northeast-trending isoclinal folds (Wood, 1978). Norris and Cooper (1977) have shown that several early fold phases recognized in the Torlesse have been obliterated beyond recognition in the higher-grade schists. The structural pattern of the schists in Otago may therefore indicate continued underthrusting after accretion, and flat-lying structures would be due to rotation of originally steeper axial surfaces during isostatic uplift following underthrusting.

Geochronology suggests that Haast schist metamorphism was completed by the end of the Triassic (Bradshaw et al., 1981), in conflict with the evidence of a New Zealand-wide unconformity of Early Cretaceous age which has traditionally been thought to represent the Rangitata Orogeny. Bradshaw et al. (1981) postulated a Late Triassic Rangitata 1 event, interpreted as marking the collision of the early Torlesse rocks with the New Zealand-Gondwanan margin. After Rangitata 1, a trench was re-established and the younger parts of the Torlesse and the Waipapa terranes were deposited.

The Early Cretaceous unconformity marks cessation of subduction and accretion along the New Zealand-Gondwanan margin. Voluminous intrusives in the Western Foreland (Separation Point Granites; Fig. 3) are of Early Cretaceous age and mainly located near the Pacific edge of the foreland. There is no evidence for overlying volcanic edifices, nor of any Early Cretaceous erosion and deposition of large volumes of detritus derived from granitic rocks. Grindley (1980) assigns F_3 flexural slip folds with conjugate strain-slip cleavages to Rangitata deformation. Otherwise Rangitata deformation of the foreland, if of any significance, is little known. The discovery of Cretaceous granulites in Fiordland (Mattinson et al., 1984) indicates that such deformation and metamorphism may be more important than hitherto assumed.

Correlation of units involved in the Rangitata event with elements in Marie Byrd Land have already been discussed, together with the model for the origin of the Torlesse terrane. The Early Cretaceous plutonic belt also can be traced into this region and continues from there into the Antarctic Peninsula (Cooper et al., 1982). In Antarctica, however, the plutonic belt is clearly associated with volcanicity. A geometric problem arises with the great width (200-500 km) of the plutonic belt in Marie Byrd Land in comparison with that in New Zealand (20-40 km). A difference in the attitude of the plate being subducted (steeply-dipping in New Zealand; more shallowly-dipping in Marie Byrd Land?) could be responsible. Alternatively, an originally narrower belt may have been repeated and doubled up by microplate movement in Marie Byrd Land, or the thinning of the belt in New Zealand may represent the end of the east Pacific zone of "Andean" plutonics.

In the Antarctic Peninsula a major event is the Late Triassic-Early Jurassic Gondwanide orogeny, with a profound unconformity separating two sequences of clastic volcanogenic rocks. One of the unsolved paleotectonic questions is the relation of this orogeny to the New Zealand Rangitata 1 event.

Correlations with Australian geology are more tenuous, although the Gimpie terrane of Queensland has been equated with the Brook Street and Maitai terranes in New Zealand (Harrington, 1983).

The basement of New Caledonia is connected with that of northern New Zealand via the Norfolk Ridge (Fig. 1). The assumption of earlier authors that the tectonic zonations in New Zealand also carries through has been challenged by Paris and Bradshaw (1977). There is no equivalent in New Caledonia of the Dun Mountain Belt of New Zealand. However, the terrigenous sequences on the west coast of New Caledonia bear a remarkable similarity to the Murihiku terrane of New Zealand. So far the Torlesse of New Zealand is not recognized in the eastern, more complex belt of pre-Senonian rocks of New Caledonia. Sedimentation in this belt appears to have been controlled by a north-facing, east-west trending subduction zone off the north end of the island and a volcanic arc near the southern end (Paris, 1981). The protolith of the "ante-Permian" of this eastern belt is not well-dated, and may be considerably younger; if so, one of the main differences between the eastern belts of New Zealand and New Caledonia would be eliminated. Metamorphism in the eastern belt is multi-phase with a possibly older low pressure/high temperature metamorphism ranging from greenschist to amphibolite facies, and a post-Tithonian pre-Senonian high-pressure/low-temperature metamorphism with lower grades in the east, and higher grades in the west. The equivalent of the New Zealand western foreland in New Caledonia is unknown because, if present, it is submerged on the Lord Howe rise.

Very little is known about the plate configuration of the Paleo-Pacific during the late Paleozoic to Early Cretaceous interval, since there are almost no magnetic anomalies preserved to reconstruct spreading ridges. However, in future, it may be possible to piece together a partial history by studies of the paleomagnetism of spilites and of the radiolaria in sediments associated with volcanics in the accreted terranes. The occurrence of exotic Permian fusu-

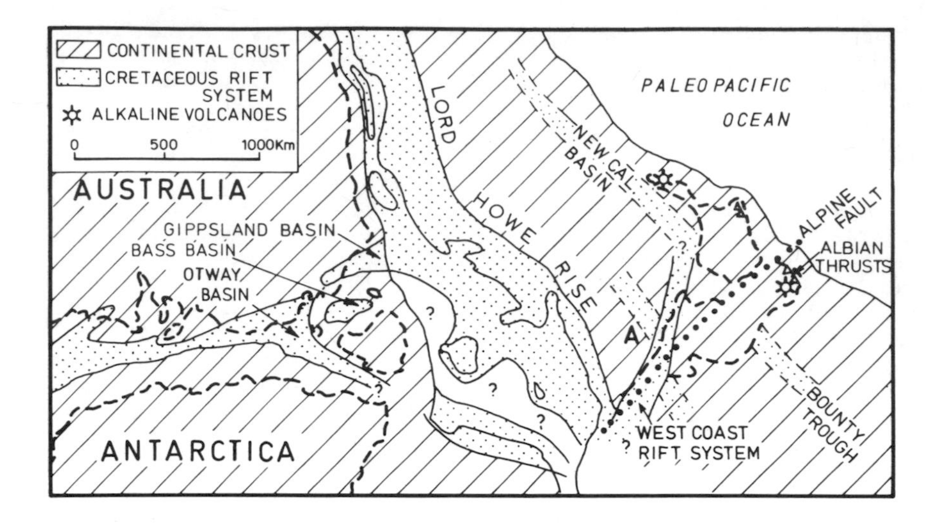

Fig. 5. Cretaceous rift systems in the New Zealand sector of Gondwana. Modified from Laird (1981). Rift with dotted boundary at A shows orientation of West Coast Rift System if Z-bending of New Zealand is straightened.

linids in oceanic assemblages in from a Tethyan (equatorial) region in now such widely dispersed areas as Japan, western North America, southern Patagonia and New Zealand (Spörli and Gregory, 1981) indicates the existence of at least one prominent R-R-R triple junction in the central part of the Pacific. Some additional constraints about plate motions may be provided by detailed structural studies in the accreted prisms, if the orientation of the slip vectors during accretion and/or collision can be determined.

Late Cretaceous-Early Tertiary Rifting Event

A major change in the tectonic regime took plate in the Cretaceous. Over most of New Zealand, the Neocomian stratigraphic hiatus is marked by a profound unconformity. The nature of the overlying mid- to Late Cretaceous deposits, which were accompanied by alkaline igneous activity (Laird, 1981), indicates their origin in an extensional tectonic setting related to the opening of the Tasman Sea from 80 to 60 Ma (Fig. 5).

In the presently northeast-trending graben system along the west coast of New Zealand, coarse continental Albian sediments of a pre-breakup sequence were deposited in fault-bounded depressions. Above the Albian strata, in part disconformably and in part with angular unconformity, are Late Cretaceous coal measures, marine sandstones, nonmarine sediments and basalts, which are cut by lamprophyre, basalt and trachyte dykes giving dates of 78-84 Ma, which are contemporaneous with the time of opening of the Tasman Sea. Major persistent faults, such as the Paparoa Fault of South Island (Fig. 6), controlled sedimentation in this "West Coast Rift System", which Laird (1981) interpreted as the northeast-trending failed arm of the rift-system known to have resulted in the separation of Australia, Antarctica and the Lord Howe Rise. It is possible that, rather than trending obliquely to the Gondwanan margin, this failed arm trended subparallel with it (Fig. 5). Such an orientation would result if the Z-bending of New Zealand is entirely post-Cretaceous.

Both the New Caledonia Basin (Eade, 1984) and the Bounty Trough (Davy, 1984) are also Cretaceous rifts, although possibly somewhat younger (Fig. 5). It is likely that they were linked, either directly or via a short transform fault which could be one of the earliest precursors of the Alpine Fault.

The situation along the Pacific margin of New Zealand is more complex. Western, landward, exposures show Aptian strata locally above an unconformity. Elsewhere, such an unconformity is difficult to observe, partly because of structural complexity. However, in some localities a prominent unconformity separates highly-deformed from less-deformed Albian rocks, which mainly consist of alternating sandstones and mudstones but also include some olistostromes. Large bodies of layered gabbro occur in the northeast of South Island and chemically similar alkaline lavas, including some boninites (Moore, 1980), are intercalated with Albian strata of the east coast of the North Island.

Complex structures beneath the intra-Albian unconformity are ascribed by Moore and Speden (1979) to sliding accompanying synsedimentary extensional faulting. However, the existence of large, closely-spaced, isoclinal, highly asymmetric folds with axial surfaces at high-angles to the unconformity surface in competent thick sandstone units, and the presence of both soft and hard rock broken formation textures (Figs. 5,

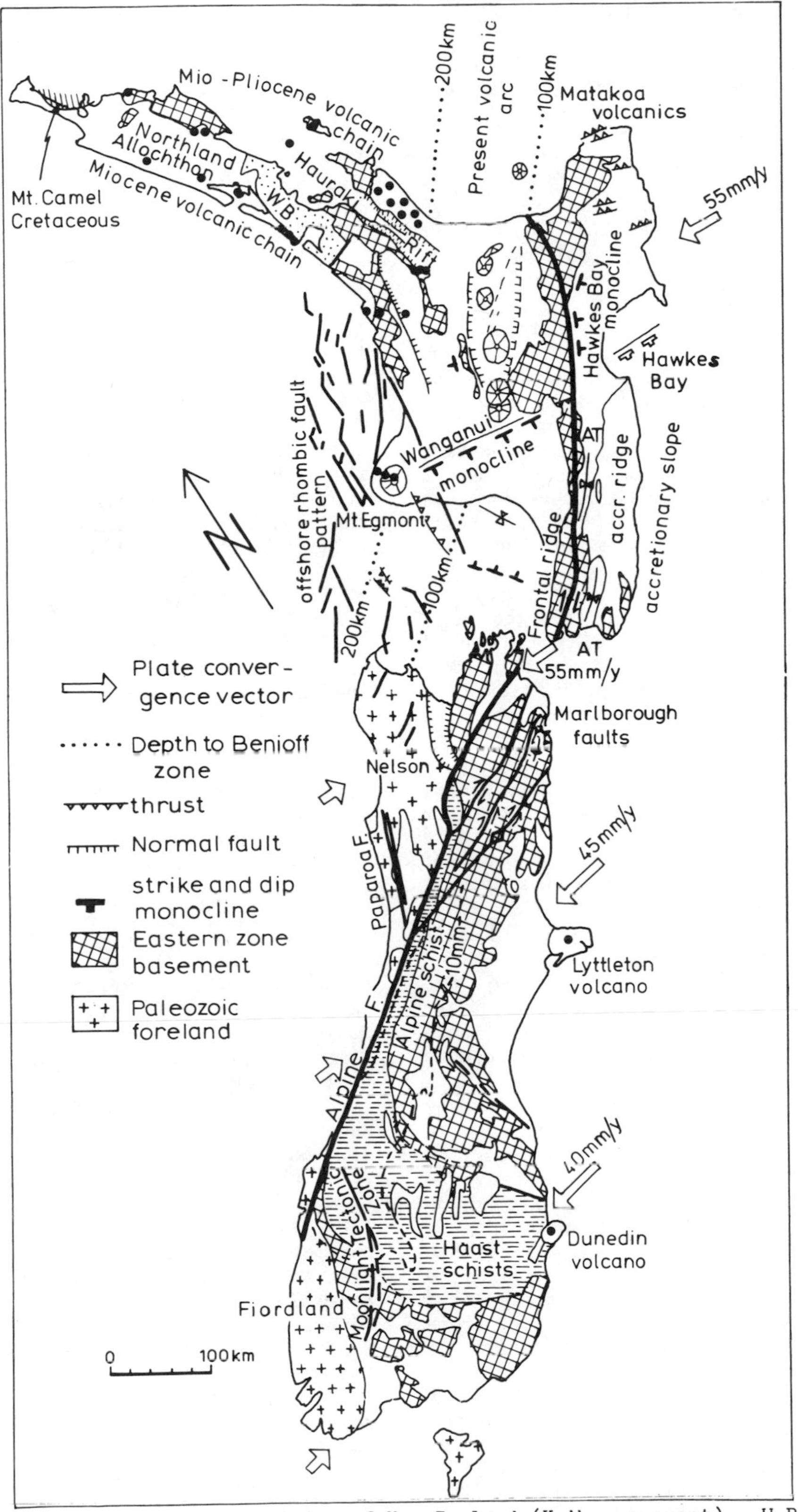

Fig. 6. Cenozoic tectonic subdivision of New Zealand (Kaikoura event). W B: Miocene Waitemata interarc basin. AT: accretionary trough. Dashed contour in South Island: uplift rate in mm/yr, after Wellman, 1979. Black dots equal late Cenozoic volcanos; volcano pattern equals dormant and active Quaternary calc-alkaline volcanoes.

8 and 9; Moore and Speden, 1979) are more consistent with imbrication, and possibly indicate that subduction processes persisted well into the Cretaceous in this region, until just before Albian rifting set in. Stratigraphic evidence for such imbrication has been presented by Feary and Pessagno (1980). Continued subduction may be supported by the existence of 92-98 Ma calc-alkaline rhyolites, andesites and dacites at the eastern edge of the Southern Alps (Mt. Somers volcanics; Oliver et al., 1979), although Crook and Feary (1982) interpret these as a surface expression of post-arc plutonism.

The question remains unanswered as to whether the highly-deformed and probably accretionary Albian sequences constitute a separate terrane which was later amalgamated with domains within which typical Cretaceous sedimentation prevails, or whether they merely record a later cessation of the Rangitata event along the Pacific margin of New Zealand. Lack of a marked suture zone may favour the second alternative but the hypothesis of the exotic Northland terrane of Crook and Feary (1982) bears further investigation.

Cretaceous of the northern promontory of the North Island (Northland) occurs in two geographically and (probably) tectonically different domains. The first is a northern domain (Mt. Camel Cretaceous, Fig. 6) containing Albian to Turonian sandstone, siltstone and conglomerate, keratophyre, and basalt, with an intra-Albian unconformity (Hay, 1975). The second, the Northland Allochthon (Fig. 6), consists of Late Cretaceous sandstone, argillite, siliceous mudstone and marl, without volcanic rocks, shows definite similarities to sequences along the East Coast of the North Island (Fig. 6; Ballance and Spörli, 1979), and no similarities to the nonmarine Pakawau Formation (Pilaar and Wakefield, 1978) off the west coast of the North Island.

Associated with the Northland Allochthon are large thrust sheets of Cretaceous-early Tertiary oceanic volcanics and associated sediments (Tangihua volcanics of Northland and Matakoa volcanics at East Cape).

Late Cretaceous to early Tertiary sedimentary sequences of New Zealand record a general time transition from coarser, nearshore facies to fine grained deep-water offshore facies, and eventually to deposition of widespread marls in the Eocene and Oligocene around the periphery of New Zealand. Alkaline igneous activity persisted into the Oligocene. This sedimentary record is compatible with drifting of New Zealand away from Gondwana, accompanied by a slow foundering, probably along a cooling trajectory governed by distance from the Tasman Sea spreading centre.

Ages of oceanic crust in the South Fiji Basin (Malahoff et al., 1982) and in the Solander Trough/Emerald Basin (Norris and Carter, 1982), indicate creation of new seafloor in the vicinity of New Zealand in Eocene to Oligocene time. Since the on-land continuation of the Solander Trough appears as the Moonlight tectonic zone of the South Island (Fig. 6), which is truncated by the Alpine Fault, a rift-transform system connecting the Moonlight Tectonic Zone with the spreading centre in the South Fiji Basin is postulated by Spörli (1980). This rift system has been more fully illustrated by Farrar and Dixon (1984) and Kamp (1984b) has proposed an Eocene to Early Miocene 100-200 km-wide rift system with all stages from intrarift subsidence to incipient seafloor spreading represented.

During the period from Cenomanian to Oligocene there is no evidence for convergent motions within New Zealand. However Eade (1984) suggests that during Late Eocene to Early Oligocene time there was a convergent margin connecting New Caledonia, Norfolk Ridge and Northland. Crook and Feary (1982) envisage a late Mesozoic to Paleogene subduction zone, facing the Australian plate along the entire Pacific margin as far as the north end of New Zealand, and attribute emplacement of the allochthonous units of Northland and East Cape (Ballance and Spörli, 1979) to collision of this arc with northern New Zealand.

Seafloor magnetic anomalies give good constraints for plate motions in the New Zealand region for the interval from about 85 Ma, when New Zealand and Marie Byrd Land separated, until the Recent. Plate configurations before 85 Ma are not so well-defined. A major problem is determining the mechanism by which the ridge between Pacific and Phoenix plates (Fig. 2) migrated past the margin towards Antarctica. One possible process is by propagation, or jumping, on a series of transforms along the margin.

Late Cenozoic "Kaikoura Event"

The still-ongoing "Kaikoura event" or orogeny led to the development of the present transform plate boundary through New Zealand. Poles of rotation derived from magnetic anomalies from the East Indian Ridge, and the Pacific Antarctic Ridge (Molnar et al., 1975; Weissel et al., 1977; Walcott, 1978a; Stock and Molnar, 1982) establish that by Anomaly 18 time (43 Ma) the pole of finite rotation of the Pacific/Indian plates had made its closest approach to New Zealand (Fig. 1). Any pre-existing zone of weakness parallel with the present Alpine Fault, such as the Eocene-Oligocene rift system, would, at this time, have come under a dextral, pure strike slip regime. Subsequently the pole moved south, away from New Zealand, leading to an increase in the rate of convergence and to a progressively more contractional regime across the Alpine Fault. Such a development is recorded by increases in the amounts of terrigenous sand, often deposited in flysch basins, and increases in the amounts of volcanic detritus (Nelson and Hume, 1977). The exact timing of the formation of the Alpine Fault as a transcurrent structure is not well determined. Kamp (1984b) postulates that it began at 23 Ma, with its time of initiation constrained by the continental rift system.

There are two schools of thought concerning the establishment of the subduction zone and the Taupo Volcanic Zone in the North Island. The first school assumes that arc magmatism began along the present northwest trends in Northland, and then migrated to its present northeasterly-trend in the Taupo Volcanic Zone and its continuation as part of the Tonga Kermadec subduction system (Cole and Lewis, 1981; Ballance et al., 1982). The second school assumes on the basis of northeast-trending radiometric age contours, K_2O index contours and other chemical index contours, that all the calc-alkaline volcanics belong to one and the same northeast-striking, down-going slab which originated 18 Ma ago as a low-angle (10° dip) subduction system, and progressively steepened to its present 50° dip, leading to progradation towards the Pacific of the accretionary prism over a distance of about 100 km (Brothers, 1984; Kamp, 1984a).

To set these problems into their proper context, the Cenozoic tectonics of crucial areas will now be discussed, proceeding from north to south (Fig. 6).

Northland

The Northland Allochthon (Fig. 6) was emplaced in latest Oligocene-earliest Miocene time (Ballance and Spörli, 1979), and contains Late Cretaceous to Oligocene strongly deformed sediments with common age reversals denoting tectonic imbrication, and zones of extreme disruption. It carries a number of exotic slabs of basic volcanics (Tangihua Volcanics), and one ultramafic massif (Bennett, 1976) which indicate obduction of seafloor from the north or northeast (Spörli, 1982). The idea of a Northland allochthon is still not universally accepted (see Brothers and Delaloye, 1982; Schofield, 1983), but is favoured by the author from field relationships. Tertiary sediments consistently on-lap onto all exposed Waipapa terrane basement, so that the widespread Cretaceous outcrop of the Northland Allochthon is always separated from the basement by a belt of Tertiary shelf strata. Drillhole evidence shows that Cretaceous (Northland) strata overlie the Tertiary shelf strata (M. Isaac, N.Z. Geol. Surv., pers. comm., 1983).

Together with analogous allochthonous sheets in the East Cape area (Ballance and Spörli, 1979), Crook and Feary (1982) regard Northland Allochthon as part of their Northland terrane and postulate that these rocks are the accretionary prism of a Late Cretaceous-early Tertiary west-facing arc which collided with and were obducted onto the northern part of New Zealand in the Late Oligocene. Obduction of the ultramafic massifs of New Caledonia (Brothers, 1974b; Paris and Lille, 1977; Paris, 1981) are part of the same process. A west-facing post-Oligocene arc, with an obducted southern part, recognized in the Three Kings Rise (Fig. 1) by Kroenke and Eade (1982) and by Kroenke and Dupont (1982) may be a remnant of this system.

Both autochthonous basement and Northland Allochthon are unconformably overlain by Miocene flysch basins, the largest of which is the Waitemata interarc basin (Ballance, 1974), which is situated between the north- and northwest-trending calc-alkaline Waitakere chain in the west, and the Coromandel chain in the east.

After Miocene time, an alkaline olivine-basalt volcanic province became established in Northland and areas to the south. It is considered to reflect transfer of the region into a behind-the-arc setting (Heming, 1980; Briggs, 1983).

Taupo Volcanic Zone

Taupo Volcanic Zone ("Present volcanic arc", on Fig. 6) is the calc-alkaline volcanic arc related to the present subduction system (Cole and Lewis, 1981). Volcanic activity dates from 1 Ma or less. Older vents appear to be oriented northwest-southeast and produced two-pyroxene andesites, and younger vents are aligned north-northeasterly and produced olivine-bearing andesite, low-Si andesite and low-Al basalts. The Recent lavas are considered to be derived from melting of subducted oceanic crust plus small amounts of sediment, at 80-100 km depth. Rhyolitic volcanism dominates in a middle segment of the zone (Wilson et al., 1984) that shows multiple caldera collapse.

Taupo Volcanic Zone occupies a graben mainly filled with rhyolitic debris 2 to 4 km in thickness. The extension rate, from geodetic measurements, is 7 ± 4 mm/yr at right angles to the graben. At the northern end of the zone, the northwest-trending, active Hauraki Rift (Ferguson et al., 1980) forms a triple junction with the Taupo Zone graben.

East Coast of North Island and Northern South Island

The Tonga-Kermadec Trench reaches the east coast of New Zealand via the Hikurangi Trough (Fig. 1; Lewis, 1980; Spörli, 1980; Cole and Lewis, 1981). Landward from the trough, a classical oblique-slip subduction complex has been developed, consisting, from southeast to northwest, of structural trench, accretionary slope, accretionary ridge, accretionary trough, frontal ridge and volcanic arc.

The structural trench is totally sediment-filled. The accretionary slope represents the Pleistocene to present day thrust regime and shows complex interaction of tectonics and sea level changes. The frontal ridge is mostly on-land and consists of older imbricated sediment, overlain by local, perched, fore-arc basins (Pettinga, 1982). Some of these are arranged in north-south trending, en-echelon synclines oriented at right angles to the vector of plate convergence. The accretionary trough is filled with Pleistocene to Recent sediment and represents the youngest forearc basin.

Whereas the accretionary prism (accretionary slope to accretionary trough) takes up the strain component mainly at a right angle to the margin, and consists mainly of young, weak, Cretaceous-Cenozoic sediments, the frontal ridge is basement-cored, and displays a record of dextral strike-slip motion in the north northeasterly direction as well as block faulting.

The entire subduction complex is segmented along strike. The frontal ridge is very narrow in the south (Fig. 6), then widens, and finally narrows again in the north. The on-land accretionary ridge and the accretionary trough are only developed in front of the southern segment. At its northern end (Fig. 6), in the Hawkes Bay area, the steep eastern thrust front is replaced along strike by the gently east-dipping, active, Hawkes Bay Monocline (Spörli, 1980). The northernmost segment is the most complex because east southeast-trending, south moving decollements, thrusts and folds interfere with the northeast-trending structure of the subduction complex (Fig. 6; Stoneley, 1968; Speden, 1976; Kenny, 1984). The east southeast-trending structures are related to thrust emplacement of the oceanic Matakoa Volcanics of Cretaceous to Paleocene age (Brothers, 1974a; Strong, 1976). Movement of the thrust sheets took place in Late Oligocene and post-Middle Miocene time. The first phase of emplacement may be equivalent to emplacement of the Northland Allochthon. The second phase is of special interest, because it correlates with persistent evidence of thrusting and shortening parallel to the strike of the subduction complex, which may be related to the lateral segmentation of the complex, and to the large cross-trending, monoclinal Wanganui-Hawkes Bay structure to be described below. Yet to be investigated is whether lateral segmentation of the subducted slab beneath the North Island (Reyners, 1983) has any influence on the overlying subduction complex, the arc and behind-arc areas.

Matakoa volcanics may continue southeastward and be recorded by isolated magnetic anomalies off the east coast, which now occur in a northeast-trending belt. If so, the formerly northwest-southeast or east southeast-trending belt of ophiolite may have been transposed and offset by dextral strike-slip movement along the accretionary complex. Part of the complex appears to have dextrally rotated at a rate of 4°/my (total rotation 65°; Walcott and Mumme, 1982), whereas some northern parts appear to have experienced no rotation, but translation south for about 70 km.

Since there is no evidence for a volcanic arc to supply the Miocene tuffs present in the accretionary prism of the east coast, Cole and Lewis (1981) and Ballance et al. (1982) have suggested that this prism was originally positioned off the northwest-trending east coast of Northland and shifted and rotated south during Alpine Fault movement in the Pleistocene.

Lateral transition from subduction into the Alpine Fault transform system occurs in the Kaikoura area of the northeastern South Island via the presently active, pure strike-slip, Marlborough faults (Freund, 1971, 1974). Strike-slip movement and southward overthrusting interfere in a complex manner in the eastern part of this area (Prebble, 1980). Although this deformation is usually assigned to the Pliocene-Quaternary, the occurrence of large boulders of folded and cleaved Eocene limestone clasts in the Early Miocene Great Marlborough conglomerate (author's personal observation, but c.f. Prebble, 1980, Fig. 4) may be evidence for significant pre-Early Miocene deformation, which then could be then interpreted as the farthest extent of deformation associated with emplacement of the Northland allochthon (Ballance and Spörli, 1979).

Western New Zealand

This region includes the Cenozoic basins of Westland and the Nelson Block in the South Island and the western part of the North Island, west of the Taupo Volcanic Zone (excluding Northland), plus adjacent offshore areas (Fig. 6). The sedimentary basins are strongly controlled by the orthogonal fault pattern developed during Cretaceous and Oligocene rifting phases. In the Miocene, the former extensional basins were compressed over almost the whole area, with reverse faulting and folding on north- to northeast-trending faults (Pilaar and Wakefield, 1978; Hunt, 1980; Knox, 1982). Probably at this time as well, the orthogonal fault pattern was changed to rhombic (Spörli, 1980), by intracrustally distributed strain resulting from onset of collisional movement on the Alpine Fault. In Westland and Nelson the typical tectonic style is of eversion of pre-existing basins (e.g. Laird, 1968).

Superimposed on the north-trending block fault pattern in the Wanganui basin is an east-west to east southeast trending major downwarp, (particularly well shown in Fig. 11 of Hunt, 1980), which is the Wanganui Monocline of Spörli, (1980). This 5 km downwarp determines the east-west outcrop pattern of Pleistocene and older strata north of Wanganui, influences the change in width of the basement ranges of the subduction complex (Fig. 6), and probably is related to structures in the northern end of the onland accretion ridge, and the indentation of Hawkes Bay. To the west it may determine the anomalous position of Mt. Egmont volcano. This structure is named herein the Wanganui Monocline-Hawke Bay cross structure. It may be caused by the northeast-southwest compressional component of oblique subduction under the North Island, and by the compressional edge-effect at the southern end of the subducted oceanic slab against the Alpine Fault transform. Similar cross structures have been described by Koons (1984) as effects of oblique subduction in the South Island.

Southern Alps

The Southern Alps are a very young product of continent-continent collision along the Alpine Fault. At present the strike slip component along the fault is 40 mm/y and the dip-slip component, at right angles to the fault movement, is 20 mm/y. Plate collision has caused 80 km of shortening across South Island in the last 15 Ma, most of which took place in the past few millions of years. Uplift in the Mt. Cook area exceeds 10 mm/y and may be as large as 17 mm/y (Wellman, 1979). Erosional products from this uplift balance volumes of sediments deposited in basins around the South Island (Adams, 1980). The sedimentary record in a basin close to the Alpine Fault is consistent with a 420 km dextral shift of its source areas in the last 11 Ma, and with strong uplift of the eastern side of the fault in the last 1.8 Ma (Cutten, 1979).

Wellman (1979) proposed a model by which the collisional shortening across South Island is converted into uplift of the Pacific Plate along a curved fault plane which originates at the crust-mantle boundary, making the Southern Alps the upturned edge of the Pacific Plate lithosphere. This upturning can be used to explain the belt of Alpine schists along the Alpine Fault with anomalously young ages of less than 10 Ma (Adams, 1979). The inferred, late out-gassing may have been aided by shear-heating along the Alpine Fault zone (Sheppard et al., 1975). Shear-heating has also been proposed for portions of the actual movement zone of the Alpine Fault (Johnston and White, 1983). Rapid differential uplift, expressed by the presence of steep reverse faults, has juxtaposed high level cataclasites and fault gouge against low-level mylonite in the Alpine Fault (Sibson et al., 1979), and sharp subordinate fault zones against narrow schist belts and zones of transposition (Findlay, 1979).

The stretching lineation in fault rocks of the Alpine Fault parallels the direction of the present plate vector (Sibson et al., 1979). Koons (1984) has separated deeper regional deformation along the Alpine Fault into an underthrusting component active since initiation of movement along the fault, and a convergent movement normal to the Alpine Fault of relatively recent origin which results in low-angle thrusting of the Pacific Plate over the Indian Plate at high crustal levels. The component of underthrusting parallel to the Alpine Fault is accommodated by large scale thrusting on northwest-trending faults in the South Island.

At the northern end of the Alpine Fault the movement is taken up by the northeast-trending Marlborough fault system, the strands of which are in the correct orientation for pure dextral strike-slip under the present plate vector. At the southern end of the Alpine Fault, lower crustal material is being uplifted in the Fiordland Block (Priestly and Davey, 1983).

Strain due to change of the originally rectangular fault grid into a rhombic pattern in the southern South Island, east and southeast of the Alpine Fault (Fig. 6), is consistent with Cenozoic movement in the Marlborough fault system and may therefore be genetically related to movement on the Alpine Fault (Norris, 1979). This concept of distributed strain has been extended in an attempt to explain the rhombic fault patterns in western North Island (Spörli, 1980) and is in accord with data from present-day deformation, which indicate a 70 to 100 km-wide belt of dextral shear (Walcott, 1978b) along the Alpine Fault. If the Z-bending of New Zealand is added to the offset of markers, the total displacement on the Alpine Fault becomes approximately 1300 km.

Southern Areas Away From the Alpine Fault

Campbell Plateau is transected by the major, dextral Campbell Fault (Fig. 1; Christoffel, 1978). Scattered Cenozoic volcanics, which include the on-land volcanic centres of Lyttleton (Christchurch) and Dunedin (Fig. 6), show an age zonation and southwestward migration interpreted by Farrar and Dixon (1984) as an effect of overriding of the Eocene-Oligocene spreading ridge in the Solander-Emerald Basin along the Puysegur-Maquarie Ridge complex subduction zone. This subduction system is dominated by strike-slip movement, with slight subduction or extension where the plate boundary mismatches the trend of the Pacific/Indian plate vector. A R-T-T triple junction joins the Pacific/Indian plates boundary with the southeast Indian ridge and the Pacific-Antarctic Ridge at latitude 61°S, longitude 161°E. The transforms cause the remarkable curvature in the Pacific-Antarctic ridge, which may represent the breakaway trace of the southern edge of the Campbell Plateau.

New Caledonia

A Cretaceous event with little evidence of orogenic movement, and probably associated with rifting as in New Zealand, is recorded by the Senonian to Early Eocene sequence in the island of New Caledonia. After this the tectonic history of New Caledonia diverges from that of New Zealand, with major deformation and emplacement of a peridotite sheet representing oceanic lithosphere from the east, occurring in the Late Eocene. This event may have produced the spectacular high-pressure metamorphic belt at the northern end of the island (Brothers, 1974b; Black, 1977). Post-Eocene deformation was complex, involved differential uplift of the island and both compression, extension, and possibly also included dextral strike slip movement along the northwest-trending "sillon" (Brothers, 1974b) or West New Caledonian Fault (Paris, 1981). The eastern edge of the New Caledonian continental mass is within tens of kilometres of the westward facing New Hebrides trench and may collide with it in 10^5 to 10^6 years, which is likely to cause major plate reorganization in the region.

Discussion

No clear picture has emerged about Paleozoic plate processes which produced the western foreland of New Zealand. There is a lack of direct evidence for accretionary processes, and it seems likely that the tectonic settings within which western foreland rocks were deposited will be determined, eventually, from formerly adjacent parts of Gondwana, such as eastern Australia, because they lack the obscuring, intense, Mesozoic and Cenozoic overprint.

Mesozoic, pre-mid-Cretaceous, tectonics appear to be those of a classical "active margin", and associated with subduction processes. It is uncertain whether deformation was continuous throughout this interval, or whether the Late Triassic Rangitata I "orogeny" reflects major plate reorganization. Some unconformities with overlying, fossiliferous, shallow water facies in the Torlesse (McKinnon, 1983) may be due rather to local emergence of ridges within an accretionary prism than to regional orogenic uplift. Eventually it may be possible to piece together the age patterns and nature of the oceanic crust subducted from the many slices of oceanic spilites, limestones and cherts associated with melanges. The Haast Schists of the South Island may represent partial subduction of the leading edge of the exotic Torlesse terrane.

At the end of the Albian Rangitata II "orogeny" practically all of the present day continental crust of the New Zealand microcontinent had been formed, with the exception of the post-Oligocene arc extending southwestward from the Three Kings Rise, the material presently accreting along the east coast of the North Island, and possibly the granitic material building up the floor of the presently active rift of the Taupo Volcanic Zone.

During Cretaceous-early Tertiary rifting, the New Zealand microcontinent became as an independent entity. It is not known where the nearest subduction system was at that time, although Crook and Feary (1982) postulate a continuously active late Mesozoic-early Tertiary, west-facing arc extending northward from Northland. The Antarctic Peninsula experienced the effects of continuous subduction during this period, which was eventually shut off by the subduction of the ridge between Phoenix and Pacific plates (Barker, 1982). It is uncertain whether the Tasman Sea, which formed during separation of New Zealand from Gondwana, originated as a back-arc basin or was generated from a ridge-transform junction after the Pacific/Phoenix plates spreading ridge had been subducted under New Zealand at the end of the Rangitata event.

Extension and upwelling of the mantle during Cretaceous-early Tertiary rifting thinned the crust of the New Zealand microcontinent, lowered its isostatic equilibrium height below sea level, and caused general submergence. Most of the areas now emergent owe their uplift to subsequent subduction/obduction, and in a few instances (Hauraki Rift?) to renewed thermal uplift. Faults associated with Cretaceous-Early Tertiary rifting seem to be mainly steep, but in future it may be interesting to search for low-angle normal faults and associated markers rotated to steep dips.

The mechanism of propagation of the present Pacific/Indian plates boundary through the New Zealand continental fragment is not clear, neither is the choice of the locus of the Alpine Fault. The fault may have followed one of the pre-existing intracontinental rift systems, especially that which connected with Oligocene spreading centres in the Emerald Basin and the South Fiji Basin (Spörli, 1980; Farrar and Dixon, 1984; Kamp, 1984b).

The mechanism for the establishment of a new volcanic arc along this plate boundary also poses considerable problems. The facts that there are clear trench and volcanic arc features visible in the Three Kings Ridge off Northland (Kroenke and Eade, 1982; Kroenke and Dupont, 1982), and that Miocene-Pliocene Northland arc volcanics do not laterally extend southwest and northeast into the pre-existing oceanic crust of New Caledonia and South Fiji basins, seem to favour a complex hypothesis (Cole and Lewis, 1981; Ballance et al., 1982) involving multiple arcs and the shifting of arcs towards the present Taupo Volcanic Zone site. Relative particle velocities across the plate boundary have increased from a low in the Late Oligocene, which is opposite to the trend required by the simple hypothesis of subduction regression (Brothers, 1984; Kamp, 1984a). The lines of equal age across the northwest-trending arcs may indicate sequential development of volcanism along the arc rather than across, as has been demonstrated for volcanism in the Antarctic Peninsula (Barker, 1982, Fig. 7).

Westward obduction of oceanic crust has been reported in Late Eocene time in New Caledonia, and Late Oligocene time in New Zealand (Brothers, 1974a; Paris, 1981). In both areas there is no firm evidence for a subduction system or arc in the vicinity of the obduction zone just prior to the obduction event. Rather, obduction appears to be the first pulse in the establishment of a new subduction system. In New Caledonia, possible Paleozoic spreading ridge material with a north-south spreading direction, and in New Zealand, possible seamount materials of Cretaceous to Paleocene ages, were obducted. In New Zealand, the migration of the volcanic arcs may have been triggered by the blocking of the Miocene subduction system during attempted underthrusting of seamounts.

Oblique subduction and irregularities imposed by the New Zealand continental crust may have caused a number of cross-trending structural features, especially the Wanganui Monocline-Hawkes Bay cross structure. Increasing intensity in the collisional component along the Alpine Fault is responsible for most of the Z-bending in New Zealand, resulting in an intracontinental zone of distributed shear. At the surface, preexisting faults have been reactivated to take up the deformation and to isolate rigid blocks of up to tens of kilometres in size. At depth, spaced zones of schistosity accommodate the movement.

One of the intriguing questions remaining is that of the nature of orogenies in New Zealand. At present little can be said about the Tuhua Orogeny because there are too many uncertainties. The Early Cretaceous Rangitata II "Orogeny" as expressed by the regional unconformity clearly represents not the peak of deformation but the shutting off of a Mesozoic subduction (and deformation) system which had been more or less continuously active at least since Middle Jurassic (or even Triassic) time. Cessation of subduction caused isostatic rebound of the crust and cooling of plutons to set their radiometric ages. In the present day, continental collision along the Alpine Fault is building the Southern Alps, by directly causing uplift, at one of the highest rates known.

References

Adams, C. J., Discovery of Precambrian rocks in New Zealand, age relations of the Greenland Group and Constant Gneiss, West Coast, South Island, Earth Planet. Sci. Lett., 281, 98-104, 1975.

Adams, C. J., Age and origin of the Southern Alps, in: Walcott, R. I., and M. M. Cresswell, (eds.), The origin of the Southern Alps, Roy. Soc. N.Z. Bull., 18: 73-78, 1979.

Adams, C. J., Contemporary uplift of the Southern Alps, New Zealand, Geol. Soc. Am. Bull., 91 (I):2-4; 91 (II):1-114, 1980.

Adams, C. J., Geochronological correlations of Pre-Cambrian and Paleozoic orogens in New Zealand, Marie Byrd Land (West Antarctica and Tasmania, in: Cresswell, M. M., and P. Vella, (eds.), Gondwana Five, Balkem Rotterdam, 191-197, 1981.

Adams, C. J., P. A. Morris, and J. M. Beggs, Age and correlation of volcanic rocks of Campbell Island and metamorphic basement of the Campbell Plateau, Southwest Pacific. N.Z. J. Geol. Geophys., 22(6):679-691, 1979.

Austin, P., Paleogeographic and paleotectonic models for New Zealand geosyncline in eastern Gondwanaland, Geol. Soc. Am. Bull., 86:1230-1234, 1975.

Ballance, P. F., An interarc basin in northern New Zealand: Waitemata group (Upper Oligocene Lower Miocene), J. Geol., 82, 432-471, 1974.

Ballance, P. F., J. R. Pettinga, and C. Webb, A model of the Cenozoic evolution of northern New Zealand and adjacent area of the Southwest Pacific, Tectonophysics, 87: 37-48, 1982.

Ballance, P. F., and K. B. Spörli, Northland Allochthon, J. Roy. Soc. N. Z., 9 (2), 259-275, 1979.

Barker, P. F., The Cenozoic history of the Pacific margin of the Antarctic Peninsula: ridge crest-trench interactions, J. Geol. Soc. London 139, 727-801, 1982.

Bennett, M. C., The ultramafic-mafic complex at North Cape, Northernmost New Zealand, Geol. Mag., 113 (1):61-76, 1976.

Black, P. M., Regional high pressure metamorphism in New Caledonia: phase equilibria in the Ouegoa district, Tectonophysics, 43, 89-107, 1977.

Bradshaw, J. D., C. J. Adams, and P. B. Andrews, Carboniferous to Cretaceous on the Pacific margin of Gondwana: The Rangitata phase of New Zealand, in: Cresswell, M. M., and P. Vella, (eds.), Gondwana Five, Balkema Rotterdam 217-221, 1981.

Bradshaw, J. D., and G. W. Grindley, S-13 Cobb: A review of the new map and the structural interpretation of the Central Belt rocks of S13 and S8 Comment and reply, N.Z. J. Geol. Geophys., 25:371-379, 1982.

Bradshaw, M. A., and B. D. Hegan, Stratigraphy and structure of the Devonian rocks of Ihungia outlier, Reefton New Zealand, N.Z. J. Geol. Geophys., 26 (4), 325-344, 1983.

Briggs, R. M., Distribution, form and structural control of the Alexandra Volcanic Group, North Island, New Zealand, N.Z. J. Geol. Geophys., 26 (1):47-55, 1983.

Brothers, R. N., Kaikoura orogeny in Northland, N.Z. J. Geol. Geophys., 17, 1-18, 1974a.

Brothers, R. N., High pressure schists in northern New Caledonia, Contrib. Min. Petrol., 46: 109-127, 1974b.

Brothers, R. N., Subduction regression and oceanward migration of volcanism, North Island, New Zealand, Nature, 309 (5970), 698-700, 1984.

Brothers, R. N., and M. Delaloye, Obducted ophiolites of North Island, New Zealand: origin, age, emplacement and tectonic implications for Tertiary and quaternary volcanicity, N.Z. J. Geol. Geophys., 25, 257- 274, 1982.

Christoffel, D. A., Interpretation of magnetic anomalies across the Campbell Plateau, south of New Zealand, Austr., Soc. Explor. Geophys. Bull., 9:143-145, 1978.

Cole, J. W., and K. B. Lewis, Evolution of the Taupo-Hikurangi subduction system, Tectonophysics, 72:1-21, 1981.

Coombs, D. S., C. A. Landis, R. J. Norris, J. M. Sinton, D. J. Borns, and D. Craw, The Dun Mountain ophiolite belt, New Zealand, its tectonic setting, constitution, and origin, with special reference to the southern portion, Am. J. Sci., 276, 561-603, 1976.

Cooper, A. F., Progressive metamorphism of metabasic rocks from the Haast schist group of Southern New Zealand, J. Petrol., 13:457-492, 1972.

Cooper, A. F., and A. Reay, Lithologies, field relationships, and structure of the Pounamu Ultramafics from the Whitcombe and Hokitika

Rivers, Westland, New Zealand, N.Z. J. Geol. Geophys., 26:359-379, 1983.
Cooper, R. A., New Zealand and South East Australia in the early Paleozoic, N.Z. J. Geol. Geophys., 18 (7):1-20, 1975.
Cooper, R. A., Lower Paleozoic rocks of New Zealand, J. Roy. Soc. N.Z., (1):29-84, 1979.
Cooper, R. A., C. A. Landis, W. C. Le Masurier, and I. G. Speden, Geological history and regional patterns in Zew Zealand and West Antarctica, their paleotectonic and paleogeographic significance, in: Craddock, C. (ed.), Antarctic Geoscience, Univ. Wisc. Press, 43-53, 1982.
Craw, D., Lithologic variations in Otago schist, Mt. Aspiring area, northwest Otago, New Zealand, N.Z. J. Geol. Geophys., 27:151-166, 1984.
Crook, K. A. W., and D. A. Feary, Development of New Zealand according to the forearc model of crustal evolution, Tectonophysics, 87, 65-107, 1982.
Cutten, H. N. C., Rappahannok Group: Late Cenozoic sedimentation and tectonics contemporaneous with Alpine Fault movement, N.Z. J. Geol. Geophys., 22 (5):535-553, 1979.
Davis, T. E., M. R. Johnston, P. C. Rankin, and R. J. Still, The Dun Mountain belt in east Nelson, New Zealand, in: Ophiolites: Proceedings of the International Ophiolite Symposium Cyprus, Cyprus Geol. Sur. Dept., 480-496, 1980.
Davy, B. W., Bounty trough, Geol. Soc. N.Z. Ann. Conf. Abst., in: Geol. Soc. N.Z. Misc. Pub., 31A, 1984.
De Wit, M. J., The evolution of the Scotia arc as a key to the reconstruction of southwestern Gondwanaland, Tectonophysics, 37, 53-81, 1977.
De Wit, M. J., Gondwana reassembly in: Cresswell, M. M., and P. Vella, (eds.), Gondwana Five, B. Ilkema, Rotterdam II, 1981.
Eade, J. V., West Norfold Ridge and the regional unconformity: Products of Eocene-Oligocene plate collision? Geol. Soc. N.Z. Ann. Conf. Abst., Geol. Soc. N.Z. Misc. Publ., 31A, 1984.
Farrar, E., and J. M. Dixon, Overriding of the Indian-Antarctic Ridge: origin of Emerald Basin and migration of late Cenozoic volcanism in Southern New Zealand and Campbell Plateau, Tectonophysics, 104, 243-256, 1984.
Feary, D. A., and E. A. Pessagno, Jr., An early Jurassic age for chert within the Early Cretaceous Oponae mélange (Torlesse supergroup), Raukumara Peninsula, New Zealand, N.Z. J. Geol. Geophys., 23:623-628, 1980.
Ferguson, S. R., M. P. Hochstein and A.C. Kibblewhite, Seismic refraction studies in the northern Hauraki Gulf, New Zealand, N.Z. J. Geol. Geophys., 23 (1):17-25, 1980.
Findlay, R. H., Summary of structural geology of Haast schist terrain Central Southern Alps, N.Z.: implications of structures for uplift and deformation within southern Alps, in: Walcott, R.I., and M. M. Cresswell, (eds.), Origin of the Southern Alps, Roy. Soc. N.Z. Bull., 18:113-120, 1979.
Freund, R., The Hope Fault: a strike slip fault in New Zealand, N.Z. Geol. Surv. Bull., n.s. 86, 49 p., 1971.
Freund, R., Kinematics of transform and transcurrent faults, Tectonophysics, 21:93-134, 1974.
Gibson, G. M., Polyphase deformation and its relation to metamorphic crystallization in rocks at Wilmot Pass, central Fiordland, N.Z. J. Geol. Geophys., 25:45-65, 1982a.
Gibson, G. M., Stratigraphy and structure of some metasediments and associated intrusive rocks from central Fiordland, N.Z. J. Geol. Geophys., 25:21-43, 1982b.
Grindley, G. W., Sheet S13 Cobb (1st ed) Geological Map of New Zealand 1:63,360 Map (1 sheet) and notes (48 p.), N.Z. Dept. Sci. Ind. Res., Wellington, 1980.
Grindley, G. W., and F. J. Davey, The reconstruction of New Zealand, Australia and Antarctica, in: Craddock, C. (ed.), Antarctic Geoscience, Univ. of Wisc. Press, 15-29, 1982.
Grindley, G. W., P. J. Oliver, and J. C. Sukroo, Lower Mesozoic position of southern New Zealand determined from paleomagnetism of the Glenham porphyry, Murihiku Terrane, Eastern Southland, in: Cresswell, M. M., and P. Vella, (eds.), Gondwana Five, Balkema, Rotterdam, 319-326, 1981.
Harrington, G. J., Correlation of the Permian and Triassic Gympic Terrane of Queensland with the Brook Street and Maitai Terranes of New Zealand, in: Permian Geology of Queensland, Geol. Soc. Austra., Queensland Div., Brisbane, 431-436, 1983.
Hatherton, T., Stokes magnetic anomaly-magnetic system or magnetic supergroup? N.Z. J. Geol. Geophys., 18 (3):519-521, 1975.
Hay, R. F., Sheet N7 Doubtless Bay (1st edition), Geological Map of New Zealand 1:63,360 Map (1 Sheet) and notes (24 pp.), Dept, Sci. Ind. Res., Wellington, 1975.
Heming, R. R., Patterns of Quaternary basaltic volcanism in the northern North Island, New Zealand, N.Z. J. Geol. Geophys., 23:335-344, 1980.
Howell, D. G., Mesozoic accretion of exotic terranes along the New Zealand segment of Gondwanaland, Geology, 8, 487-491, 1980.
Hume, B. J., The relation between Charleston Metamorphic Group and the Greenland Group in the Central Paparoa Range, South Island, New Zealand, J. Roy. Soc. N.Z., 7 (3):379-392, 1977.
Hunt, T., Basement structure of the Wanganui Basin, onshore interpreted from gravity data, N.Z. J. Geol. Geophys., 23 (1):1-16, 1980.
Hunter, H. WW., Geology of the Cobb intrusives, Takaka Valley, Northwest Nelson, New Zealand, N.Z. J. Geol. Geophys., 20 (3):469-501, 1977.
Ireland, T. R., A. Reay, and A. F. Cooper, The Pounamou ultramafic belt in the Diedrich Range, Westland, New Zealand, N.Z. J. Geol. Geophys., 27, 247-256, 1984.

Johnston, D. W., and S. H. White, Shear heating associated with the Alpine Fault, Tectonophysics, 92, (1-3), 241-252, 1983.
Kamp, P. J. J., Neogene and Quaternary extent and geometry of the subducted plate beneath North Island, New Zealand: Implication for Kaikoura tectonics, Tectonophysics, 108:241-266, 1984a.
Kamp, P. J. J., The extent, continuity and geotectonic development of a mid-Cenozoic continental rift system through western New Zealand, and implications of age of Alpine Fault inception, Geo. Soc. N.Z. Ann. Conf. Abst., Geol. Soc. N.Z. Misc. Publ., 31A, 1984b.
Kenny, J., Stratigraphy, sedimentology and structure of the Ihungia décollement Raukamara Peninsula, North Island, New Zealand, N.Z. J. Geol. Geophys., 27, 1-19, 1984.
Knox, G. J., Taranaki basin, structural style and tectonic setting, N.Z. J. Geol. Geophys., 25 (2), 125-140, 1982.
Koons, P. O., Implications of continental collision in southern New Zealand, Geol. Soc. N.Z. Ann. Conf. Abst., Geol. Soc. N.Z. Misc. Publ., 31A, 1984.
Kroenke, L. W., and J. Dupont, Subduction-obduction: A possible north-south transition along the west flank of the Three Kings Ridge, Geo-Marine Lett., 2:11-16, 1982.
Kroenke, L. W. and Y. V. Eade, Three Kings Ridge: A west facing arc, Geo-Marine Lett., 2:5-10, 1982.
Laird, M., The Paparoa tectonic zone, N.Z. J. Geol. Geophys., 11, 435-453, 1968.
Laird, M. G., The late Mesozoic fragmentation of the New Zealand fragment of Gondwana, in: Cresswell, M. M., and P. Vella, (eds.), Gondwana Five, Balkema, Rotterdam: 311-318, 1981.
Landis, C. A., and D. G. Bishop, Plate tectonics and regional stratigraphic-metamorphic relations in the southern part of the New Zealand geosyncline, Geol. Soc. Am. Bull., 83:2267-2284, 1972.
Landis, C. A., and D. S. Coombs, Metamorphic belts and orogenesis in southern New Zealand, Tectonophysics, 4:501-518, 1967.
Lewis, K. B., Quaternary sedimentation on the Hikurangi oblique subduction and transform margins New Zealand, in: Ballance, P. F., and H. G. Reading, (eds.), Sedimentation in oblique slip mobile zones, Spec. Pub. Int. Assoc. Sediment., 4, 171-189, 1980.
McKinnon, T. C., Origin of the Torlesse Terrane and coeval rocks, South Island, New Zealand, Geol. Soc. Am. Bull., 94:967-985, 1983.
Malahoff, A., R. Feden, and H. Fleming, Magnetic anomalies and tectonic fabric of marginal basins north of New Zealand, J. Geophys. Res., 87:4109-4125, 1982.
Mattinson, J. M., D. I. Kimbrough, and J. J., Bradshaw, Zircon and apatite U-Pb age constraints and initial Pb and Sr isotopic compositions of Cretaceous granulites and gabbronorites from Fiordland, southwest New Zealand, Abs., Geol. Soc. N.Z. Ann. Conf., Geol. Soc. N.Z. Misc. Publ., 31A: 1984.
Molnar, P., T. Atwater, T. Mammerickx, and S. M. Smith, Magnetic anomalies, Bathymetry and tectonic evolution of the South Pacific since the Late Cretaceous, Roy. Astron. Soc. Geophys. J., 60:383-420, 1975.
Moore, P. R., Late Cretaceous-Tertiary stratigraphy, structure and tectonic history of the area between Whareama and Ngahape, eastern Wairarapa, New Zealand, N.Z. J. Geol. Geophys., 23 (2):167-177, 1980.
Moore, P. R., and I. Speden, Stratigraphy, structure and inferred environment of deposition of the Early Cretaceous sequence, eastern Wairarapa New Zealand, N.Z. J. Geol. Geophys., 22 (4):417-433, 1979.
Nelson, C. S., and T. M. Hume, Relative intensity of tectonic events revealed by the Tertiary sedimentary record in the North Wanganui Basin and adjacent area, New Zealand, N.Z. J. Geol. Geophys., 20 (2):369-392, 1977.
Nelson, K. D., A suggestion for the origin of mesoscopic fabric in accretionary mélange based on feature observed in the Chrystalls Beach complex, South Island, New Zealand, Geol. Soc. Am. Bull., 93:625-634, 1982.
Norris, R. J., A geometrical study of finite strain and bending in the South Island, in: Walcott, R. J., and M. M. Cresswell, (eds.), Origin of the Southern Alps, Roy. Soc. N.Z. Bull., 18:21-28, 1979.
Norris, R. J., and R. M. Carter, Fault bounded blocks and their role in localizing sedimentation and deformation adjacent to the Alpine Fault Southern New Zealand, Tectonophysics, 87: 11-23, 1982.
Norris, R. J., and A. F. Cooper, Tour C: Structure and metamorphism of the Haast Schist in the Kawarau Gorge and Lake Hawea areas, Field trip guides Queenstown Conference, Geol. Soc. N.Z., C1, 1977.
Oliver, G. J. H., Geology of the granulite and amphibolite gneisses of Doubtful Sound, Fiordland, New Zealand, N.Z. J. Geol. Geophys., 23: 27-41, 1980.
Oliver, G. J. H. and J. H. Coggon, Crustal structure of Fiordland, New Zealand, Tectonophysics, 54:253-292, 1979.
Oliver, P. J., T. C. Mumme, G. W. Grindley, and P. Vella, Palcomagnetism of the Upper Cretaceous Mount Somers Volcanics, Canterbury, New Zealand, N.Z. J. Geol. Geophys., 22 (2):199-212, 1979.
Paris, J. P., Geologie de la Nouvelle Caledonie, un essai de synthese, Mem. BRGM, 112, 278 p., 1981.
Paris, J. P., and J. D. Bradshaw, Paleogeography and geotectonics of New Caledonia and New Zealand in the Triassic and Jurassic, Int. Symp. Geody., Noumea 1976, Editions Technip, Paris, 209-216, 1977.
Paris, J. P., and R. Lille, New Caledonia: Evolution from Permian to Miocene, Mapping data and hypotheses about geotectonics, Int. Symp. Geody SW Pacific Noumea 1976, Editions Technip. Paris 195-208, 1977.

Pettinga, J. R., Upper Cenozoic structural history, coastal southern Hawkes Bay, New Zealand, N.Z. J. Geol. Geophys., 25, (2): 149-191, 1982.

Pilaar, W. T. W., and L. L. Wakefield, Structural and stratigraphic evolution of the Taranaki Basin offshore New Zealand, APEA J. 1978, 93-102, 1978.

Prebble, W., Late Cenozoic sedimentation and tectonics of the East Coast Deformed Belt in Marlborough, New Zealand, in: Ballance, P F., and H. G. Reading, (eds.), Sedimentation in oblique slip mobile zones, Spec. Publ. Int. Assoc. Sediment., 4:217-228, 1980.

Priestley, K., and F. J. Davey, Crustal structure of Fiordland, southwestern New Zealand from seismic refraction measurements, Geology, 11: 660-663, 1983.

Reyners, J., Lateral segmentation of the subduction plate at the Hikurangi Margin, New Zealand seismological evidence, Tectonophysics, 96:203-223, 1983.

Schofield, J. C., Northland Cretaceous-Oligocene strata-allochthonous or autochthonous? N.Z. J. Geol. Geophys., 26:155-162, 1983.

Shelley, D., Temperature and metamorphism during cleavage and fold formation of the Greenland Group, north of Greymouth, J. Roy. Soc. N.Z., 5 (1):65-75, 1975.

Shelley, D., Takaka River recumbent fold complex Nelson, New Zealand, N.Z. J. Geol. Geophys., 27:139-149, 1984.

Sheppard, D. S., J. D. Adams, and C. W. Bird, Age of metamorphism and uplift in the Alpine Schist Belt, New Zealand, Geol. Soc. Am. Bull., 86: 1147-1153, 1975.

Sibson, R. H., S. H. White, and B. Atkinson, Fault rock distribution and structure within the Alpine Fault zone: A preliminary account, in: Walcott, R. I. and M. M. Cresswell, (eds.), Origin of the Southern Alps, Roy. Soc. N.Z. Bull., 18:55-65, 1979.

Speden, I. G., Geology of Mt. Taitai, Tapuaeroa Valley, Raukumara Peninsula, N.Z. J. Geol. Geophys., 19:71-114, 1976.

Spörli, K. B., Mesozoic tectonics, North Island, New Zealand, Geol. Soc. Am. Bull., 89:412-425, 1978.

Spörli, K. B., New Zealand and oblique slip margins: Tectonic development up to and during the Cenozoic, in: Ballance, P. F., and H. G. Reading, (eds.), Sedimentation in oblique slip mobile zones, Spec. Publ. Int. Assoc. Sediment., 4:140-170, 1980.

Spörli, K. B., Review of paleostress/strain directions in Northland and of the structure of the Northland Allochthon, Tectonophysics, 87, 25-36, 1982.

Spörli, K. B., and A. B. Bell, Torlesse mélange and coherent sequences, eastern Ruahine Range, North Island, New Zealand, N.Z. J. Geol. Geophys., 19(4), 427-447, 1976.

Spörli, K. B., and M. R. Gregory, Significance of Tethyan fusulinid limestones of New Zealand, in: Cresswell, M. M., and P. Vella, (eds.), Gondwana Five, Balkema, Rotterdam, 223-229, 1981.

Spörli, K. B., and J. A. Grant-Mackie, Upper Jurassic fossils from the Waipapa Group of Tawharanui Peninsula, North Auckland, New Zealand, N.Z. J. Geol. Geophys., 19, (1), 21-34, 1976.

Stock, J., and P. Molnar, Uncertainties in the relative positions of Australia, Antarctica, Lord Howe, and Pacific Plates since Late Cretaceous, J. Geophys. Res., 87 (BG):4697-4714, 1982.

Stoneley, R., A lower Tertiary décollement on the East Coast, North Island, New Zealand, N.Z. J. Geol. Geophys., 11:128-156, 1968.

Strong, C. P., Cretaceous foraminifera of the Matakaoa Volcanic Group, N.Z. J. Geol. Geophys. 19:140-143, 945-947, 1976.

Turnbull, I. M., Structure of the Caples terrane of the Thomson Mountains, Northern Southland, New Zealand, N.Z. J. Geol. Geophys., 23 (7): 43-62, 1980.

Walcott, R. I., Present tectonics and late Cenozoic evolution of New Zealand, Geol. J. Roy. Astron. Soc., 52:137-164, 1978a.

Walcott, R. I., Geodetic strains and large earthquakes in the Axial Tectonic Belt of North Island, New Zealand, J. Geophys. Res., 83:4419-4429, 1978b.

Walcott, R. I., and T. C. Mumme, Paleomagnetic study of Tertiary sedimentary rocks from the East Coast of the North Island, New Zealand, Geophys. Div., Dept. Scient. Ind. Res., Report 189, 1982.

Weissel, J. K., D. E. Hayes, and E. M. Herron, Plate tectonic synthesis: the displacements between Australia, New Zealand and Antarctica since the Late Cretaceous, Marine Geol., 25, 231-277, 1977.

Wellman, H. W., An uplift map for the South Island of New Zealand, and a model for uplift of the Southern Alps, in: Walcott, R. I., and M. M. Cresswell, (eds.), The origin of the Southern Alps, Roy. Soc. N.Z. Bull., 18:13-20, 1979.

Williams, J. C., and I. E. M. Smith, Geochemical evidence for paired arcs in the Permian volcanics of Southern New Zealand, Contr. Mineral. Petr., 68:285-291, 1979.

Wilson, C. J. N., A. M. Rogan, I. E. M. Smith, D. J., Northey, I. A. Nairn, and B. F. Houghton, Caldera volcanoes of the Taupo Volcanic Zone, New Zealand, J. Geophys. Res., 89 (B10), 8463-8484, 1984.

Wodzicki, A., Geology of the pre-Cenozoic basement of the Taranaki-Cook Strait-Westland area, New Zealand, based on recent drillhole data, N.Z. J. Geol. Geophys., 17 (4):747-57, 1974.

Wood, B. L., The Otago schist megaculmination and the tectonic significance in the Rangitata Orogen of New Zealand, Tectonophysics, 47:339-368, 1978.

PALEOZOIC TECTONIC DEVELOPMENT OF EASTERN AUSTRALIA IN RELATION TO THE PACIFIC REGION

Erwin Scheibner

Geological Survey of New South Wales, G.P.O. Box 5288, Sydney 2001

Abstract. Eastern Australia consists mainly of the composite Tasman Fold Belt system, or Tasmanides, which evolved from latest Precambrian to Triassic time. The region was part of eastern Gondwanaland, and developed within the tectonically complex southwestern Pacific region, which has been the site of episodic orogenic activity throughout the Phanerozoic. Pre-Mesozoic activity apparently resulted from interaction of eastern Gondwanaland with hypothetical Paleo-Pacific plates. The Tasman Fold Belt system evolved in four main stages: latest Precambrian, early Paleozoic, Silurian to Early Carboniferous, and Late Carboniferous to Triassic. Stage 1: Upper Proterozoic sediments and volcanics of the Australian Craton, west of the Tasmanides, are deduced to have been deposited in intraplate and passive margin settings. By earliest Cambrian time a well-developed, complex west-Pacific-type active plate margin apparently existed in eastern Australia. Accordingly, an episode of continental break-up involving sea-floor spreading, coupled with separation and dispersion of microcontinents and possibly some plate convergence, is inferred to have occurred during latest Precambrian time, and preceded the distinctive episode of extension close to the Cambrian/Precambrian boundary. Precambrian complexes form the basement and internal massifs of the Tasmanides. Orogenic episodes in Paleozoic rocks are inferred to have been caused by changes in style of subduction (Mariana versus Chilean), and collisions of microcontinents and volcanic arcs. Extensional volcanic rifts, marginal basins, and other basins were closed and inverted. Syn-kinematic S-type granites were emplaced in high temperature metamorphic belts, while late and post-kinematic S- and I-type granites intruded inverted belts and earlier stabilized blocks. A-type granites intruded volcanic piles associated with volcanic rifting. Stage 2: The lower Paleozoic Kanmantoo Fold Belt, and older parts of the Thomson Fold Belt, which lie on the inner, western side of the Tasmanides, are believed to result from collisions of microcontinents. During the Late Ordovician to Early Silurian Benambran Orogeny, the Molong volcanic arc and its microcontinental basement collided with the Victorian microcontinent to the east. The transmitted stress caused easterly thrusting in the Stawell-Bendigo Foreland Fold and Thrust Belt still farther east. Stage 3: The Lachlan Fold Belt (east of the Kanmantoo Fold Belt), Thomson Fold Belt to its north, and Hodgkinson-Broken River Fold Belt in northeastern Queensland, were stabilized during the Carboniferous Kanimblan Orogeny. These fold belts comprise early Paleozoic structural elements, together with middle Paleozoic elements which resulted from closure of a complex back-arc region. Stage 4: The New England Fold Belt, which was stabilized during mid-Permian to Late Triassic time, displays composite early to late Paleozoic volcanic arc-fore-arc basin and accretionary prism complexes, which were intruded by mainly post-kinematic granitoids. This fold belt is thrust westward over its foredeep, the Sydney-Bowen Basin. Mesozoic rifting, break-up, and separation were clearly connected with the general dispersion of parts of Gondwanaland. However, development of eastern Gondwanaland was influenced also by the Pacific active plate margin, the orogenic development of which is still continuing.

Published in 1987 by the American Geophysical Union.

Introduction

This paper was presented at the 27th International Geological Congress (August 1984) Symposium: "Circumpacific Orogenic Belts and Evolution of the Pacific Basin" organized by Working Group 2 of the Inter-Union Commission on the Lithosphere. It includes ideas regarding suspect tectonostratigraphic terranes in eastern Australia (Scheibner, 1983b, 1985). Its main purpose is to review data from the Tasmanides supporting the existence of a Paleo-Pacific plate. All arguments are based on the assumption that actualistic, plate tectonic models can be used to decipher the geologic past back at least as far as one billion years ago.

Modified Plate Tectonics

The basic tenets of plate tectonics (Le Pichon et al., 1973) are expanded herein to include the following tectonic models:

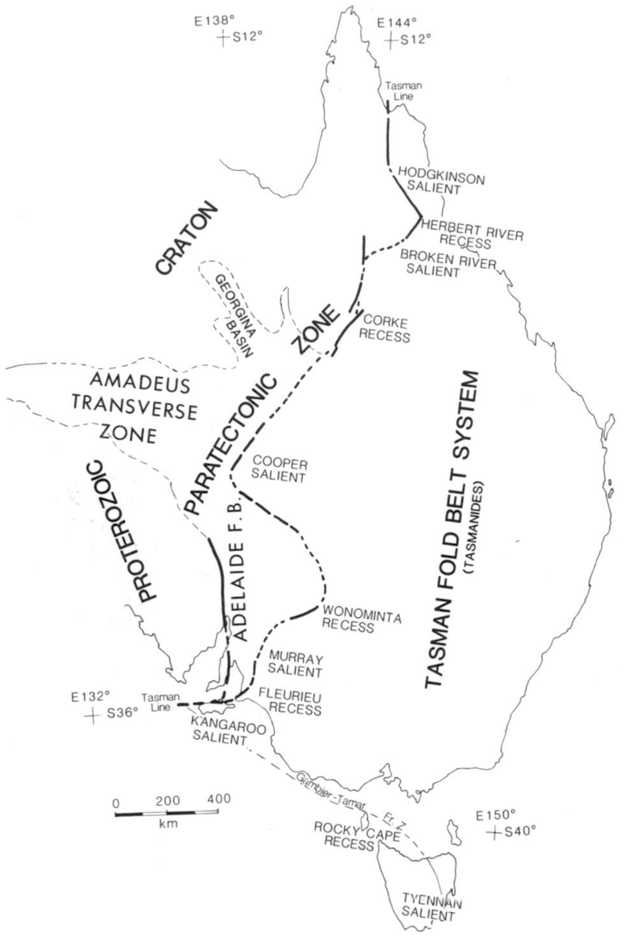

Fig. 1. Framework of eastern Australia. Salients and recesses in the Tasman Fold Belt System and the distribution of the paratectonic zone.

1. A- and B-type subduction.
2. Two different modes of B-type subduction: Chilean and Mariana types.
3. A subduction-related volcanic arc develops only if an asthenospheric wedge is between lower and upper plates.
4. Subduction and growth of an accretionary wedge can occur in association with a flat-dipping, Chilean-type subduction zone without formation of a related volcanic arc.
5. If decoupling of convergent plates is completely effective, the accretionary wedge may be absent.
6. Tectonic erosion of the upper plate can occur during subduction.
7. Lateral tectonic segmentation in plate convergent (orogenic) belts is caused by major, along-strike variations in dip of subducting plates, and by varying lithospheric properties of plates.
8. Development of A-type subduction and foreland fold and thrust belts is related to changes in the mode of B-type subduction or to plate collisions. Foreland fold and thrust belt structures are strongly influenced by the geometry of B-type subduction and by the inherited structure of the upper plate.
9. The complex relationships observed in areas of plate interaction are the rule rather than the exception. Examples are oblique plate movements causing transtension or transpression, leaking transforms, and jumps in spreading centres.
10. During plate convergence allochthonous lithospheric fragments, (or tectonostratigraphic terranes), may be accreted.
11. Besides continuous deformation in the accretionary wedge, episodes of orogenic deformation are associated with changes in the mode and style of subduction, coupling of converging plates, docking and collision of tectonostratigraphic terranes, and other types of collisional events.
12. The various types of orogenic granites; M-, I-, A- and S-types, characterize certain types of tectonic regimes. M-types are mostly developed in ensimatic volcanic arcs, but also can be associated with major lineaments (fracture zones) which penetrate the whole lithosphere; I-types occur in volcanic arcs and continental margin magmatic arcs; A-types are developed in rifts and volcano-tectonic depressions; and S-types occur in high-temperature metamorphic belts and in belts where large-scale overthrusting of continental crust has occurred.

Sources of Data

Basic stratigraphic and other geologic data for the Tasmanides are contained in the following recent publications: Cas (1983), Collins et al. (1982), Cooper and Grindley (1982), Crook (1980), Crook and Powell (1976), Flint and Parker (1982), Geological Society of Australia (1971), Queensland Geological Survey (1975), Henderson and Stephenson (1980), Korsch (1977), Korsch and Harrington (1981), Leitch (1974), Owen and Wyborn (1979), Packham (1969, 1973), Parkin (1969), Pickett (1982), Plumb (1979), Pogson (1982), Powell (1983a,b, 1984a,b), Roberts et al. (1972), Rutland (1976), Scheibner (1976, 1978, 1985), VandenBerg (1978), von der Borch (1980), Veevers (1984), Webby (1978), White and Chappell (1983), Williams (1978).

Structural Framework

This paper is limited to the Late Proterozoic to Mesozoic orogenic development of eastern Australia, prior to Mesozoic breakup of eastern Gondwanaland.

The geology of eastern Australia comprises the following three elements:

1. The composite, Early to Middle Proterozoic Australian Craton on the west, with associated epicratonic basins and paratectonic fold belts;

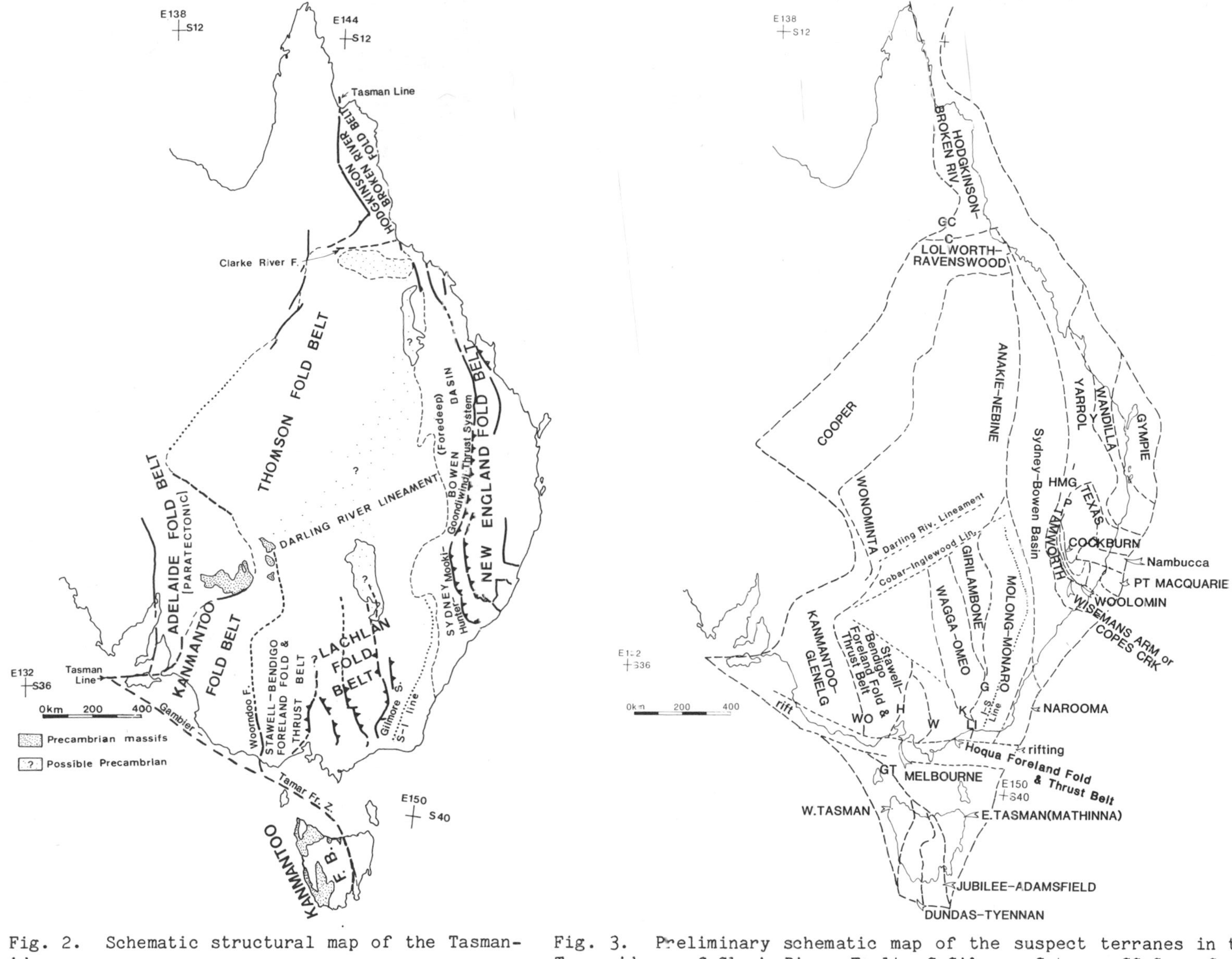

Fig. 2. Schematic structural map of the Tasmanides.

Fig. 3. Preliminary schematic map of the suspect terranes in the Tasmanides. C-Clark River Fault; G-Gilmore Suture; GC-Gray Creek Fault Zone; GT-Gambier-Tamar Fracture Zone; H-Heathcote Greenstone Belt and Mount William Fault; HMG-Hunter-Mooki-Goondiwindi Thrust System; K-Kiewa Thrust (Suture); LI-Long Plain-Indi Fault Zone; P-Peel Fault System (Thrust); W-Mount Wellington Greenstone Belt; WO-Woorndoo Fault; Y-Yarrol Fault System (thrust); rift-denotes zones of Meso-Cenozoic rifting.

2. The composite Late Proterozoic to Mesozoic Tasman Fold Belt System on the east;

3. The widespread, Late Carboniferous to Holocene platform cover of sedimentary and volcanic rocks, some of which are related to post-Gondwanaland break-up and separation.

The Early to Middle Proterozoic Australian Craton, subdivided into major structural units exposed in shield-like blocks (Geological Society of Australia, 1971; Plumb, 1979; Rutland, 1976), does not concern us here. However, the Late Proterozoic to Paleozoic sedimentary cover of the craton is related to development of the Tasmanides and thus of importance. This is the "Tasman Paratectonic Zone" of (Fig. 1; Rutland, 1976), which was characterized by Late Proterozoic rifting. Veevers (1984) suggests that such rifting also occurred in the Georgina Basin area (Fig. 1). The southern part of the paratectonic belt was deformed during the Middle Cambrian to Early Ordovician (Fig. 1; Thomson, in Parkin, 1969; Rutland et al., 1981). The Amadeus Transverse Zone was deformed during the Carboniferous Alice Springs-Kanimblan Orogeny (Powell, 1984a).

The boundary between the Proterozoic craton and the Tasmanides follows the Tasman Line of Hill (1951) (Fig. 1; cf. Harrington, 1974; Scheibner, 1974b, 1978). Veevers (1984) places this line in a similar position with some variations.

The Tasman Fold Belt System (TFBS) (Scheibner, 1974b, 1978) comprises five orogenic belts, a foreland fold and thrust belt, and a foredeep basin (Fig. 2). These elements are summarized below, generally from west to east, and differ in timing of their stages of tectonic evolution, from pre-cratonic (cratonic is used in the sense of normal continental crust) featuring flyschoid sedimentation, through transitional, with molassic sedimentation, to cratonic (or neocratonic) (cf. Scheibner, 1976).

The Kanmantoo Fold Belt (KFB) (Fig. 2) occupies western Tasmania, west of the Tamar Fault Zone (Williams, 1978), western Victoria, west of the Woorndoo Fault (VandenBerg, 1978), the adjoining southeastern part of South Australia east of the paratectonic Adelaide Fold Belt (Flint and Parker, 1982; Parkin, 1969), and an area in northwestern New South Wales (N.S.W.), northwest of the Darling Depression and east of the Adelaide Fold Belt (Scheibner, 1972a). Large tracts of this orogenic belt are concealed below the Cenozoic Murray Basin, but aeromagnetic anomalies (Tucker and Hone, 1984) that are thought to reflect extensions of the Stavely Greenstone Belt (VandenBerg and Wilkinson, in Cooper and Grindley, 1982), appear to link up beneath the basin with equivalent exposures in northwestern N.S.W. and thus delineate the eastern margin of the KFB.

The KFB developed from a Late Proterozoic to early Paleozoic pre-cratonic province containing deep-water turbiditic strata, some volcanics, with shallow-water sediments deposited close to basement inliers. The pre-cratonic development was terminated by the diachronous Middle Cambrian to Early Ordovician Delamerian Orogeny (Thomson in Parkin, 1969). Transitional tectonic, shallow marine to continental sedimentation started locally in the Middle Cambrian (Powell et al., 1982) and terminated in the Early Ordovician. Sedimentation and volcanism associated with the development of the Lachlan Fold Belt to the east (LFB) subsequently affected the region of the KFB. During the Early Devonian (?latest Silurian), continental to shallow-water sedimentation occurred in grabens in Victoria (Grampians; VandenBerg, 1978) where it was preceded by extrusion of rhyolites, while in N.S.W. (Cootawundy beds; Webby, in Cooper and Grindley, 1982) it was associated with mafic andesitic to felsic volcanism (G. Neef, pers. comm., 1984).

Middle Devonian to Early Carboniferous continental clastics were part of a larger molassic sheet of sediments which probably connected with similar accumulations further east. After continent-wide Carboniferous Alice Springs-Kanimblan Orogeny the KFB behaved as a neocraton.

The Kanmantoo pre-cratonic province probably continued further north into Queensland, but this area is nearly completely concealed by the Late Carboniferous to Cenozoic Trans-Australian Platform Cover and it is difficult to separate the early Palaeozoic fold belt from the middle Paleozoic one. For this reason, an all-embracing name, the Thomson Fold Belt (TFB), was introduced by Kirkegaard (1974).

The southern boundary of the TFB is taken along the Darling River Lineament (Fig. 2) or at the major Bouguer gravity change in northwestern N.S.W., the western boundary is the Tasman Line, the northern boundary is the Clarke River Fault, and the eastern boundary forms the western edge of the Sydney-Bowen Basin.

Pre-cratonic development of the TFB was terminated during a Middle to Late Ordovician episode of deformation. The Devonian to Early Carboniferous Burdekin, Drummond, and Adavale basins represent the transitional tectonic stage (Murray and Kirkegaard, 1978). The presence of mainly felsic volcanic, locally andesitic volcanics, as in the KFB, is of particular interest. After the Kanimblan Orogeny the TFB changed into a neocraton.

The east-verging Stawell-Bendigo Foreland Fold and Thrust Belt of Cox et al. (1983) occurs to the east of the KFB in Victoria. In a stratotectonic sense it is the foreland basin of the KFB, although in respect to the timing of its deformation it is linked to early deformation in the Lachlan Fold Belt (LFB). This fold and thrust belt, with its thin-skinned structure (Cox et al., 1983), comprises Cambrian to Ordovician turbidites (VandenBerg, 1978; Cas et al., 1983), with the narrow Heathcote Greenstone Belt at its front and possibly also its sole. The greenstones include Cambrian calc-alkaline andesite, boninite, and MORB tholeiite (Crawford, 1983). Emplacement of this belt could have occurred during the Late Ordovician-Early Silurian Benambran Orogeny and was structurally related to the development of the LFB.

The Lachlan Fold Belt (LFB) (Fig. 2) is a complex orogenic belt whose early Paleozoic history culminated in a Late Ordovician to Early Silurian collisional belt during the Benambran Orogeny (Pogson, 1982; Scheibner, 1982). The Melbourne Trough (VandenBerg, 1978) saw turbiditic sedimentation during this event, and sedimentation lasted into the Middle Devonian. However, in the region west of the Gilmore Suture (Fig. 2), the Silurian is generally absent, and basin formation with some volcanic rifting occurred during the Early Devonian. From the Middle Silurian to the Middle Devonian the region east of the Gilmore Suture became a complex back-arc area. This development was terminated during the Middle Devonian Tabberabberan Orogeny which diminished in intensity from southeast to northwest (Powell, 1983a). Subsequent diachronous molassic sedimentation lasted from late Early Devonian to Early Carboniferous time, and was terminated during the Carboniferous Kanimblan Orogeny which converted the LFB into a neocraton.

The Hodgkinson-Broken River Fold Belt (HBFB) (Fig. 2) in northeast Queensland contains Ordovician to Early Carboniferous rocks which accumulated partly on Precambrian basement (Day et al., 1983; Henderson and Stephenson, 1980). The main process of cratonization started in the Middle Devonian and continued into the Carboniferous. Post-kinematic felsic igneous activity continued into the Permian.

The easternmost and youngest fold belt of the Tasmanides is the New England Fold Belt (NEFB) (Fig. 2). It is separated from the other orogenic belts to the west by its foredeep, the Sydney-Bowen Basin. The pre-cratonic development was complex and started in the Cambrian (Cawood, 1976). The NEFB comprises composite early to late Paleozoic volcanic arc-fore-arc basin-accretionary prism complexes which have been intruded by syn-, but mainly post-kinematic granitoids. Its development was punctuated by Middle Devonian and Middle to Late Carboniferous deformations, with the terminal Hunter-Bowen Orogeny spread over the period between mid-Permian to Late Triassic (Day et al., 1978, 1983; Korsch and Harrington, 1981; Leitch, 1974). Cratonization progressed from south to north. Post-kinematic igneous activity lasted into the Mesozoic.

The latest Carboniferous to Triassic Sydney-Bowen Basin (Fig. 2) is a foredeep of the NEFB, and represents platform cover on older fold belts to the west. During the later stages of terminal deformation, the NEFB was thrust eastwards over the Sydney-Bowen Basin along the Hunter-Mooki-Goondiwindi Thrust System.

Probably one of the most characteristic features of the Tasmanides is the widespread orogenic plutonic activity (White and Chappell, 1983; Richards in Henderson and Stephenson, 1980). While some granites can be related to hypothetical B-subduction processes, a large proportion cannot be explained this way. It has been proposed by Pogson (1982) and Scheibner (1982) that some of these granites are related to thickening of marginal basin fill by thrusting, with consequent crustal melting, and others are related to melting associated with rifting (Collins et al., 1982; Barron et al., 1982).

The Tasman Fold Belt System (TFBS) is cut by major lineaments, which appear to be old fracture zones. These lineaments subdivide the TFBS into blocks or subplates, which were active from time to time (Scheibner, 1974b). Today these subplates are welded into the Australian continent.

Continental Accretion and Accretion of Allochthonous Terranes

The eastern Australian region has been considered by some worker to be a classic example of lateral accretion of continental crust (David and Browne, 1950; Packham, 1960, 1969). Progressive eastward stepwise accretion to the old Australian craton is indicated by general eastward younging of orogenic deformation and granitic intrusions. However, this process was not simple peripheral eastward growth of the Australian craton, because crosscutting extensional features in the form of rifts, basins and marginal seas were created episodically between separated continental blocks or microcontinents. The extensional features, formed on thinned continental crust or on newly created oceanic crust, were filled by sedimentary rocks, deformed, metamorphosed and eventually cratonized. This process resulted in a complex mosaic of older and younger crustal structures. A further complication has been introduced by possible accretion of allochthonous terranes.

In the other orogenic belts for which active plate margin settings are recognized (e.g. North American Cordillera, Appalachians) all terranes beyond the autochthonous miogeoclinal (or paratectonic) zone fringing the cratonal foreland are considered "suspect" (Coney et al., 1980; Williams and Hatcher, 1982). They are "suspect" in that their original paleogeographic settings are uncertain, and for some terranes there is paleontological and/or paleomagnetic evidence which shows they originated in regions far removed from where they now occur. The possible presence of such terranes in the Tasmanides has been only recently suggested (Fig. 3; Cawood, 1983; Flood and Fergusson, 1982; Harrington, 1983; Powell, 1983a,b, 1984b; Scheibner, 1982, 1983b, 1985). Some of these terranes will be mentioned below.

Episodes of terrane accretion and orogenic activity appear to be closely related. While continuous deformation occurred in arc-trench gap areas, episodes of stronger deformation affecting limited areas (Powell, 1983a,b) punctuated the tectonic development. These distinct, often diachronous, orogenic episodes may result from changes in the mode of B-subduction (from Mariana to Chilean-type), with resulting closure of mar-

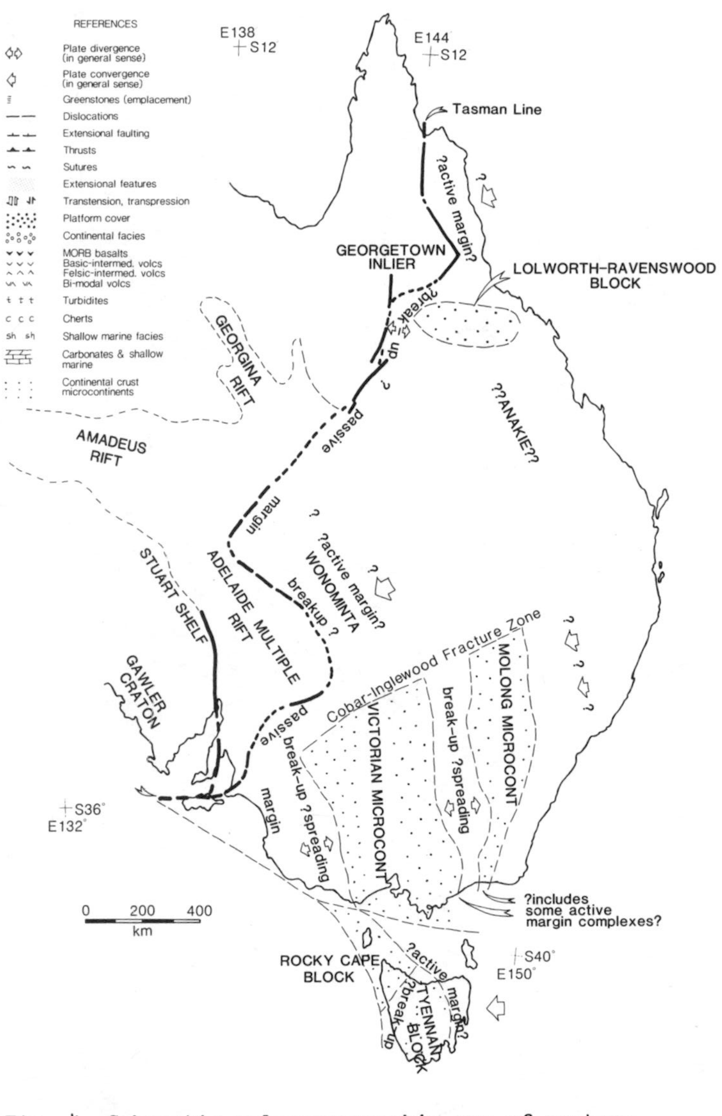

Fig. 4. Schematic palaeogeographic map of eastern Australia for the late Proterozoic.

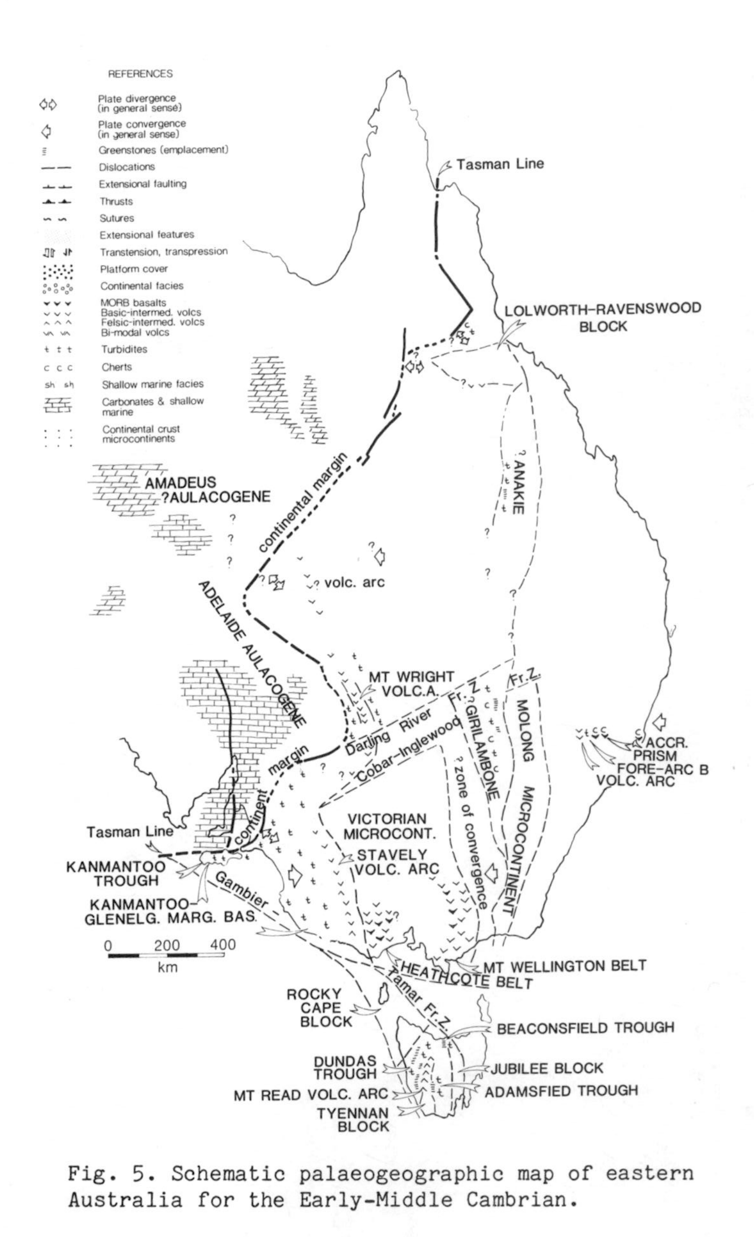

Fig. 5. Schematic palaeogeographic map of eastern Australia for the Early-Middle Cambrian.

ginal basins, and collision of volcanic arcs and microcontinents. Major orogenic episodes were followed by rearrangements of stratotectonic units or formation of new ones at the active plate margin, and the local changes presumably were the result of rearrangement of major plate movements and interactions.

The following seven orogenic episodes or events were important during development of the Tasmanides: Delamerian (Middle Cambrian to Early Ordovician), Benambran (Late Ordovician to Early Silurian), Quindongan (Middle Silurian), Bowning (Early Devonian), Tabberabberan (Middle Devonian), Kanimblan (Carboniferous), and Hunter-Bowen (mid-Permian to Triassic).

Late Precambrian Development

In tectonic analyses and syntheses of the Tasmanides, usually only the Paleozoic development is considered. However, the presence of flysch facies, ophiolites, and arc volcanics indicates that by earliest Cambrian time a well-developed active plate margin existed (Oversby, 1971; Scheibner, 1972a,b; Solomon and Griffiths, 1972; Crawford and Keays, 1978; Crook, 1980; Powell, 1984b). Hence break-up, and sea-floor spreading associated with the separation and dispersion of microcontinents may have occurred in Precambrian time, prior to the distinctive episode of extension at the Cambrian/Precambrian boundary (Plumb, 1979; Veevers, 1984; von der Borch, 1980; and others).

The foundation for this statement stems from the following deductions. Stratigraphic and geochemical data (cf. Cook, 1982; Cooper and Grindley, 1982; Crawford and Keays, 1978; Crawford, 1983) suggest the existence of Early Cambrian volcanic arcs in the southwestern part of the present Tasmanides (see Fig. 5). The volcanic arc in New South Wales has been called the Mount Wright Volcanic Arc (Scheibner, 1972a), and may be linked, using aeromagnetic data, (Tucker and Hone, 1984) with the Stavely Volcanic Arc (Greenstone Belt) in Victoria (Crawford and Keays, 1978; Crook, 1980; Crawford, 1983). This volcanic arc occurred west of the Victorian Microcontinent (Figs. 4,5), which had a width of over 500 km. The allochthonous Mount Wellington Greenstone Belt was possibly derived from a second volcanic arc (Crawford and Keays, 1978; Crawford, 1983; Cook, 1980), located on the eastern side of the Victorian Microcontinent (Fig. 5). If these Early Cambrian volcanic arcs were related to subduction of oceanic crust, such oceanic crust may have been older by several tens of millions of years, as today it is unusual for subduction of young oceanic crust, or even spreading centers, to take place. The disposition of these volcanic arcs requires the existence of two subduction zones, one on each side of the Victorian Microcontinent. This in turn means that Precambrian oceanic crust must have existed both east and west of the Victorian Microcontinent and that such crust formed in a passive or active plate margin setting. The Late Proterozoic rocks of the Australian Craton should reflect the history of such Precambrian development.

Late Proterozoic Rifts

During the Late Proterozoic (Fig. 4), the Australian Craton was covered by strata of the Central Australian Platform Cover (Geological Society of Australia, 1971; Plumb, 1979). Extensive shallow-water to continental sediments and less important intraplate volcanics accumulated in basins (Adelaide, Amadeus, Officer) that were associated either with intraplate rifting or perhaps compression (Lambeck, 1983). Very thick accumulations were localized over the most recently cratonized Precambrian mobile belts (Plumb, 1979). At a later stage, especially during the latest Proterozoic glaciations, extensive platform cover rocks extended from western Tasmania to Western Australia (Plumb, 1979).

Major Late Proterozoic rift zones are discordant to the eastern margin of the craton and the subsequent Paleozoic Tasman active plate margin (Fig. 4). The remains of these rifts are located opposite salients in the Tasmanides (Fig. 1), indicating deeper crustal relationships. The Adelaide Fold Belt is opposite the Murray Salient and the Amadeus Transverse Zone (Rutland, 1976) is opposite the Cooper Salient (Figs. 1,4), although here the Permian Pedirka Basin conceals the immediate contact area. Thomas (1983) and other workers have suggested that salients of orogenic belts develop at sites of embayments formed in rifted continental margins, which act as forelands to the orogens, and conversely, recesses in orogenic belts may be related to former continental margin promontories. Another possibility is that salients are the result of accretion of microcontinental terranes. Thus the Murray Salient may either have formed on the site of a continental margin embayment or reflects the collision of the Victorian Microcontinent.

Controversy exists regarding the origin and the tectonic development of the paratectonic Adelaide Fold Belt. An earlier interpretation of Sprigg (1952) suggested deposition of the Adelaidean sediments in a miogeosynclinal (miogeoclinal) continental terrace, implying the existence of a continental margin immediately to the southeast. In many respects the Late Proterozoic rift zones are similar to aulacogenes (Fig. 5; Olenin, 1967; Rutland, 1976; Scheibner, 1972b, 1974a; von der Borch, 1980), although Rutland et al. (1981) argued that the Adelaide Rift (Fig. 4) developed as a protracted multiple rifted arch system in a passive continental margin (cf. Veevers and Cotterill, 1976). Von der Borch (1980) modified the earlier general aulacogene model and also the multiple rift model, by suggesting a Cambrian triple junction between an intracratonic Central Flinders Zone (aulacogene) and the ancient continental margin to the south.

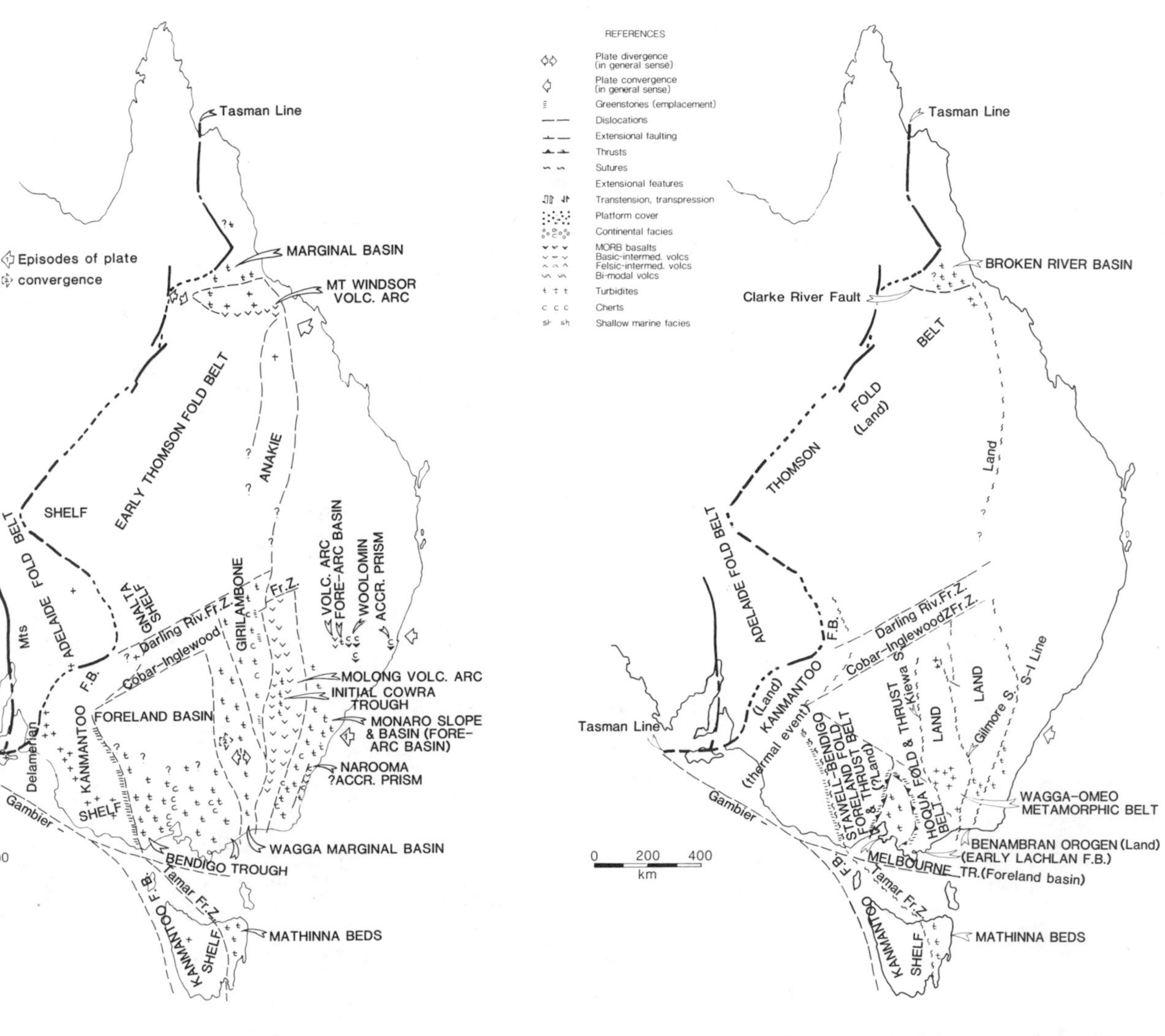

Fig. 6. Schematic palaeogeographic map of eastern Australia for the Late Cambrian to Ordovician; + indicate granitoids.

Fig. 7. Schematic palaeogeographic map of eastern Australia for the Early Silurian; + indicate granitoids.

Recently Preiss (1983) suggested that the early Adelaidean Callana and Burra groups represent pre- and syn-rift deposits and the Umberatana and Wilpena groups are post-rift sediments. Webb et al. (1983) date the Beda volcanics, which are correlatives of the Wooltana volcanics of the Callana Group within the rift, at 1076 ± 33 Ma, the Burra Group as postdating a metamorphic event at 849 ± 32 Ma, and the start of the Umberatana and Wilpena groups at about 724 Ma. Thus, passive margin formation and associated break-up could have occurred east of the Tasmanides from about 1000 to 724 Ma. Any oceanic crust generated could be as old as 250 Ma by Early Cambrian time, and thus gravitationally quite unstable.

If we accept the modified continental margin and aulacogene model of von der Borch (1980), we can speculate that the prolonged Late Proterozoic rifting and basin formation (Rutland et al., 1981; von der Borch, 1980) were related to continental break-up further east than the subsequent Cambrian break-up, which localized the ensimatic Kanmantoo Trough at the continental margin (Figs. 4,5).

Late Proterozoic Active Margin?

Precambrian rocks form scattered outcrops within the Tasmanides (Fig. 2) but their subsurface nature and extent is not known and controversial. Alternative views are that the Tasmanides developed across either an originally ensialic basement (Rutland, 1976), or an interspersed ensialic and ensimatic basement (Scheibner, 1974a, 1976), or a mainly ensimatic basement (Crook, 1980).

The Kanmantoo Fold Belt contains inliers of variably metamorphosed sedimentary and volcanic rocks (Fig. 2). The Wonominta beds of New South Wales, which according to K. Mills (pers. comm., 1985) comprise older, pre-Late Proterozoic metamorphics, as well as Late Proterozoic turbidites and minor mafic volcanics metamorphosed to the greenschist facies. In some areas the younger rocks appear to be conformable beneath early Paleozoic sediments. In Tasmania, the Tyennan Block (Fig. 4) (Williams, 1978; Cooper and Grindley, 1982) comprises Precambrian rocks, predominantly of sedimentary origin and metamorphosed to greenschist facies, together with minor amphibolite and eclogite derived from basic to intermediate igneous rocks. The main metamorphic event is dated at 800 Ma (Raheim and Compston, 1977). In northwestern Tasmania, the Rocky Cape Block comprises comparatively unmetamorphosed sediments and mafic volcanics. On the western side of King Island, strongly deformed metasediments have been intruded by syntectonic granites (735-725 Ma). On the east side of the Rocky Cape Block, the Arthur Lineament, a narrow belt less than 10 km wide, is formed by greenschists with amphibolite dykes. Further east, on the north coast, there is a thick, complexly folded sequence (Burnie Formation of Gee, 1977) of turbidites, mudstones, and some mafic pillow lavas. Deformation of all these rocks occurred during the so-called Penguin Orogeny (670-725 Ma) (Williams, 1978; Cooper and Grindley, 1982).

The evidence is permissive that rocks of Precambrian inliers in the eastern Tasmanides formed in an active plate margin setting; this would support the idea of a late Proterozoic onset of plate interactions in eastern Australia.

We can thus speculate that during Late Proterozoic, the break-up of the Australian Craton proceeded from west to east, and resulted in separation of microcontinental blocks of unknown size, separated by one or more marginal seas possibly floored by oceanic crust. With time, such oceanic crust became gravitationally unstable and would have led to the formation of an active plate margin. Unfortunately little is known about the Precambrian rocks which could have formed at this active plate margin. They were incorporated into the Tasmanides and are only exposed locally.

Early Paleozoic Development

By earliest Cambrian time a well-developed west Pacific-type active plate margin may have existed in eastern Australia.

Recently Veevers and Powell (Veevers, 1984) summarized data from Australia which suggest that Early Cambrian extension affected the whole continent and resulted in break-up along the western limit of the Tasmanides, the so-called Tasman Line (Figs. 2,5; Hill, 1951; Scheibner, 1974b). This extension can be related to a strain field defined by a pole of rotation (Scheibner, 1974b; Veevers, 1984) fitted to lines which appear to represent transform faults (Fig. 5: cf. Veevers, 1984, Fig. 182B).

Cambrian Volcanism

In many areas of Australia, the Cambrian succession starts with intraplate, mainly tholeiitic basalts. Volcanism elsewhere was of a different character, and Veevers and Powell (Veevers, 1984) postulated four main tectonic settings for Early Cambrian volcanics in Australia: 1) newly generated oceanic lithosphere; 2) eruptives along the newly-formed continental margin (Truro Volcanics in the Adelaide Fold Belt); 3) volcanic rifts or grabens near the continental margin (the Mount Read Volcanics in Tasmania (cf. Williams, 1978; Brown et al., 1980)); 4) platform volcanics; tholeiites with minor agglomerate (Antrim Plateau Volcanics in Western Australia and the Table Hill Volcanics of the Officer Basin). An additional (5) tectonic setting had volcanic arc character and included the Stavely, Heathcote and Mount Wellington greenstone belts in Victoria (Crawford and Keays, 1978; Crawford, 1983; Edwards, 1979; Scheibner, 1972a,b, 1974a).

Kanmantoo Trough

According to von der Borch (1980), sedimentological evidence indicates that separation had occurred in the Kanmantoo Trough area by earliest Cambrian time (Fig. 5). The ensimatic character of this trough had been suggested previously (Scheibner, 1972a,b, 1974a), on the basis of the large thickness, over 6 km (Daily and Milnes, 1973), of rapidly deposited, partly turbiditic sediments. Direct evidence for oceanic crust is missing. In contrast to an earlier model (Scheibner, 1972a,b), the writer feels that the Kanmantoo Trough formed part of a larger marginal sea (Kanmantoo-Glenelg Marginal Basin), floored partly with Late Proterozoic oceanic crust. The protolith of the Glenelg River Metamorphic Complex in Victoria (VandenBerg, 1978) also accumulated in this marginal sea.

Victorian Microcontinent

To the east of this marginal sea lay the Victorian Microcontinent (Figs. 4,5), which was nearly 500 km wide. Part of this microcontinent consists of Early Proterozoic rocks, and part of Late Proterozoic active plate margin complexes like the Wonominta beds in N.S.W., noted above, and similar rocks in Tasmania.

The eastern limit of the Victorian Microcontinent is the concealed, eastern root region of the Kiewa Suture or Thrust (Figs. 6,7). The western limit, at the surface, is the Woorndoo Fault (Fig. 2; VandenBerg, 1978; VandenBerg and Wilkinson in Cooper and Grindley, 1982). In the subsurface the western limit is concealed by the allochthonous Glenelg River Metamorphic Complex, the Stavely Greenstone Belt, and the Stawell-Bendigo Foreland Fold and Thrust Belt (Figs. 2,5, 7). The autochthonous cover of the Victorian Microcontinent is the Early Ordovician to Middle Devonian sediments deposited in the Melbourne Trough (Fig. 7; VandenBerg, 1978). The Early Cambrian intermediate volcanics in the Mount Eastern area are also autochthonous (cf. VandenBerg and Wilkinson in Cooper and Grindley, 1982). Other outcropping sediments and volcanics in the area of the Victorian Microcontinent appear to be allochthonous. The Heathcote Greenstone Belt may form the sole and front of the west-verging Hoqua Foreland Fold and Thrust Belt (Fig. 7). Emplacement of opposing allochthonous foreland fold and thrust belts (Fig. 7), which originally were deposited as cover on the Victorian Microcontinent, occurred during the latest Ordovician-Early Silurian Benambran Orogeny. Sedimentation in the intervening foreland basin (Melbourne Trough) was continuous (Fig. 7), probably because of the rigid basement (Victorian Microcontinent). We will return to this problem later.

Molong Microcontinent

As well be discussed later, the ?Late Cambrian-Ordovician Molong Volcanic Arc (Fig. 6) is underlain by at least 20 km of older, continental-type crust. Possibly this basement was a microcontinent, the so-called Molong Microcontinent (Figs. 4,5). In the area west of Parkes, the Girilambone Group may be the basement of the Molong Volcanic Arc (Fig. 6). If this is correct, then the Girilambone Group and its correlatives in Queensland, the Anakie Metamorphics (Fig. 5) are pre-Late Cambrian.

The eastern limit of the Molong Microcontinent coincides with the S-I line of White and Chappell (Fig. 7; 1977), which according to these authors indicates the eastern limit of the Australian Proterozoic Craton. The western limit of the Molong Microcontinent was the Gilmore Suture (Fig. 7; Scheibner, 1982). During ?Late Cambrian and Ordovician time the Molong Microcontinent with its superposed volcanic arc was separated by the Wagga Marginal Basin (Fig. 6) from the Victorian Microcontinent.

Cambrian Plate Interactions

To explain the presence of Early Cambrian volcanic arc rocks of the Mount Wellington Greenstone Belt (Fig. 5) on the eastern side of the Victorian Microcontinent, (Crawford and Keays, 1978; Crawford, 1983; Crook, 1980), it is necessary to assume west-dipping subduction in the area east of the Victorian Microcontinent, and an adequate volume of oceanic crust to sustain subduction. This points towards Late Proterozoic separation of the Victorian and Molong microcontinents. Continental margin andesites of the Stavely Greenstone Belt (Fig. 5) on the western side of the Victorian Microcontinent (Crawford and Keays, 1978) could represent a volcanic arc which originated by east-dipping subduction of the Kanmantoo-Glenelg Marginal Basin west of the Victorian Microcontinent.

The Stavely Greenstone belt may link up with Mount Wright volcanics to the north northwest. The strong positive aeromagnetic expression of the Stavely Greenstone Belt (VandenBerg and Wilkinson in Cooper and Grindley, 1982; D. Tucker pers. comm.) can be followed north-northwesterly and appears to be terminated by the Cobar-Inglewood Lineament (Fig. 5). A less intensive, sinistrally offset aeromagnetic high, continues towards the northeast, and joins the wide and complex magnetic high which is the expression of the outcropping Kanmantoo Fold Belt in northwestern N.S.W. The Mount Wright Volcanic Arc volcanics and younger intrusions (Fig. 5) are some of the sources of these anomalies.

The Heathcote and Mount Wellington Greenstone belts (Fig. 5) contain both low-Ti lavas (boninites) and apparently superposed MORB lavas (Crawford, 1983). Crawford (1983) suggested that subduction during Early Cambrian time of a back-arc basin spreading centre beneath an intra-oceanic volcanic arc caused formation of these rocks. This active spreading centre provided the heat source to generate low-Ti lavas from hydrous shallow, sub-back-arc basin sediments and from upper mantle beneath the arc. Continued activity of the subducted spreading centre provided the superposed MORB lavas.

Arrangement of Cambrian Stratotectonic Units

The arrangement of stratotectonic units in the area of present-day Victoria and southern N.S.W. during the Early Cambrian is shown in Figure 5. The Tasmanian segment was probably separated by the Gambier-Tamar (earlier name Gambier-Beaconsfield) Fracture Zone of Crawford and Campbell (1973) and Harrington et al. (1973). No direct southern continuation of the Kanmantoo-Glenelg Marginal Basin can be identified. Instead, remnants of several narrow rift-like troughs between basement blocks can be recognized (Corbett in Cooper and Grindley, 1982). The Smithton Trough, containing latest Proterozoic to Cambrian epicontinental sediments and tholeiitic basalts, developed in the western Rocky Cap Block. The other troughs (Dundas, Dial Range, Fossey Mountains, Beaconsfield and Adamsfield troughs) were situated between the basement blocks (Rocky Cape, Tyennan, Forth, Jubilee, and Beaconsfield). Most of these troughs are represented by sequences containing tholeiitic basaltic lavas and disrupted ultramafic-mafic complexes of somewhat ophiolitic character, but no true ophiolites (Brown et al., 1980). The Dundas Trough complex interfingers on the east with the Mount Read volcanics which mostly rest on the Tyennan Block. The Mount Read Volcanics either represent remnants of an Early Cambrian continental margin volcanic arc (Solomon and Griffiths, 1972) or a volcanic rift (Brown et al., 1980).

To the north, the Kanmantoo-Glenelg Marginal Sea probably interconnected across the Darling River and Cobar-Inglewood Lineaments (fracture zones) with the narrow Bancannia Trough (Scheibner, 1972a). This trough was in the area of the Wonominta Recess (Fig. 4) which possibly developed in response to a continental margin promontory. To the east of the Bancannia Trough was the Mount Wright Volcanic Arc (Fig. 5; cf. Packham, 1969; Scheibner, 1972a; Edwards, 1979). Further east was an unstable shelf or deeper terrace where the Copper Mine Range beds (Pogson and Scheibner, 1971) accumulated. Relationships of these rocks to the Girilambone Group (Figs. 5,6) to the east, which is a flysch-like complex containing possible dismembered ophiolites, remains uncertain.

In the area north of the Bancannia Trough and the Mount Arrowsmith area, Wopfner (1972) suggested Early Cambrian sedimentation in a so-called "Circum-Denison Arc". Wopfner also mentioned Late Proterozoic to Early Cambrian tuffs and volcanics in the Warburton basin, and these could indicate further northerly continuation of the Mount Wright Volcanic Arc (Fig. 5; cf. Cook, 1982).

Further north, data are very sparse (Murray and Kirkegaard, 1978). In the Lolworth-Ravenswood Block (Fig. 5), the basement of which could be a small Precambrian microcontinent, the Cape River beds, Mount Windson Volcanics, Argentine Metamorphics, Charters Towers Metamorphics, and similar complexes (Queensland Geological Survey, 1975; Day et al., 1983) are thought to be of Late Cambrian to Early Ordovician ages. These rocks may have been deposited in a Late Cambrian-Early Ordovician volcanic arc-back-arc setting (Murray and Kirkegaard, 1978; Day et al., 1983) which is younger than comparable associations to the south.

Cambrian Orogenic Deformations

At the Australian active plate margin, diachronous orogenic deformations preceded by mild episodic tectonic activity started in Middle Cambrian time and continued into the Ordovician. This orogenic deformation was named the Delamerian Orogeny in South Australia (Thomson in Parkin, 1969), and the same name has been used in Victoria and N.S.W. (Scheibner, 1972a; VandenBerg, 1978). Early to Late Ordovician deformation in the Thomson Fold Belt to the north is unnamed (Kirkegaard, 1974; Murray and Kirkegaard, 1978). Recently, Powell (1984) related Ordovician deformation in Queensland to dextral transpression along a transform boundary, which in N.S.W. changed to oblique subduction. This is a possible but not yet proven relationship.

During this orogeny, inner marginal basins were closed and inverted by collisional movements of the inner microcontinents with their attached volcanic arcs, which is similar to accretion of allochthonous, suspect terranes (Scheibner, 1983b, 1985). Metamorphism affected the Precambrian craton to the east, as shown by the 520 ± 40 Ma thermal pulse in the Broken Hill Block of Harrison and McDougall (1981). The paratectonic cratonic cover was deformed, and resulted in the Adelaide Fold Belt in South Australia (Fig. 7; Thomson in Parkin, 1969; Rutland et al., 1981; Parker, 1983). Late Cambrian to Early Ordovician orogenic granites (cf. Milnes et al., 1977; Richards and Singleton, 1981) were emplaced in the inverted Kanmantoo-Glenelg Marginal Basin (Kanmantoo-Glenelg Belt), and also the old craton (Fig. 6; cf. Thomson in Parkin, 1969). The orogenic belt which resulted from deformation of the Early Cambrian active plate margin has been referred to as the Kanmantoo Fold Belt (Fig. 7; Scheibner, 1972a,b; 1978). In Queensland the slightly younger complexes were cratonized by the Middle Ordovician and formed the core of the Thomson Fold Belt (Fig. 7; Murray and Kirkegaard, 1978; Day et al., 1983).

The late orogenic, transitional tectonic, Middle Cambrian to Early Ordovician sediments of the Kanmantoo Fold Belt were deposited in mostly shallow marine to continental settings (cf. Cooper and Grindley, 1982). On the site of the former Victorian Microcontinent a wide foreland

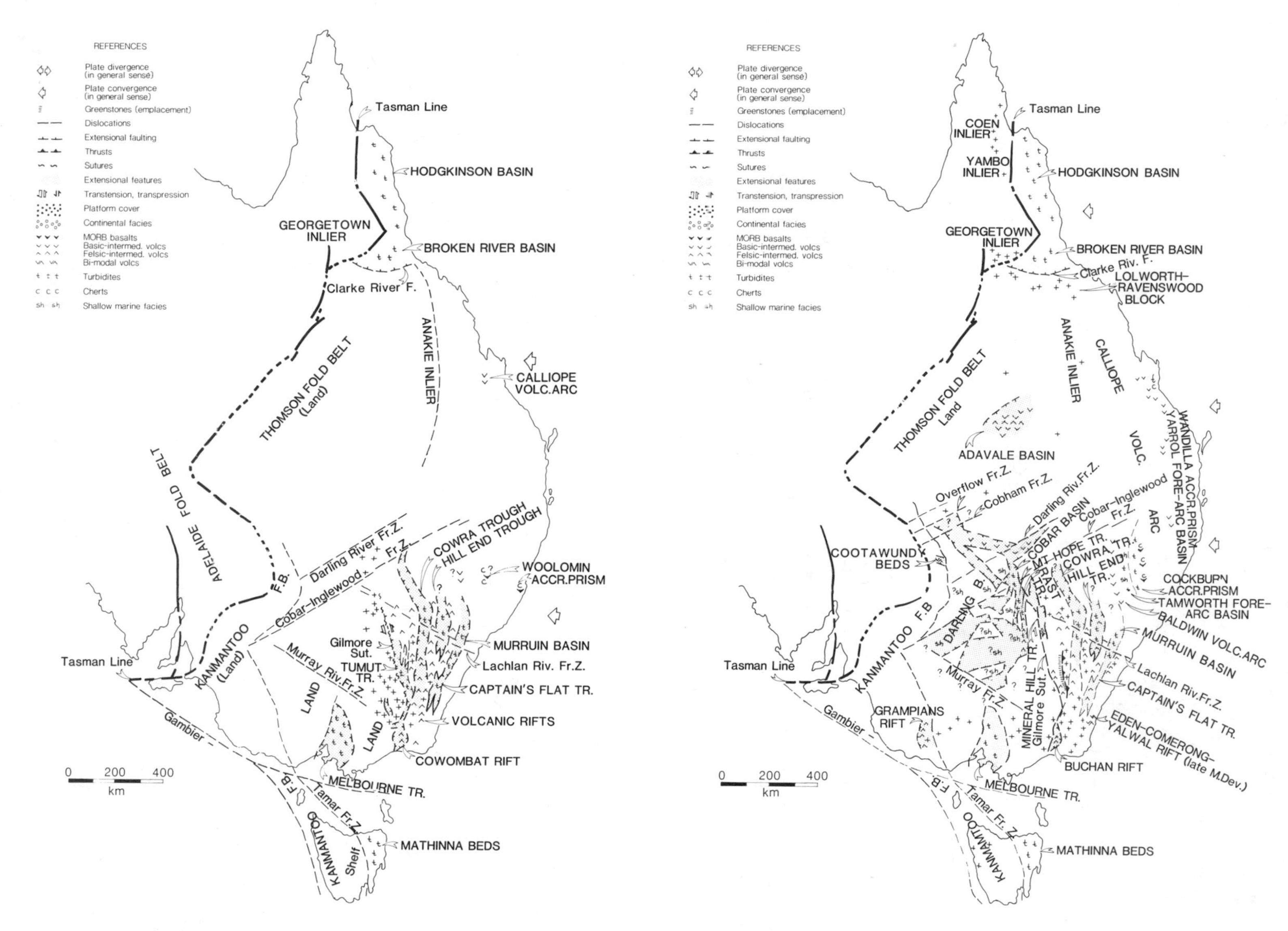

Fig. 8. Schematic palaeogeographic map for eastern Australia for the middle-Late Silurian; + indicate granitoids.

Fig. 9. Schematic palaeogeographic map for eastern Australia for the Early-Middle Devonian; + indicate granitoids.

basin developed (Cox et al., 1983), which included the Stawell, but mainly the Ballarat (Bendigo) and Melbourne troughs (Figs. 6,7) (VandenBerg, 1978; VandenBerg in Cooper and Grindley, 1982).

Ordovician Stratotectonic Units in the Lachlan Fold Belt

Possibly in the latest Cambrian, but definitely by earliest Ordovician time (Sherwin, 1979), the Molong Volcanic Arc developed east of the Victorian Microcontinent. It can be traced from the N.S.W./Victorian border northward to the Cobar-Inglewood fracture zone or lineament (Fig. 6). Its relationship to the Mount Windsor Volcanic Arc of Queensland (Day et al., 1983) is not known. Most authors (Oversby, 1971; Packham, 1973; Scheibner, 1972b, 1974a; Cas et al., 1980; Crook, 1980; Powell, 1983a,b, 1984b) agree on a model in which Ordovician volcanics in New South Wales formed a volcanic island arc which was bordered on the west by a marginal sea represented by the Wagga Marginal Basin (Fig. 6), and on the east by open ocean, represented by Monaro Slope and Basin (Fig. 6).

Molong Volcanic Arc

Geochemical data (Wyborn, 1977; Owen and Wyborn, 1979; I. Clarke, pers. comm., 1985) from rocks of the Molong Volcanic Arc indicate that the volcanism has dominantly shoshonitic character, with minor tholeiitic basalts and island arc tholeiites in the southern part of the arc (Wyborn, 1977; Owen and Wyborn, 1979), and minor K-rich calc-alkaline andesites and basalts, and minor trachytes in the northwestern part (I. Clarke, pers. comm.).

Uncertainty about the origin of modern shoshonitic suites (cf. Johnson et al., 1978) has led to reservations about the validity of a subduction model for the Molong Volcanic Arc (cf. Owen and Wyborn, 1979), and to the suggestion that these volcanics were related to Ordovician rifting and fracturing of continental crust in a marginal plateau environment (Wyborn, 1977). In the southeast Pacific, modern shoshonites occur mostly in areas where there is older continental-type crust. Coulon and Thorpe (1981) came to the conclusion that shoshonites form where the continental crustal thickness exceeds 20 km. It seems that it is necessary to conclude that there was older continental-type crust in the areas of Ordovician shoshonitic volcanism of the Molong Volcanic Arc. This crust could have originated during the previous orogenic episode, or could be the Molong Microcontinent, discussed earlier.

The writer favours the concept of evolution of the Molong Volcanic Arc in a convergent plate setting. Most authors envisage west-dipping subduction beneath the arc, but uncertainty is introduced by the apparent eastward-younging of the arc, with oldest rocks on the west and youngest on the east. If the conclusion of Dickinson (1973) about progressive migration with time of arcs away from the trench can be applied here, then subduction could have been towards the east. Wyborn (1977) in the southern part of the volcanic arc noted an eastward progression of volcanics from tholeiitic through island-arc tholeiitic to shoshonitic. This would support a model of eastward subduction.

The Molong Volcanic Arc occupies a crustal block which is bounded on the west by the Gilmore Fault Zone or Suture (Fig. 7; Scheibner, 1982). The fault zone features reverse faults modified from westerly-dipping thrusts by later deformation. The Molong Volcanic Arc and its basement had underthrust the basin to the west. The Gilmore Suture, and its northern continuation, has many Late Silurian to Devonian intermediate, basic, and ultrabasic intrusions associated with it, and it is an important gold-bearing metallogenic feature (Suppel and Degeling, 1982). North of the Lachlan River Fault Zone (Fig. 8) the Gilmore Suture may have two strands, the eastern one defined by Alaskan-type intrusions within the Girilambone terrane.

Wagga Marginal Basin

To the west of the Molong Volcanic Arc a region of turbiditic basinal sedimentation is called the Wagga Marginal Basin (Fig. 6; Scheibner, 1972b). New palaeontological data (Kilpatrick and Fleming, 1980) indicate that sedimentation started in earliest Ordovician time. Earlier tectonic models have to be amended to take this new date into account.

If westward subduction under the Molong Volcanic Arc did occur, then the Wagga Marginal Basin had a back-arc basin setting and as such could be expected to be partly floored by oceanic crust. However, there is no direct evidence for such crust, although it has been argued above that some Precambrian oceanic crust could have been present between Victorian and Molong microcontinents.

The fill of the basin was converted into the Wagga-Omeo Metamorphic Belt during the Late Ordovician-Early Silurian Benambran Orogeny. The belt is bounded by the Gilmore Suture on the east and by the Kiewa Thrust or Suture on the west. Along the east-dipping Kiewa Thrust, high-temperature/low-pressure Ordovician metamorphics, intruded by synkinematic, anatectic, S-type granites, are in contact with little-metamorphosed rocks of the Tabberabbera subzone of VandenBerg (1978). This subzone and the Hoqua zone of VandenBerg (1978) in turn were thrust westward, and are classified as the west-verging Hoqua Foreland Fold and Thrust Belt (Fig. 7; in earlier papers the name Tabberabbera Foreland Fold and Thrust Belt was used (Scheibner, 1983b), but is not employed here, "Tabberabbera" is used for the Middle Devonian Orogeny). The Mount Wellington

Greenstone Belt is at the western front and sole of this belt.

Fore-arc Basin and Accretionary Wedge

Using the calculations of median volcanic arc size values of Dickinson (1973), and accepting that the minimum duration of the Molong Volcanic Arc was 65 million years, the arc-trench gap should have been about 130-150 km wide, and the width of the arc over 80 km. This means that in the model of westward subduction the Monaro Slope and Basin sediments (Scheibner, 1974a) would mainly represent fore-arc basin fill, which supports earlier suggestions by Crook (1980). Only in the South Coast region of N.S.W. might we expect to find accretionary wedge rocks, and possibly the Wadonga Beds, which contain some basic volcanics, fulfill this role (Scheibner, 1974a, 1982). In the South Coast region, Powell (1983a,b) recently recognized Ordovician beds showing stripey-cleavage which he interpreted as characteristic of an outer arc slope, accretionary prism setting.

Tectonic Model for Ordovician-Early Silurian Time

The following model is used to explain the evolution of this region. During the earliest Ordovician, and probably latest Cambrian time, west-facing Mariana-type, B-subduction developed east of the Molong Microcontinent. The Molong Volcanic Arc formed on this block. The back-arc basin (Wagga Marginal Basin) was partly floored by oceanic crust which formed during Late Cambrian to Early Ordovician subduction, and/or was a remnant from an earlier marginal sea between Victorian and Molong microcontinents. To the east of the volcanic arc was the fore-arc basin, Monaro Slope and Basin, and further east, the accretionary prism (Cas et al., 1980; Powell, 1982b, 1984b). During the later part of the Ordovician, the Mariana-type boundary changed into Chilean-type, and the Wagga Marginal Basin was closed. The volcanic arc underthrust the marginal basin on the Gilmore Suture, and overthrust the Victorian Microcontinent on the Kiewa Suture. The deformation and thrust pile-up of the Wagga Marginal Basin fill resulted in high-temperature/low-pressure metamorphism, with areas of ultra-metamorphism which led to the formation of anatectic S-type granites (Pogson, 1982; Scheibner, 1982). Subsequent granitic plutonism lasted through the Silurian for about 30 million years (Fagan, 1979). These effects record the Benambran Orogeny of the Lachlan Fold Belt.

In the area west of the Kiewa Suture the rocks of the Hoqua subzone and the Wellington Greenstone Belt (cf. VandenBerg and Wilkinson in Cooper and Grindley, 1982), which represent the cover of the Victorian Microcontinent, were thrust westward. They form a structure which best could be described as the Hoqua Foreland Fold and Thrust Belt. There was continuous sedimentation in the foreland basin, the Melbourne Trough, which was protected by the rigid basement. Collisional stresses, however, were transmitted westward and the east-verging Stawell-Bendigo Foreland Fold and Thrust Belt developed at this time (Cox et al., 1983). In the rest of the Kanmantoo Fold Belt this collision caused a weak metamorphic event (Milnes et al., 1977). Rapid uplift and erosion is indicated by presence of Early Silurian granite clasts in Silurian conglomerates in Victoria. The Wagga-Omeo Metamorphic Belt remained mostly emergent during the Silurian.

To the south, Tasmania was hardly affected by the Benambran Orogeny, which suggests the Gambier-Tamar Fracture Zone (Harrington et al., 1973) acted so as to isolate the Tasmanian segment. Shallow-water sedimentation existed in the western part of the island while the flysch-like Mathinna beds (Ordovician to Early Devonian) were deposited east of the Tamar Zone (Williams, 1978; Corbett in Cooper and Grindley, 1982).

Early Silurian orogenic deformation and subsequent granite emplacement occurred in the "Western Belt" of the Tuhua Orogen in New Zealand (cf. Cooper and Grindley, 1982; Sporli, this volume).

The above plate tectonic model for the Ordovician of Victoria and New South Wales seems to be restricted to the area south of the Darling River Lineament. It is difficult to propose a meaningful tectonic model for the Ordovician development of the Thomson Fold Belt because of the scarcity of data (cf. Murray and Kirkegaard, 1978; Henderson in Henderson and Stephenson, 1980), and therein lies the weakness in attempts to make a general Ordovician plate model analogous with that of the present Andaman Basin (Cas et al., 1980; Powell, 1983a,b). There seem to be similarities in lithology and in intensity deformation, between the Girilambone Group and the Anakie Metamorphics. Possible physical connection between these complexes is supported by the presence of Girilambone-type rocks north of the Darling River Lineament in the N.S.W. State border area (cf. Scheibner, 1976), and possibly in the Nebine Ridge further north, but these rocks are of uncertain age. If they represent Late Proterozoic-Early to Middle Cambrian active plate margin complexes, they have no bearing on models of Ordovician plate interactions. If they represent Ordovician back-arc basin complexes (Pogson, 1982), there should have been a volcanic arc to the east. Until now, no Ordovician volcanic arc has been recognized immediately to the east of the Anakie Metamorphics (cf. Murray and Kirkegaard, 1978).

In the New England area, Cawood (1976, 1980) described early Paleozoic strata unconformably beneath an Early Devonian fore-arc sequence (Tamworth Group). The oldest rocks, of late Middle Cambrian to early Late Cambrian age, were derived from a volcanic arc. The well-documented accretionary prism complex (Woolomin Formation, Fig. 6; Cawood, 1980) is pre-Early Devonian (Leitch and Cawood, 1980), and recently M.A. Lanphere (pers. comm., 1983) obtained Late Ordovician (444 Ma) K-Ar ages on micas from blueschists in this complex (Barron et al., 1976). These data point to an early Paleozoic convergent plate setting (Leitch, 1982), but it is difficult to tie it to the Ordovician model of the Lachlan Fold Belt. At present it will be useful to treat the early Paleozoic rocks in the Peel Fault System (Fig. 11), as an additional suspect terrane (Copes Creek Terrane of Fig. 3; Howell et al., 1985; Scheibner, 1983b, 1985).

In the southwestern part of the Thomson Fold Belt, in the Warburton Basin, Early to early Late Ordovician sediments, including graptolitic shales, overlie Cambrian carbonates and indicate deepening of the basin (Kirkegaard, 1974). Seismic data indicate a large thickness of sediments in the western part of the Thomson Fold Belt (Kirkegaard, 1974).

Silurian to Early Carboniferous Development

The Silurian record indicates that a major rearrangement of the active plate margin of eastern Gondwanaland (Fig. 8) followed the Late Ordovician-Early Silurian Benambran Orogeny.

Tasmanian Segment During the Mid- to Late Silurian

The Tasmanian segment (cf. Williams, 1978; Cooper and Grindley, 1982) shows no evidence of major tectonic change. Shallow-water, platformal sedimentation continued west of the Tamar fracture zone. To the east, the Mathinna beds, a sequence of deep water, mainly turbiditic sediments, may include Silurian strata.

New South Wales-Victorian Segment During the Mid- to Late Silurian

Major changes occurred in the region between the Gambier-Beaconsfield and Darling River lineaments or fracture zones, with the most significant changes between the Darling River and Murray River fracture zones (Fig. 8).

In the region west of the Gilmore Suture, Early to early Late Silurian sediments are absent, except from the Melbourne Trough. In the Melbourne Trough south of the Murray River Lineament, sedimentation between the Ordovician and Silurian was continuous and similar to that of Eastern Tasmania.

To the east of the Gilmore Suture, during the mid- and Late Silurian, several extensional volcanic rifts and troughs formed (Fig. 8). At least one of these troughs, the Tumut Trough, was partly floored by ophiolites (Ashley et al., 1979). The large thickness of sediments in the Hill End and Cowra troughs (Cas, 1983) suggests that the crust there was extensively thinned. Initially, the troughs had the character of volcanic rifts, with bimodal but dominantly felsic volcanism, and exhibit Kuroko-type massive sulphide mineralization (Gilligan et al., 1979). Subsequently, the troughs were filled predominantly by turbidites which initially were quartz-rich and later became lithic and volcaniclastic (Packham, 1968; Cas, 1978a). Locally, deep-water emplacement of silicic lavas has been documented (Cas, 1978b). Marginal trough facies include olistostromes and slump deposits (Crook and Powell, 1976).

The ridges, rises, or highs between the troughs were sites of shallow-water sedimentation, with some carbonates, and submarine to subaerial volcanism. Some of these volcanic piles of S- and I-type character (Owen and Wyborn, 1979; Wyborn et al., 1981) were intruded by comagmatic granites. On the rises which had previously been the site of Ordovician arc volcanism, Siluro-Devonian porphyry-type mineralization was associated with felsic igneous activity (Bowman et al., 1983). The stratigraphy and lithology of these Silurian rocks has recently been described by Pickett (1982) and Cas (1983).

Segments North of the Darling River Lineament During the Mid- to Late Silurian

No sedimentary rocks of Silurian age are known from the Thomson Fold Belt (cf. Day et al., 1983). However, further north, in the Broken River and Hodgkinson basins (Fig. 8), sedimentation from Late Silurian time on was controlled by major fault zones. These fault zones mark the hinge line between the craton (Georgetown and Yambo inliers; Fig. 9) to the west and the subsiding basins to the east and southeast (de Keyser, 1963). Shallow marine facies in the Broken River Basin have been described from the western part (Arnold and Henderson, 1976; Day et al., 1983). Limestone blocks and lenses in the generally deep-water eastern part represent either exotic blocks or deposits on transient volcanic highs (Arnold and Fawckner, 1980). In the Hodgkinson Basin (Fig. 8) shallow marine facies were deposited in the west, while turbidites dominated in the east. These sediments contain a large proportion of silicic to intermediate detritus, possibly derived from a Siluro-Devonian, Andean-type volcanic chain which lay to the west, on the edge of the old craton. This volcanic chain has been eroded, leaving only its roots preserved as a line of plutons of Devonian age (Fig. 9). If this interpretation is correct, a relatively simple convergent plate margin existed here, with the described basinal sediments partly representing a fore-arc basin (cf. Henderson in Henderson and Stephenson, 1980).

New England Fold Belt Region During the Mid- to Late Silurian

In the region further east, in the present New England Fold Belt in Queensland, Late Silurian calc-alkaline volcanics, volcanic sediments, limestone, and chert crop out in a series of isolated fault blocks in the N.S.W. border region

(Day et al., 1983). According to Marsden (1972) these rocks represent remnants of a volcanic island arc which has been called the Calliope Island Arc by Day et al. (1978), and is referred to here as Calliope Volcanic Arc (Fig. 8). The main development of this arc occurred during the Devonian. It is possible that the arc continued further south into N.S.W., and because of subsequent displacement, is correlative with rocks west of the Tamworth belt (Fig. 9). In N.S.W., east of the probable continuation of the volcanic arc, is the Woolomin Formation, a sequence dominated by chert and basic volcanics, with fault-bounded slices of limestone and serpentinite, which has been interpreted by most authors as a subduction complex, and herein is shown as the Woolomin accretionary prism (Fig. 8). Detailed descriptions are provided by Cawood (1980). It has been suggested (Marsden, 1972) that the Siluro-Devonian Calliope Volcanic Arc was separated by a marginal sea from the regions to the west (Anakie Inlier in Queensland). A similar marginal basin (Murruin Basin) possibly existed in the south (Scheibner, 1974a).

Tectonic Models for the Mid- to Late Silurian

It is perhaps premature to suggest a tectonic model which would satisfactorily explain all different tectonic settings in eastern Australia as deduced from scattered and often poor quality data.

Recently Powell (1983b, 1984b) proposed a simple and constant plate geometry, which puts southeastern Australia into regional dextral shear, to explain the transition from Late Ordovician to Silurian palaeogeography. He envisaged southwestward-dipping subduction of an eastern oceanic plate after Early Ordovician time, and ascribed cessation of andesitic volcanism and onset of deformation during the Middle Ordovician in Queensland, to oblique convergence. The Late Ordovician Benambran Orogeny was considered to be the result of a change from Mariana- to Chilean-type subduction. During the Silurian, dextral transtension operated generally south of the Darling River Lineament, with westward-dipping subduction in the Broken River-Hodgkinson region (Powell, 1984b, Fig. 216). This plate interaction was compared with that of Neogene western North America, which features the Basin and Range structural province. There are some problems with this model. The Basin and Range province is characterized by relatively thin lithosphere, and has continental facies and bimodal volcanics, whereas supposed Australian Paleozoic analogues had marine sedimentation, with over 5 km of turbiditic fill in most troughs. This indicates crustal/lithospheric differences between the two regions, but may not be a serious problem.

According to Powell (1983, 1984b) the Tumut Trough (Fig. 8), a pull-apart basin, opened and subsequently closed by dextral transtension and transpression respectively. Wyborn (1977) conversely suggested that this trough opened as a sinistral pull-apart basin. Closure occurred by dextral transpression during latest Silurian time.

According to Scheibner (1972b, 1974a, 1976) major palaeogeographic changes south of the Darling River Lineament were the consequence of stepping-out of the subduction zone and the development of a new zone to the east. The problem is that the first evidence for this new subduction arrangement, the Calliope Volcanic Arc, commenced in the Late Silurian, while the earliest extensional basins in the Lachlan region had started to form in the mid-Silurian. (The older sediments in the area of the Cowra Trough were related to an earlier inherited paleogeography (cf. Pickett, 1982)).

A model in which the Calliope Volcanic Arc was the frontal volcanic arc would put all the Lachlan region in the back-arc position. If subduction was oblique, besides extension due to rollback of the subduction zone, a transtensional component would be generated, and such a tectonic regime is indicated by available data for the region east of the Gilmore Suture. Insufficient structural data, and later, Devonian to Carboniferous, deformations which have strongly distorted the original shape of the mid- to Late Silurian extensional troughs, make it difficult to deduce the sense of transtension.

It is not known if the subduction zone in the Hodgkinson-Broken River region was in continuity with that in the New England region, and the two were perhaps only separated by a transform fracture zone (in Henderson and Stephenson, 1980), or if there were two different zones. Subduction in the Hodgkinson-Broken River region could be related to subduction of a marginal sea west of the New England region, and this sea could have had a continuation further south. However, I cannot offer convincing evidence to support this suggestion. Such subduction would help to explain the wide zone of igneous activity in the Lachlan region. Although part of the igneous activity there was clearly associated with rifting, it is difficult to explain the widespread S- and I-type granitoids emplaced during the Late Silurian and Devonian. One would expect crustal melting in the troughs, which obviously had the highest heat flow, yet granitoids were emplaced in the highs and not in the troughs until their closure and the subsequent metamorphism of the fill.

Tasmanian Segment During the Early to Middle Devonian

In Tasmania, the Silurian tectonic setting continued into the Early Devonian. Sedimentation west of the Tamar fracture zone was shallow marine to continental, while turbiditic sedimentation characterized the Mathinna beds to the east (Fig. 9; Williams, 1978).

During the Middle Devonian there was a major

tectonic event that can be correlated with the Tabberabberan Orogeny on the mainland. This event is quite well constrained by stratigraphic evidence supported by isotopic dating (cf. Williams, 1978), and occurred between middle Early Devonian and the late Middle Devonian time. Perhaps the most important aspect of this deformation is the direction of tectonic transport, being from the northeast, west of the Tamar fracture zone, but from the southwest in the area to the east. This area east of the Tamar fracture zone is one of the probable allochthonous terranes in the Tasmanides (Fig. 3; Scheibner, 1983b Baillie, 1984). Some foliated granitoids were emplaced during this tectogenic event, but post-kinematic granitoids of Late Devonian and Early Carboniferous age were more widespread.

Early to Middle Devonian Between the Gambier-Tamar Fracture Zone and the Clark River Fault

Major changes occurred in this region during the earliest Devonian (compare Figs. 8,9). Immediately to the east of the Gilmore Suture, the Tumut Trough in N.S.W. (Basden, 1982; Crook and Powell, 1976) and the Cowombat Rift in Victoria (VandenBerg and Wilkinson in Cooper and Grindley, 1982) were closed and inverted, their fill was deformed and metamorphosed, and the Silurian granites were unroofed during the latest Silurian to earliest Devonian Bowning event (cf. Packham, 1969).

At about the same time, to the west of the Gilmore Suture, formation of the Mount Hope trough (Fig. 9) was accompanied by rift-related volcanism along much of its eastern margin and in the adjacent Mineral Hill Trough. Further west, in the region of the Kanmantoo Fold Belt in northwestern N.S.W., the rift-related volcanism of the Cootawundy beds developed (Fig. 9). Possible rift-related volcanism (rhyolite) also occurs in the Grampians in western Victoria (Fig. 9). The Adavale Basin, and the eastern side of the Anakie Inlier in the Thomson Fold Belt in Queensland, also contain Early Devonian volcanics.

South of the Darling River Fracture Zone felsic volcanics predominate. Minor basic rocks in the Mount Hope Trough reflect bimodal rift volcanism. To the north of the Darling River Lineament, mafic intermediate volcanics (basaltic andesites) occur, as well as felsic rocks in the Cootawundy beds (G. Neef, pers. comm., 1984). Mafic andesites present in the Louth block of the Darling Basin around Louth and Bourke (encountered during mineral exploration drilling) have been previously ascribed to the Ordovician; however, an indeterminate bryozoan fauna of younger aspect is associated with them (J.W. Pickett and D.W. Suppel, pers. comm., 1984). Continental mafic andesitic and felsic volcanics (Gumbardo Formation, (Galloway, 1970)) also occur in the Adavale Basin and on the eastern side of the Anakie Inlier (cf. Day et al., 1983).

Close to the Darling River Fracture Zone the rifts appear to be symmetrically arranged, trending northwest-southeast to the north and northeast-southwest in the south (Fig. 9). The north-northwest trending Mineral Hill Trough, which lies east of the Gilmore Suture, is an exception to this pattern.

During Early Devonian time, basin formation in the Mount Hope Trough, rifts, and pull-apart structures appear to be associated with dextral master faults, while a sinistral sense is indicated for the Mineral Hill Trough (Scheibner, 1983a; R.A. Glen, D.J. Pogson and others, in prep.). Intensive volcanism in the Darling Basin occurred south of the Lachlan River Fracture Zone (Fig. 8), (Scheibner and Stevens, 1974) and to the north of the Darling River Fracture Zone. By way of contrast, volcanism in the Mineral Hill Trough occurred between these lineaments. In the eastern part of the Darling Basin, mainly in the Cobar Basin, and less so in the Mount Hope and Rast troughs, large thicknesses (locally over 7 km) of mainly turbiditic sediments accumulated during the Early Devonian, while shallow-water facies dominated elevated blocks and areas to the west (Pogson and Felton, 1978; Glen, 1982b, in prep.).

The Cobar Basin and Mount Hope and Rast troughs (Fig. 9), despite large thicknesses of sedimentary fill, show generally positive Bouguer anomalies, perhaps reflecting crustal thinning during basin formation, and underplating by denser material (Scheibner, 1983a). This interpretation has been used to extend the suspected rifts towards the southwest (Fig. 9) into areas concealed beneath the Late Carboniferous to Cenozoic platform cover.

The Grampian Graben or Rift was filled by over 6 km of molassic fluviatile-lacustrine sediments with thin shallow marine intercalations. These sediments rest on the Wickliffe and Rocklands Rhyolites (Spencer-Jones, 1965), which are assumed to be related to rifting and graben formation. Early to Middle Devonian granitoids (Richards and Singleton, 1981) intrude the sequence, which was folded during the Early or Middle Devonian time.

The Cootawundy beds in northwestern N.S.W. comprise continental sediments with shallow marine intercalations containing plant fragments, trilobite tracks, and other trace fossils. These rocks are intruded by dykes and sills of andesite, rhyolite, and quartz porphyry. A few andesitic lava flows have also recently been mapped by G. Neef (pers. comm., 1984).

Small outcrops of Early to Middle Devonian volcanics (andesites predominate over minor rhyolites) and shallow marine sediments occur east of the Anakie Inlier (cf. Day et al., 1983). Similarly, in the Adavale Basin, Early Devonian shallow marine sediments conformably overlie volcanics of the Gumbardo Formation (C.G. Murray, pers. comm., 1985). Marine transgression continued into the Middle Devonian, followed by late

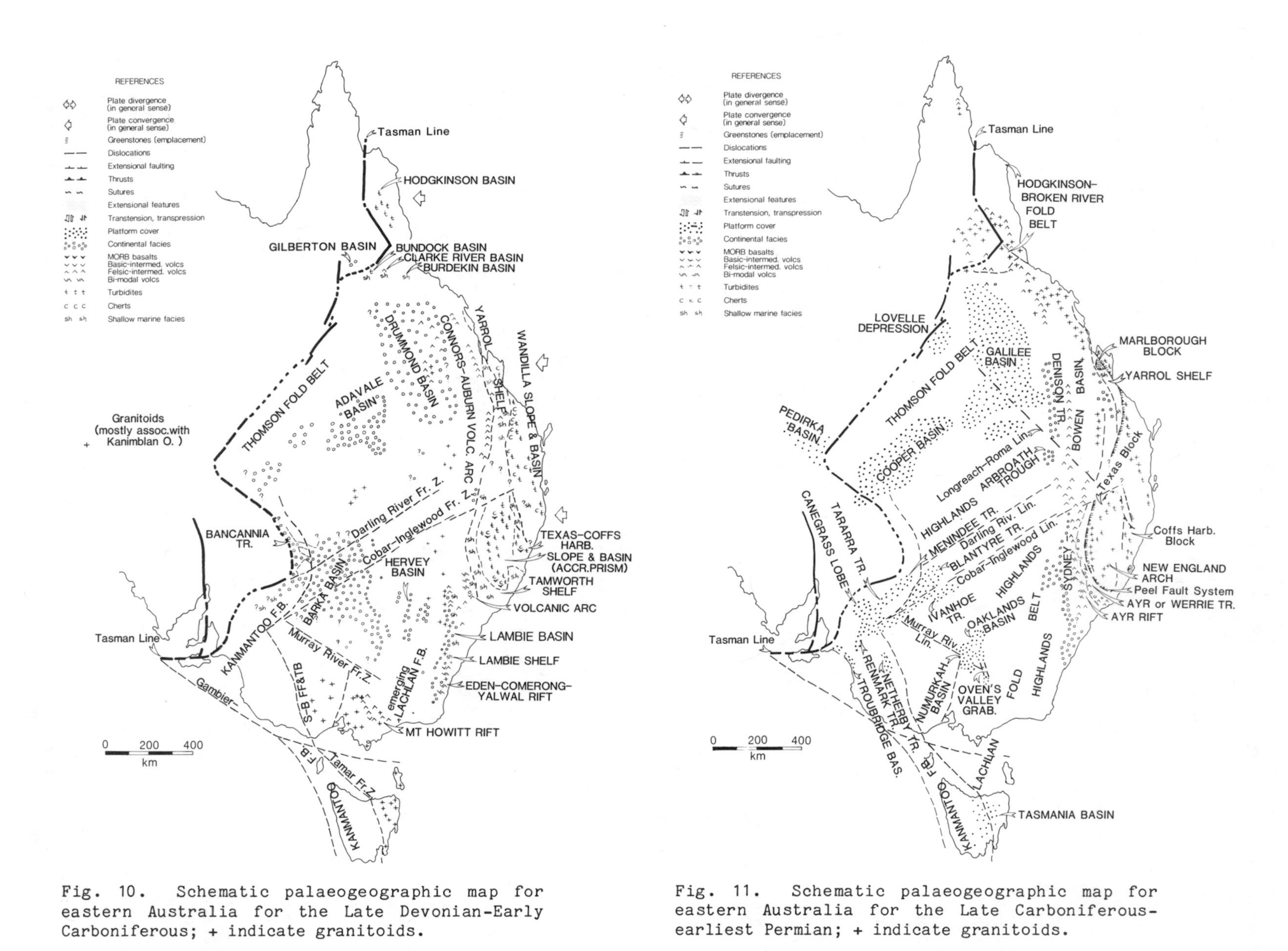

Fig. 10. Schematic palaeogeographic map for eastern Australia for the Late Devonian-Early Carboniferous; + indicate granitoids.

Fig. 11. Schematic palaeogeographic map for eastern Australia for the Late Carboniferous-earliest Permian; + indicate granitoids.

Middle Devonian regression (cf. Day et al., 1983). Both these areas were folded during the late Middle Devonian Tabberabberan event, and this folding was accompanied by granite emplacement in the Anakie Inlier region.

Early Devonian granites comagmatic with the rift volcanics occur in the Darling Basin (Mount Hope Trough (cf. Barron et al., 1982)), while the tectonic setting of the Tibooburra Granite in northwestern N.S.W. (410 Ma) (S. Shaw in Cooper and Grindley, 1982) is unknown.

South of the Murray River Lineament in the Melbourne Trough, facies changes are the only reflection of the Bowning event, sedimentation there being continuous from Silurian to Devonian time. Marine facies were replaced by fluviatile facies in Middle Devonian. The total thickness of Ordovician to Middle Devonian fill is about 10 km. Tight folding during the Tabberabberan event occurred in the Middle Devonian (VandenBerg, 1978).

To the east of the Gilmore Suture, the Silurian palaeogeography continued without much change into the Early Devonian and lasted generally into the Emsian. Plutonic activity increased during the Devonian, with emplacement of large I-type batholiths east of the S-I line during the Middle Devonian (White and Chappell, 1983). The southern part of the Lachlan region, as well as experiencing plutonism, was also strongly affected by folding and strike-slip faulting in the Middle Devonian Tabberabberan event (Packham, 1969), giving rise to the Tabberabberan Highlands which became the source of molassic clastics (Powell, 1984b). While this event had diminished effects towards the north and west in N.W.S. (Powell and Edgecombe, 1978; Powell and Fergusson, 1979; Powell et al., 1980; Glen, 1982a), pre-cratonic facies were replaced by transitional tectonic molassic facies everywhere in the Lachlan region.

Early to Middle Devonian in the Hodgkinson-Broken River Region

Generally, the palaeogeography of Silurian time continued into the Devonian in the Hodgkinson-Broken River region. Turbiditic sedimentation continued into Late Devonian time in the Hodgkinson Basin, while in the Broken River Basin, as in the Lachlan region, marine sedimentation was replaced by dominantly continental sedimentation with brief marine incursions after the Tabberabberan event (Day et al., 1983). Large granitoid batholiths were emplaced in adjacent cratonic areas, in the Coen, Yambo and Georgetown inliers and in the Lolworth-Ravenswood Block (Fig. 9; Richards, 1980).

Early to Middle Devonian in the New England Region

The palaeogeography seen in the Silurian was similar in the Devonian in the New England region but, in addition to the Calliope Volcanic Arc and its probable continuation in N.S.W., the Baldwin Volcanic Arc (named by Veevers et al., 1982), sedimentation characterized by volcanic detritus occurred in the Yarrol and Tamworth fore-arc basins (Fig. 9; Crook, 1980; Day et al., 1978; Leitch, 1974, 1975). There appears to be a change from intermediate to felsic rocks with time (Crook 1964), probably reflecting the maturing of the volcanic arc.

Pelagic and turbiditic sediments and basic volcanics of ophiolitic character to the east of the fore-arc basins are interpreted as the Wandilla and Cockburn Accretionary Prisms (Fig. 9; Wandilla Slope and Basin of Day et al., 1978; former Woolamin Slope and Basin of Scheibner, 1972b, 1974b).

A late Middle Devonian event, which correlates with the Tabberabberan Orogeny in the Lachlan region, affected the Calliope Volcanic Arc. The marginal sea which separated the New England region from the rest of the Tasmanides, in the west, was closed not only in Queensland but also in N.S.W. (Murruin Basin). Some granitoids were emplaced in the Calliope Volcanic Arc. Local unconformities and disconformities (cf. Korsch and Harrington, 1981) are the expression of this event in the Tamworth belt.

Tectonic Models for the Early to Middle Devonian

The tectonic models previously discussed for Late Silurian, time also apply to Early to Middle Devonian time. The Middle Devonian Tabberabberan Orogeny (Tuhua event in New Zealand; Sporli, this volume) may have been caused by collision of the submerged Lord Howe block, which is possibly Precambrian, with regions to the east (Solomon and Griffiths, 1972). The collision resulted in the progressive emergence of the Lachlan Fold Belt, and gave rise to widespread subsequent molassic deposits in late orogenic transitional tectonic basins. The onset of molassic sedimentation was diachronous, and progressed eastwards from the foreland towards the zone of collision, suggesting that the convergent movement of the Australian plate played an active role in the collisional deformation.

The convergent margin in the New England region continued into the Late Devonian, but this region ceased to act independently until Early Permian time, as it was accreted to the rest of the Tasmanides.

Powell (1984b) suggests that an eastward-migrating heat source under the Lachlan region can explain the distribution of orogenic granitoids. However, while the general easterly younging is valid for the eastern part of the Lachlan region, opposing trends are present in Central and Western Victoria and elsewhere. Increase in crustal thickness due to the possible Tabberabberan collision probably contributed to the associated plutonism, but no widely accepted model for the genesis of igneous activity has been put forward.

Late Devonian-Early Carboniferous in the Kanmantoo and Lachlan Fold Belts

Late Devonian to Early Carboniferous post-kinematic granitoids in Tasmania truncate the Tabberabberan folds, and no sediments of this age are known from Tasmania. Similarly, no sediments or igneous rocks of this age have been described from the Kanmantoo Fold Belt in western Victoria or southeastern South Australia. However, in northwestern N.S.W. and eastern South Australia, major subsidence occurred and resulted in accumulations of over 4 km of molassic continental clastics, including red beds (Bancannia Trough) (Fig. 10). These continental clastics are similar to the coeval clastics in the Lachlan Fold Belt (Packham, 1969; Evans, 1977).

Across the Lachlan Fold Belt and to the west, Late Devonian to Early Carboniferous molassic rocks, mainly continental clastics with some red beds, and minor marine rocks in the east, are preserved in superposed synclinoria (Fig. 10; Conolly, 1969; Scheibner, 1976). Controversy exists as to whether they formerly constituted a continuous sedimentary blanket (Powell, 1984b) or were isolated intramontane basins (Conolly 1969a, b; Cas, 1983). Onset of sedimentation was diachronous, and this to some extent defines the following three main basinal areas (Fig. 10): the Barka Basin in the west (formerly the Ravendale Terrestrial Basin of Scheibner (1974a), (Glen et al., in prep.), the Hervey Basin in the centre (Conolly, 1965a,b), and the Lambie Basin further southeast (Conolly, 1969a,b). The greatest thickness of sediment, 4.7 km, has volcaniclastic rocks in its upper part which were derived from a volcanic source to the east, possibly the frontal arc in New England (Powell and Fergusson, 1979).

Sedimentation in the Barka Basin started in latest Early to early Middle Devonian time (Glen, 1982a), while in the east sedimentation started in the Late Devonian (Frasnian) (Conolly, 1969a, b; Roberts et al., 1972). The upper limit of the sequence is Early Carboniferous (Evans, 1977). Sedimentation was preceded on the South Coast of N.S.W. during late Givetian or early Frasnian time by rifting associated with volcanism which resulted in the Eden-Comerong-Yalwal Rift (McIlveen, 1974). The bimodal volcanics in it are intercalated with continental and marine strata (Steiner, 1975; Fergusson et al., 1979). Associated with the rifting are A-type granitoids (Collins et al., 1982). Another major area of volcanic rifting was in central-eastern Victoria (Marsden, 1976), especially in the Mount Howitt Trough or Rift, which is a graben structure in which over 4 km of bimodal, dominantly felsic, volcanics and continental sediments accumulated. A wide region in Victoria featured igneous activity, with cauldron subsidence features described by Hills (1959).

The Kanimblan Orogeny during the Early Carboniferous caused metamorphism, folding, and faulting, and converted the Lachlan Fold Belt into a neocraton (Packham, 1969; Powell, 1976).

Late Devonian-Early Carboniferous in the Thomson Fold Belt

The Middle Devonian Tabberabberan Orogeny in the Thomson Fold Belt was followed by widespread deposition in transitional tectonic downwarps. Volcanics, rare granites, and continental minor marine sediments occur in the Drummond Basin (Olgers, 1972). Continental sediments, including red beds, conformably overlie Middle Devonian rocks in the Adavale Basin.

The Burdekin basin in northeastern Queensland was the site of oscillating shallow-marine and continental sedimentation followed by gentle warping and emplacement of granite in Middle Carboniferous time.

In the subsurface below the Bowen Basin (Fig. 13), in the Roma area of southern Queensland, possible Late Devonian sediments are intruded by Late Devonian granitoids, and similar granites occur at Eulo in southwestern Queensland (Day et al., 1983). The tectonic setting of these rocks has not been explained yet.

The Kanimblan event converted the Thomson Fold Belt into a neocraton during the Early to Middle Carboniferous (Murray and Kirkegaard, 1978).

Late Devonian-Early Carboniferous in the Hodgkinson-Broken River Region

Major changes occurred in the Hodgkinson-Broken River region. In the Broken River region two new basins developed on the former Broken River Basin. The Clarke River Basin contains dominantly continental sediments with shallow marine strata of Early Carboniferous age which unconformably overlie Siluro-Devonian sediments of the Broken River Basin. The Bundock Basin, contains dominantly continental clastics with marine intercalations of Early Carboniferous age in the higher part of the sequence, which disconformably overlie the Siluro-Devonian sediments of the Broken River Basin (cf. Day et al., 1983). Sediments of both these basins were deformed during the late Early Carboniferous (Henderson and Stephenson, 1980).

In the Hodgkinson Basin (Fig. 10) turbiditic facies sedimentation continued into the Late Devonian, but at the end of the Devonian strong folding and fault movement began. This event continued into the Carboniferous causing multiple deformation and metamorphism (cf. Henderson and Stephenson, 1980) and was followed by the emplacement of post-kinematic granitoids of mid-Carboniferous and younger age (Richards, 1980).

Late Devonian-Early Carboniferous in the New England Region

Subsequent to the Middle Devonian Tabberabberan Orogeny, an Andean-type continental volcanic arc developed along the western margin of the New England region. In Queensland it has been named the Connors-Auburn Volcanic Arc (Fig. 10; Day et al., 1978), while its continuation in N.S.W., displaced by subsequent dextral transcurrent movement, remains unnamed and concealed beneath the Sydney-Bowen Basin. As noted above Powell, (1983a, 1984b) suggests that this arc continued further south, just east of the Lambie Basin, and was the source of the volcanolithic detritus for the upper part of the sequence.

In northern parts of the arc, massive andesitic flows predominate over felsic rocks and locally abundant basalts. In the southern part, felsic rocks predominate, with locally abundant andesites (Marsden, 1972; Day et al., 1978). Volcaniclastic detritus derived from the N.S.W. portion of the arc indicates that there it was of andesitic to dacitic character (Crook, 1964; Powell, 1984b).

To the east of this volcanic arc, the fore-arc basin had the character of an unstable shelf. In Queensland, the name Yarrol Shelf, the same name as that used for the Early-Middle Devonian fore-arc basin, has been employed (Day et al., 1978). This shelf developed over deformed rocks of the Late Silurian-Middle Devonian Calliope Volcanic Arc. In N.S.W., this fore-arc basin, again displaced by subsequent faulting, has been called the Mandowa Unstable Shelf (Scheibner, 1972b, 1974a), but most authors retain the name Tamworth Shelf from the earlier episode, as has been done in Queensland. In N.S.W., the new fore-arc basin was at least partly built over the earlier volcanic arc, and the gravity high in the Tamworth structural belt could be an expression of the earlier arc.

The sediments in the fore-arc basin are dominated by volcaniclastics derived from the arc, together with interbedded primary volcanics (Crook, 1964; Marsden, 1972). During the widespread marine transgression in Early Carboniferous time, characteristic oolitic limestone formed locally.

Eastward, the fore-arc basin passed into a continental slope and ocean basin environment which in Queensland is referred to as the Wandilla Slope and Basin. Day et al. (1978) used the name Woolomin Slope and Basin for its extension in N.S.W. However, as mentioned before, the name Woolomin is not appropriate and should be replaced by Texas-Coffs Harbour Slope and Basin to include the most typical region.

The distinctive rocks of this zone are cherts and metabasalts, but the sequence is mostly turbiditic volcaniclastic and quartz arenites, interbedded with massive argillites. Slump deposits are common. Fergusson (1982a,b) has described melanges and imbricate structures from the Coffs Harbour Block in N.S.W., and provided evidence that these rocks represent an accretionary prism. Redeposited oolites and Early Carboniferous fossil fragments from the fore-arc basin occur in some parts of this accretionary prism and help in stratigraphic correlations (Fleming et al., 1974).

The Early Carboniferous Kanimblan event, which so profoundly affected the Tasmanides, had a mild expression in the New England region, but major changes occurred during the Late Carboniferous. In the Yarrol Shelf in the north, in late Early Carboniferous time, a marine regression occurred, the area undergoing sedimentation decreased, and there was a change in provenance of the detritus from volcanic to plutonic (Roberts and Engel, 1980; Day et al., 1978). In N.S.W. felsic volcanism with rare andesites continued. In the fore-arc basin in N.S.W. shallow marine sedimentation was limited to the eastern part of the shelf, while elsewhere continental sedimentation predominated (Roberts and Engel, 1980).

Tectonic Models for the Late Devonian-Early Carboniferous

Most of the published tectonic models for the Late Devonian to Early Carboniferous (Day et al., 1983; Leitch, 1982; and Powell, 1984b), envisage a convergent margin in the New England region. The continental Connors-Auburn arc developed across the region, and the presence of a fore-arc basin and accretionary prism to the east of it indicates westward-dipping subduction.

The rest of the Tasmanides was characterized by transitional tectonism. There was widespread molassic sedimentation of predominantly continental facies with marine incursions from the east, and rift-related volcanic (bimodal) and plutonic (A-type) activity. This development occurred in a partly stabilized, wide, back-arc region which was dominated by stresses generated during convergence of the hypothetical Pacific and East Gondwanaland (or Australian) plates. It terminated during the late Early and Middle Carboniferous Kanimblan Orogeny. Structures formed during this event indicate general east-west compression, perhaps with some oblique component. However, thrusts in the Amadeus Transverse Zone (Fig. 1; Rutland, 1976) in the centre of the Australian continent, which formed in the generally coeval Alice Springs event, indicate north-south compression. Powell (1984a) describes major kink zones from the Tasmanides that apparently postdate east-west compressional structures. He interpretes the kink zones to be a result of continent-wide north-south compression, the same as that which affected the Amadeus Transverse Zone.

Late Carboniferous to Triassic Development

Late Carboniferous to Triassic Platform Cover of the Tasmanides

The Kanimblan Orogeny was followed in early Late Carboniferous time by continental and alpine glaciation (Powell and Veevers in Veevers, 1984). Subsequent deposition in the Kanmantoo, Lachlan, and Thomson Fold Belt regions occurred in plat-

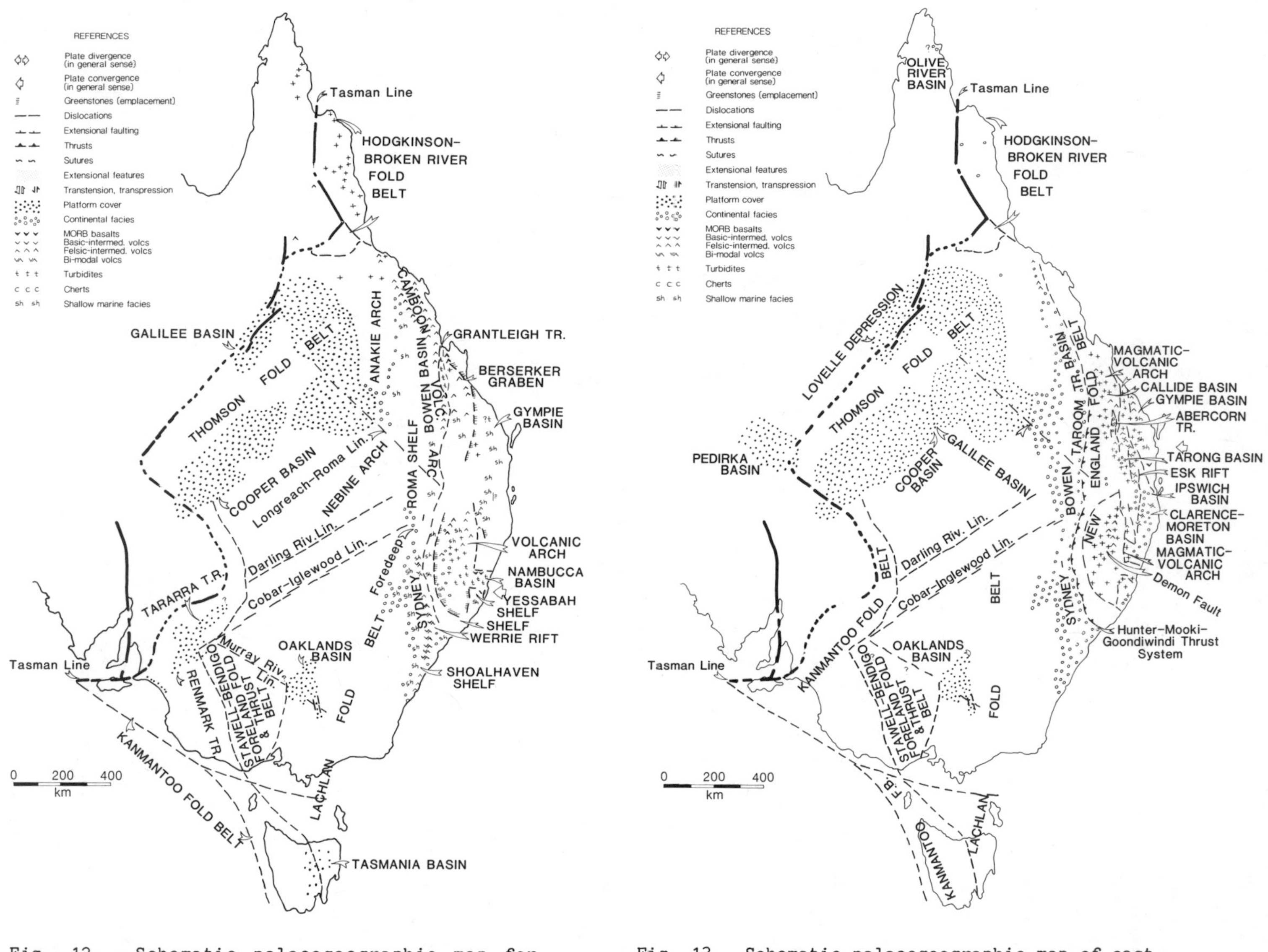

Fig. 12. Schematic palaeogeographic map for eastern Australia for the Early-middle Permian; + indicate granitoids.

Fig. 13. Schematic palaeogeographic map of eastern Australia for the middle Permian-Triassic; + indicate granitoids.

formal epicratonic basins during Late Carboniferous-Early Permian and Permian-Triassic time (Veevers, 1984). Continental to marine glacigenic sediments occur in the Galilee and Cooper basins and Lovelle Depression in Queensland (Fig. 11), and in basins mostly concealed beneath the Cenozoic Murray Basin in the three southeastern states. Some outcrops of glacigene sediments occur in South Australia and Victoria, and in the Tasmania Basin (cf. Doutch and Nicholas, 1978; Veevers, 1984).

Late Carboniferous-Triassic Post-kinematic Igneous Activity in the Tasmanides

Subsequent to the Kanimblan event, post-kinematic I-type granitoids (Bathurst-type of Vallance (1969)) were emplaced in the north-eastern part of the Lachlan Fold Belt. Locally some volcanics (Rylstone Tuff) appear to have been associated with them. Aeromagnetic data indicate that these granites occur in the sub-surface beneath the Meso-Cenozoic cover further north towards the Cobar-Inglewood Lineament (A. Agostini, pers. comm., 1984).

An even larger area of post-kinematic igneous activity, comprising large granitic batholiths and comagmatic eruptives, is present in north-eastern Queensland. These rocks occur in a wide belt extending from the northern part of the Thomson Fold Belt, through the Hodgkinson-Broken River Fold Belt and its western foreland, which comprises the Georgetown, Coen, and Cape York-Oriomo inliers (Richards, 1980; Day et al. 1983). Large cauldron subsidence areas typically developed, accompanied by continental volcaniclastic sediments (Branch, 1966; Oversby et al., 1980). The tectonic setting of these igneous rocks may be similar to those of the Lachlan Fold Belt or the New England region.

Sydney-Bowen Basin

During the latest Carboniferous and earliest Permian a foreland basin developed, that in Queensland is called the Bowen Basin, in northern N.S.W. the Gunnedah Basin, and further south the Sydney Basin (Fig. 11). These aligned basins are collectively referred to as the Sydney-Bowen Basin (Figs. 11,12) (Bembrick et al., 1973).

Based on differing styles of deposition and deformation, the Bowen Basin has been divided into a western and eastern area separated by a structural high (Hawthorne, 1975). The western area developed over the Thomson Fold Belt and is characterized by dominant continental sedimentary rocks and subordinate paralic and shallow-marine sedimentary rocks, which subsequently have been gently folded. By contrast, the basement of the eastern area is not known, and there sediments and volcanics, over 7 km thick, have been tightly folded and overthrust by the New England Fold Belt.

In the north, the oldest basin deposits are of latest Carboniferous age. The eastern part of the basin is dominated by andesitic volcanics and associated continental sedimentary rocks. These persist into the Early Permian. The tectonic setting of the volcanics of the Bowen Basin has not been established, but they could be related to dextral rifting during basin formation (Evans and Roberts, 1980). Continental, mainly glaciogenic, sediments filled the rapidly subsiding western part of the Bowen Basin (Day et al., 1983).

In N.S.W., the relationship between the foredeep and the New England orogen is complicated. The foredeep was superposed on the frontal volcanic chain, and a close relationship is suggested with the fore-arc basin area. In the Gunnedah Basin, volcanism was widespread during the latest Carboniferous and earliest Permian, and there is good evidence for rift-related bimodal volcanism (Boggabri Volcanics and Werrie Basalt) (Runnegar, 1970).

In the northeastern part of the Sydney Basin, a northwest-trending depression developed during the latest Carboniferous, and fluvio-glacial sediments were followed by marginal marine sediments and interbedded volcanics (Mayne et al., 1974). In the southern part of the basin, glaciogene and fluvio-glacial clastics were deposited in broad valleys, probably draining glaciated highlands to the west (Herbert, 1972).

During the Early Permian a major transgression occurred which, according to Evans and Roberts (1980), was related to eustatic sea level rise. This transgression resulted in widespread marine sedimentation in the Sydney-Bowen Basin. At the basin margin, coal measures and other continental facies interfingered with marine facies.

In the Bowen Basin the axis of deposition shifted from the west (Denison Trough) to the east with time (Malone, 1964; Dickins and Malone, 1973). In the Sydney Basin the rate of sedimentation increased rapidly east of a north-south trending hinge line (cf. Bembrick et al., 1973; Herbert, 1980).

During mid-Early Permian time (Fig. 12) a major orogenic event occurred to the east in the New England Fold Belt, and was expressed as a regression in the Sydney-Bowen Basin (Evans and Roberts, 1980). Connection between the two basins was interrupted (Runnegar, 1970), and widespread coal measures accumulated. This regression was followed by another transgression during late Early Permian time which reflected another eustatic sea level rise (Evans and Roberts, 1980). During the Late Permian, marine regression was followed by deposition of coal measures, and there was deposition of continental (mainly fluviatile) to marginal marine facies during the Triassic. In the Bowen Basin the main subsidence area was in the southeast (Taroom Trough, Fig. 13), while in N.S.W. the Sydney Basin was the main depocenter (Herbert and Helby, 1980; Mayne et al., 1974).

The Sydney-Bowen Basin was deformed during Late Triassic time. Structures indicate an increase in intensity of deformation eastwards towards the orogen, which overthrust its foredeep along the Hunter-Mooki-Goondiwindi Thrust System (Fig. 13; cf. Bembrick et al., 1973), the northern continuation of which is called the Burunga-Leichhardt Thrust (Exon, 1974; Thomas et al., 1982).

Late Carboniferous-earliest Permian in the New England Region

In Queensland during the late Early Carboniferous, volcanism waned in the Connors-Auburn Volcanic Arc. During the latter part of the Late Carboniferous, extensive granitoid batholiths were emplaced (Fig. 11) (cf. Richards, 1980). Sedimentation in the fore-arc basin (Yarrol Shelf) decreased and the provenance of detritus changed from volcanic to terrigenous. According to Evans and Roberts (1980) the whole region was affected by dextral regional shear, and the area of the former Wandilla Slope and Basin was uplifted in the north, while further south some areas were depressed and became the site of Late Carboniferous deposition of argillites and cherts interbedded with basic volcanics. This shearing was followed in southeastern Queensland by the Late Carboniferous event, which caused deformation, metamorphism, and emplacement of granitoids (Green, 1973; Day et al., 1978).

During the early Late Carboniferous in N.S.W., volcanism continued in the area of the previous volcanic arc, but changed character. Widespread ignimbrites, are interpreted (Scheibner, 1974a, 1976) as reflecting volcanic rifting (Ayr Volcanic Rift). Towards the east, close to the volcanic chain, sedimentation comprised continental and marginal marine facies, while elsewhere marine facies dominated (Roberts and Engel 1980). Deposits of mainly continental facies during the Namurian (mid-Carboniferous) accumulated in a narrow trough called the Ayr Basin (Fig. 11; Scheibner, 1974a, 1976) or Werrie Basin (Evans and Roberts, 1980).

According to Roberts and Engel (1980), during the Namurian the central part of the New England region was uplifted to form the New England Arch of Campbell (1969). It is possible that the Peel Fault (Fig. 11) developed at the western margin of the composite accretionary prism during this time, or, alternatively, during the later part of Late Carboniferous, when the folding and metamorphism of the accretionary prism in the Woolomin-Texas and Coffs Harbour Blocks occurred.

The steeply east-dipping Peel Fault System had a complex history, including multiple upthrust and strike-slip movements (Crook, 1963). It is probably an old suture, as indicated by the diverse slices of early to late Paleozoic rocks which occur along it, and, it separates the composite accretionary prism and fore-arc basin complexes. Multiple (Carboniferous and Permian) emplacement of serpentinites and other ultramafic and mafic rocks occurred along this structure. The Carboniferous age for early serpentinite emplacement originally suggested by Benson (1913) is supported by dating of nephrite associated with serpentinite south of Tamworth at 279 and 286 Ma, recalculated to new constants (Lanphere and Hockley, 1976).

Following Late Carboniferous deformation, S-type granitoid plutons of the Bundarra (286 Ma), and Hillgrove (289 Ma) suites were emplaced into the deformed accretionary prism rocks (Shaw and Flood, 1981). The Bundarra suite forms a belt parallel with the Peel Fault System and is probably related to thrusting along this suture (Shaw and Flood, 1981).

According to Flood and Fergusson (1982), the structures and facing of the accretionary prism complexes in the Coffs Harbour and Texas Blocks indicate large-scale oroclinal bending, which probably was caused by southeastward dextral displacement of the southern part of the Queensland segment of the New England Fold Belt (Murray and Whitaker, 1982). This bending was accentuated by subsequent Late Permian and Triassic dextral displacement along the Demon Fault (Fig. 13; Korsch et al., 1978).

The timing of this deformational event as Late Carboniferous to earliest Permian is further supported by the presence of unconformities at the base of the Permian sediments (Olgers and Flood, 1970; Runnegar, 1970; and others). Either immediately succeeding, or perhaps contemporaneously with this orogenic event, an extensional feature, the Nambucca Basin, started to form.

Early Permian in the New England Fold Belt

The NEFB underwent major modifications during the Early Permian (Fig. 12).

The volcanism which ranged from basaltic to felsic in character and which earlier was limited to the Bowen Basin, spread eastward over the Carboniferous Connors-Auburn Volcanic Arc and Yarrol Shelf. Day et al. (1978) referred to this volcanic chain as the Camboon Volcanic Arc. The tectonic setting is uncertain and could be either rift or subduction related. An elongate, deep marine trough (Grantleigh Trough) developed to the east of the central part of this volcanic chain, and was filled by 6.5 km of flysch-like sediments and spilitic pillow basalts (Malone et al., 1969). Further east, the former frontal-arc basin received volcanogenic sediments, and its eastern edge collapsed to form a graben (Berserker Graben), which was filled with 3 km of volcanics and volcanogenic sediments with a rich marine fauna (Kirkegaard et al., 1970). Further southeast, in the area of the present Gympie Block, an anomalous area of relatively shallow water fossiliferous Permian sedimentation occurs (Gympie Basin). The basal Highbury Volcanics of this basin have been correlated by Harrington

(1983) with volcanic arc rocks of the Brook Street Group in New Zealand.

Marine sedimentation of shallow marine facies encroached on all sides of the New England Arch (Fig. 11; Runnegar, 1970). To the east of the arch, flysch-like sediments and argillites and interbedded rare pillow basalts, were deposited in the Nambucca Basin. Some felsic volcanics, possibly rift related, occur in the western part of this basin.

During the middle Early Permian, large areas of the NEFB were affected by the Hunter event (Carey and Brown, 1938). In N.S.W., its strongest effects can be observed east of the Peel Fault System. Deformation there was accompanied by shearing and regional metamorphism which locally reached amphibolite facies (Binns, 1966). Multiple deformations affected the fill of the Nambucca Basin (Leitch, 1978). Contemporaneous emplacement of serpentinites occurred along the Peel Fault System and other major dislocations. In the area west of the Peel Fault System, the Tamworth Synclinorial Zone or Belt developed during deformation of the fore-arc basin. It is characterized by west-verging folds and thrusts (Voisey, 1959), and by burial metamorphism. The whole belt was thrust over the Sydney-Bowen Basin (Carey, 1934).

In Queensland a large region between the Bowen and Gympie Basin was uplifted. The Grantleigh Trough (Fig.12) was deformed and overthrust westward to form the Gogango Overfold Zone (Olgers et al., 1964). Less severe deformation occurred further north, which resulted in the Strathmuir Synclinorium. Along the Yarrol Fault System, which occupies a position similar to the Peel Fault System, an ophiolitic complex (Marlborough Block), which is comprised predominantly of serpentinite, was thrust westward over the fore-arc basin complexes (Yarrol Block) and the Gogango Overfold Zone (Murray, 1974). The area west of the Yarrol Fault System, displays open folds and high-angle reverse faults or small thrusts similar in style to those west of the Peel Fault System in N.S.W., and the Berserker Graben was also folded. There is however, no evidence of folding in the Gympie Basin during this event, and marine sedimentation was continuous here in the Permian (Day et al., 1978).

The Hunter event was followed by emplacement of granites dated at 260-269 Ma. Remnants of late Early Permian marine sediments in the NEFB indicate that during the marine transgression which effected the foredeep most of the fold belt remained emergent. These marine sediments occur in southeastern Queensland and northeastern N.S.W.

Late Permian-Late Triassic in the New England Fold Belt

Sedimentation during the Late Permian (Fig. 13) was limited to the Gympie Basin, while widespread igneous activity occurred throughout the remainder of the NEFB. This igneous activity forms a Late Permian to Triassic magmatic arc.

In N.S.W., calc-alkaline, mainly felsic, volcanics are often associated with cauldron subsidence structures (McPhie, 1982) which are intruded by subvolcanic granitoids of post-kinematic A- and I-type character (Shaw and Flood, 1981; Korsch and Harrington, 1981). Similar granitoids of Late Permian to Triassic and occasionally Early Jurassic ages occur throughout the NEFB and even further north in the Hodgkinson-Broken River Fold Belt (Richards, 1980).

During the Late Permian, deformation affected large areas of the NEFB and is well documented in an area of marine sedimentation in the Queensland-N.S.W. border region (Thomson, 1973; Murray et al., 1981). Deformation again affected the Gogango Overfold Zone more stongly than elsewhere. The effects in the Gympie Basin are uncertain; according to Runnegar and Fergusson (1979) Permian and Triassic sediments are conformable, while an unconformable relationship between them has been suggested by Harrington (1983).

During the Early Triassic, marine sedimentation in the NEFB was limited to the Gympie region and even here it ceased during the late Early Triassic (Runnegar and Fergusson, 1969).

Rifting affected the NEFB during the Early-Middle Triassic, and the Esk Rift and Abercorn Trough (Fig. 13) formed, accompanied by volcanism. A large thickness (5 km) of volcanogenic sediments was deposited in the Esk Rift, and a lesser thickness (1.5 km) in the Abercorn Trough. At the end of the Middle Triassic, sedimentation in these rifts ceased and new basins formed: the Tarong, Callide, and Ipswich or Clarence-Moreton basins (Day et al., 1974). The Clarence-Moreton (Ipswich) Basin was filled by early rift volcanics, followed by deposition of coal measures (Day et al., 1974).

The magmatic arc continued to develop, mainly in Queensland (Richards, 1980). Further displacement of crustal blocks occurred mainly along north-south, strike-slip faults, one of the best documented being the dextral Demon Fault (Fig. 13) in northeastern N.S.W. (Korsch et al., 1978).

During the Late Triassic, the Bowen orogenic events occurred (Carey and Brown, 1938). Sedimentation terminated in the foredeep (Sydney-Bowen Basin), as well as in the NEFB, and the Gympie Basin was deformed (Day et al., 1974). After the Bowen event the NEFB behaved as a neocraton and subsequent sedimentation has platformal character.

Tectonic Models for Late Carboniferous to Late Triassic Development

After the Kanimblan (Alice Springs) Orogeny, most of eastern Australia behaved as a neocraton, with widespread platformal sedimentation commencing after the retreat of the continental and alpine glaciation (Veevers, 1984). An active plate margin setting was still present in the New England region, and post-kinematic igneous activity occurred along nearly the whole length of the present eastern coastal region.

Not enough attention has been devoted to the problem of basin formation in the platformal region of eastern Australia, and no consistent kinematic model has been proposed for the whole region. Some basins, like the Renmark and Tararra Troughs (Fig. 12), are parallel to major resurgent fracture systems (Darling River and Cobar-Inglewood Lineaments). Some, like the Oaklands Basin-Ovens Valley Graben, appear to be framed by basement faults. Others, like the Cooper Basin, appear to follow the structural grain in the basement which is well expressed in the gravity data. Evans and Roberts (1980) linked the formation of the platformal Galilee Basin and Lovelle Depression, the sub-basins in the foreland basin (Sydney-Bowen Basin), and the basins in the orogenic New England region (Ayr or Werrie Trough, Grantleigh Trough, Berserker Graben, and others) to a dextral shear couple. This shear couple produced different effects on either side of an inferred northwest-trending transform fracture zone, which is at present expressed as the Longreach-Roma gravity lineament.

Most of the published tectonic models for the development of the NEFB envisage an active plate margin setting, but interpretations differ in detail. Usually an Andean-type convergent margin is suggested for Carboniferous time, in contrast with an intra-oceanic arc for Late Silurian-Devonian time. Volcanism became rift-like in N.S.W., but waned in Queensland, to be followed there by widespread post-kinematic plutonism in both the arc and further north in the already consolidated Hodgkinson-Broken River Fold Belt and its foreland. It is not clear if this long-lasting plutonism was related to subduction or not.

During latest Carboniferous to earliest Permian time, a widespread deformational event resulted in the formation of the core of the New England Fold Belt. The composite accretionary prism (comprising early, middle and late Paleozoic complexes that contain high-pressure/low-temperature metamorphic rocks was thrust, on the Peel-Yarrol Fault System, westward over the fore-arc basin (Scheibner and Glen, 1972; Day et al., 1978). Nephrite associated with serpentinite emplaced along the Peel Fault Zone has been dated as earliest Permian (Lanphere and Hockley, 1976). In Queensland a flat-lying thrust sheet of oceanic crust and mantle forming the Marlborough Block (Fig. 11), was thrust westwards over the fore-arc basin (Murray, 1974). Some authors (Leitch, 1975; Crook, 1980) have compared this thrust to the Coast Range Thrust in California. However, a closer analogy exists with the obduction of the accretionary prism over the volcanic arc suggested by Kroenke and Dupont (1982) for the Three Kings Rise. Obduction in the Three Kings Rise region appears to have been caused by an oceanic plateau which collided with the rise and choked the subduction zone. If a similar collision occurred in the NEFB, the colliding block has not yet been identified.

The Bundarra Suite S-type granitoids are parallel to the Peel Fault System and appear to be related to thrusting (Shaw and Flood, 1981). Some authors (Cawood, 1980; Crook, 1980) have suggested that the formation of syn-kinematic granitoids in the accretionary prism rocks is due to subduction of a spreading centre causing high heat flow and partial melting. This problem remains unresolved.

Major displacement of crustal blocks bounded by strike-slip faults yielded the well-developed block structure of the New England Fold Belt. The southeastern Queensland region was dextrally displaced in a southeasterly direction (Murray and Whitaker, 1982) causing major oroclinal bending of the accretionary prism complexes, as described by Flood and Fergusson (1982) from the Texas and Coffs Harbour Blocks.

Subsequently or contemporaneous with the latest Carboniferous-earliest Permian event, the ensimatic Nambucca Basin (Fig. 12) formed in N.S.W. Tectonic interpretations of this basin are contradictory. Evans and Roberts (1980) suggested basin formation in association with dextral shear, while Cawood (1982, 1983) suggested a sinistral transtensional origin during an episode of terrane dispersion. Scheibner (1974a, 1976) interpreted the Nambucca Basin as a marginal sea formed by stepping out of the subduction zone, and suggested that deformation of the fill of this basin occurred during its closure by subduction. These rocks show multiple intensive deformations together with burial metamorphism (Leitch, 1978). Leitch (1982), however, argued that the rocks do not show the imbricate structure typical of accretionary prisms and the large-scale structure remains unresolved.

The short-lived Grantleigh Trough (Fig. 12) represents a similar, possibly ensimatic, basin in Queensland (Day et al., 1978). However, this basin developed on the western side of the NEFB, close to the foreland basin or foredeep. In Queensland another basin developed in an intermediate position in the NEFB, the Berserker Graben. During this time the Gympie Basin appears to have undergone independent development, and it has been considered to be an allochthonous terrane (Harrington, 1983).

The tectonic setting of the Permian Camboon Volcanic Arc (Day et al., 1978) in Queensland is uncertain. Initially volcanism developed in association with the rift formation of the foreland basin (Sydney-Bowen Basin). Subsequently this volcanism spread over the western part of the NEFB in Queensland, while retaining its rift character in N.S.W.

During the mid-Early Permian Hunter event the structure of the NEFB was further developed. The strongest deformation, including regional metamorphism, occurred in the central part of the NEFB. Generally, east-west compression resulted in further thrust movement on the Peel-Yarrol Fault System and in new emplacement of serpentinites along this and other major faults. The former fore-arc basin (Tamworth-Yarrol Belt) shows evidence of burial metamorphism that increases towards the inner part of the NEFB. The NEFB was thrust westwards over the basin along the Hunter-Mooki-Goondiwindi Thrust System (Fig. 13). Again interpretation of the general sense of displacements is controversial. The above event could have resulted from terrane collision and accretion in the region further east. Recently Cawood (1984) tried to correlate the tectonic development of the NEFB and the Rangitata Orogen in New Zealand.

Subsequent tectonic development is characterized by the formation of a magmatic arc and by disruptive volcanic rifting and basin formation typical of transitional tectonism. The magmatic arc appears to show zoning of associated metallogenic provinces (Weber and Scheibner, 1977) similar to that described from subduction-related arcs, and Cawood (1984) argued for a close relationship with plate convergence in the New Zealand orogenic region. The problem needs further assessment and analysis, and more data.

The igneous activity and deformations appear to migrate northwards with time, suggesting an unstable migrating plate configuration, but no clear explanation is at hand. The Permian to Early Triassic clastic, volcanic and carbonate Gympie terrane was accreted possibly during the Early Triassic, and the terminal deformation in the NEFB and its foreland occurred in the Late Triassic. Obviously the NEFB is not a complete orogen, and parts essential for tectonic interpretation remain hidden in the marginal plateaux and microcontinents of the southwest Pacific, produced during the Mesozoic breakup of this region (Sporli, this volume).

Concluding Remarks

Evidence from eastern Australia indicates that oceanic Pacific plates existed east of the continental margin in pre-Mesozoic time.

As far as direct evidence is concerned, well-documented ophiolites in eastern Australia are rare. The Coolac Ophiolite Suite (Ashley et al., 1979) has marginal sea character, and there are some other similar occurrences. MORB basalts from the NEFB appear to have formed close to a continental margin (Scheibner and Pearce, 1978), but it cannot be proven that they represent the remains of a major oceanic plate.

Indirect evidence points to the existence of an active plate margin in eastern Australia starting probably in the late Proterozoic. The region appears to have varied from a complex back-arc region similar to the present southwest Pacific, but on smaller scale, through an oblique zone of convergence perhaps like the present Sunda region, to fairly simple Andean-type continental margin. The eastern Australian active plate margin could have developed only by interaction with other major oceanic plates. The presence of allochthonous tectonostratigraphic terranes in eastern Australia needs to be proven by paleomagnetic and/or palaeontological evidence.

The magnitude and longevity of subduction-related igneous chains (intra-oceanic island arcs, continental margin arcs, and magmatic arcs or arches) should be proportional to the amount of subducted oceanic lithosphere. Eastern Australia has an abundance of igneous rocks, but it is impossible to prove which are related directly to B-subduction. For example, some syn-kinematic anatectic granites possibly formed as a result of thrust pile-up (A-subduction), and in zones of high-temperature/low-pressure metamorphism. Some igneous activity is related to rifting far away from possible coeval B-subduction, and because of a thin lithosphere, bimodal magmatic rocks may be calc-alkaline and not alkaline as in rift-valleys on old shields. During some intervals, like the Siluro-Devonian, orogenic granitoids were emplaced all over the Tasmanides, and it is unlikely that all of these granitoids were B-subduction related. The same applies to many post-kinematic granitoids and to those associated with the transitional or late-orogenic tectonism. These igneous rocks might be connected with lithospheric faults, along which large displacements could lead to emplacement of upper mantle melts at high levels in the neighbouring crustal blocks, causing crustal melting (cf. Leake, 1978; Pitcher, 1982). Some M-type intrusions, like the Alaskan-type, could also have been emplaced along deep fractures away from B-subduction.

All this evidence leads to the conclusion that, based on igneous activity in the Tasmanides, we have problems in estimating the amount of subducted oceanic lithosphere, but in spite of this the writer feels it is possible to reasonably assume large-scale B-subduction during Early Cambrian, Ordovician, Late Silurian-Devonian, and late Paleozoic times.

Acknowledgements. I would like to thank Dr. Jim Monger, Chairman of the ICL Working Group 2, for his suggestion and encouragement to produce this paper. The ideas presented evolved over a considerable time and many colleagues contributed and stimulated these ideas. I am particularly thankful to Jeannet Adrian and Helena Basden, Dennis Pogson, and Richard Glen for their help with the early draft of this paper.

Permission to publish this paper was given by the Secretary of the NSW Department of Mineral Resources, Sydney.

References

Arnold, G.O., and J.F. Fawckner, The Broken River and Hodgkinson province, in: Henderson, R.A., and Stephenson, P.J. (eds), The Geology and

Geophysics of Northeastern Australia. Geol. Soc. Aust., Qld. Div., 175-189, 1980.
Arnold, G.O., and R.A. Henderson,, Lower Palaeozoic history of the south-west Broken River Province, north Queensland, Geol. Soc. Aust. J., 23, 73-93, 1976.
Ashley, P.M., P.F. Brown, B.J. Franklin, A.S. Ray and E. Scheibner, Field and geochemical characteristics of the Coolac Ophiolite Suite and its possible origin in a marginal sea, Geol. Soc. Aust. J., 26, 45-60, 1979.
Baillie, P.A., Palaeozoic suspect terrane in Southeastern Australia and North Victoria Land, Antarctica, Geol. Soc. Aust., Abstr., 12, 43 p., 1984.
Barron, B.J., E. Scheibner, and E. Slansky, A dismembered ophiolite suite at Port Macquarie, New South Wales, N.S.W. Geol. Surv., Records, 18, 69-102, 1976.
Barron, L.M., E. Scheibner, and D.W. Suppel, The Mount Hope Group and its comagmatic granites on the Mount Allen 1:100,000 Sheet, New South Wales, N.S.W. Geol. Surv., Q. Notes, 47, 1-17, 1982.
Basden, H., Preliminary report on the geology of the Tumut 1:100,000 Sheet area, southern New South Wales, N.S.W. Geol. Surv., Q. Notes, 46, 1-18, 1982.
Bembrick, C.S., C. Herbert, E. Scheibner, and J. Stuntz, Structural subdivision of the New South Wales portion of the Sydney-Bowen Basin, N.S.W. Geol. Surv., Q. Notes, 11, 1-13, 1973.
Benson, W.N., The geology and petrology of the Great Serpentine Belt of New South Wales, Part I, Linn. Soc. N.S.W., Proc., 38, 490-517, 1913.
Binns, R.A., Granitic intrusions and regional metamorphic rocks of Permian age from the Wongwibinda district, northeastern New South Wales, R. Soc. N.S.W., J. Proc., 99, 5-36, 1966.
Bowman, H.N., S.J. Richardson, and I.B.L. Paterson, eds., Palaeozoic island arc and arch deposits in the central west of New South Wales, N.S.W. Geol. Surv., Records, 21(2), 329-406, microfiche M1-M93, 1983.
Branch, C.D., Volcanic cauldrons, ring complexes, and associated granites of the Georgetown Inlier, Queensland, Aust. Bur. Min. Resour. Geol. Geophys., Bull., 76, 1966.
Brown, A.V., M.J. Rubenach, and R. Varne, Geological environment, petrology and tectonic significance of the Tasman Cambrian ophiolitic and ultramafic-mafic complexes, Int. Ophiol. Symp. Cyprus, Cyprus Geol. Surv. Dept., 649-659, 1980.
Campbell, K.S.W., Carboniferous System, in: The Geology of New South Wales, G.H. Packham (ed), Geol. Soc. Aust. J., 16, 245-261, 1969.
Carey, S.W., The geological structure of the Werrie Basin, Proc. Linn. Soc. N.S.W., 59, 351-374, 1934.
Carey, S.W., and W.R. Browne, Review of the Carboniferous stratigraphy, tectonics and palaeogeography of New South Wales and Queensland, J. Proc. Roy. Soc. N.S.W., 71, 591-614, 1938.
Cas, R.A.F., Basin characteristics of the Early Devonian part of the Hill End Trough based on stratigraphic analysis of the Merrions Tuff, Geol. Soc. Aust., J., 24, 381-401, 1978a.
Cas, R.A.F., Silicic lavas in Palaeozoic flysch-like deposits in New South Wales, Australia: behaviour of deep subaqueous silicic flows, Geol. Soc. Am. Bull., 89, 1708-1714, 1978b.
Cas, R.A.F., A review of the facies patterns, palaeogeographic development and tectonic context of the Palaeozoic Lachlan Fold Belt of southeastern Australia, Geol. Soc. Aust., Spec. Publ., 10, 104 p., 1983.
Cas, R.A.F., S.F. Cox, L. Bieser, B.E. Clifford, R.L. Hammond, G. McNamara, and I. Stewart, Lower Ordovician turbidites of Central Victoria: Submarine fan or basin plain, and tectonic significance, Geol. Soc. Aust., Abstr. 9, 200-201, 1983.
Cas, R.A.F., C.McA. Powell, and K.A.W. Crook, Ordovician palaeogeography of the Lachlan Fold Belt: a modern analogue and tectonic constraints, Geol. Soc. Austr., J., 27, 19-32, 1980.
Cawood, P.A., Cambro-Ordovician strata, northern New South Wales, Search, 7, 317-318, 1976.
Cawood, P.A., Structural relations in the subduction complex of the New England Fold Belt, eastern Australia, J. Geol., 90, 381-392, 1982.
Cawood, P.A., The geological development of the Palaeozoic New England Fold Belt, Ph.D Thesis, Univ. of Sydnoy, N.S.W., 429 p., 1980.
Cawood, P.A., Accretionary tectonics and terrane dispersal within the New England Fold Belt, Eastern Australia, Proc. Circum-Pacif. Terrane Conf., Stanford Univ. Publ., Geol. Ser., XVIII, 50-52, 1983.
Cawood, P.A., The development of the SW Pacific margin of Gondwana: Correlations between Rangitata and New England Orogens, Tectonics, 3, 539-553, 1984.
Collins, W.J., Beams, S.D., White, A.J.R., and Chappell, B.W., Nature and origin of A-type granites with particular reference to southeastern Australia, Contrib. Mineral. Petrol., 80, 189-200, 1982.
Coney, P.J., D.L. Jones, and J.W.H. Monger, Cordilleran suspect terranes, Nature, 288, 329-333, 1980.
Conolly, J.R., The stratigraphy of the Hervey Group in central New South Wales, R. Soc. N.S.W., J. Proc., 98, 37-83, 1965a.
Conolly, J.R., Petrology and origin of the Hervey Group, Upper Devonian, central New South Wales, Geol. Soc. Aust., J., 12, 123-166, 1965b.
Conolly, J.R., Southern and Central Highlands Fold Belt: II. Upper Devonian Series, Geol. Soc. Aust., J., 16, 150-178, 1969a.
Conolly, J.R., Southern and Central Highlands Fold Belt: Late Devonian sedimentation, Geol. Soc. Aust., J., 16, 224-226, 1969b.
Cook, P.J., The Cambrian palaeogeography of Australia and opportunities for petroleum exploration, APEA, J., 22, 42-64, 1982.
Cooper, R.A., and G.W. Grindley, eds., Late Pro-

terozoic to Devonian sequences of southeastern Australia, Antarctica and New Zealand and their correlation, Geol. Soc. Aust., Spec. Publ., 9, 103 p., 1982.

Coulon, C., and R.S. Thorpe, Role of continental crust in petrogenesis of orogenic volcanic associations, Tectonophysics, 77, 79-93, 1981.

Cox, S.F., J. Ceplecha, V.J. Wall, M.A. Etheridge R.A.F., Cas, R. Hammond, and C. Willman, Lower Ordovician Bendigo Trough sequence, Castlemaine area, Victoria - Deformational style and implications for the tectonic evolution of the Lachlan Fold Belt, Geol. Soc. Aust., Abstr., 9, 41-42, 1983.

Crawford, A.J., Tectonic development of the Lachlan Fold Belt and construction of the continental crust of southeastern Australia, Geol. Soc. Aust., Abstr., 9, 30-32, 1983.

Crawford, A.J., and R.R. Keays, Cambrian greenstone belts in Victoria: marginal sea-crust slices in the Lachlan Fold Belt of southeastern Australia, Earth Planet. Sci. Lett., 41, 197-208, 1978.

Crawford, A.R., and K.S.W. Campbell, Large-scale horizontal displacement within Australo-Antarctica in the Ordovician, Nature Phys. Sci., 241, 11-14, 1973.

Crook, K.A.W., Structural geology of part of the Tamworth Trough, Linn. Soc. N.S.W., Proc., 87, 397-409, 1963.

Crook, K.A.W., Depositional environment and provenance of Devonian and Carboniferous sediments in the Tamworth Trough, New South Wales, R. Soc. N.S.W., J. Proc., 97, 41-53, 1964.

Crook, K.A.W., Forearc evolution in the Tasman Geosyncline: the origin of southeast Australian continental crust, Geol. Soc. Austr. J., 27, 215-232, 1980.

Crook, K.A.W., J. Bein, R.J. Hughes, and P.A. Scott, Ordovician and Silurian history of the southeastern part of the Lachlan Geosyncline, Geol. Soc. Austr. J., 20, 113-138, 1973.

Crook, K.A.W., and C.McA. Powell, The evolution of the southeastern part of the Tasman Geosyncline, 25th Int. Geol. Congr. Excursion Guide, 17A, 1976.

Daily, B., and A.R. Milnes, Stratigraphy, structure and metamorphism of the Kanmantoo Group (Cambrian) in its type section east of Tuakalilla Beach, South Australia, Roy. Soc. S. Aust., Trans., 97, 213-251, 1973.

David, T.W.E., and W.R. Browne, The Geology of the Commonwealth of Australia, Arnold, London, 1950.

Day, R.W., L.C. Cranfield and H. Schwarzbock, Stratigraphic and structural setting of Mesozoic basins in southeastern Queensland and northeastern New South Wales, in: The Tasman Geosyncline, (A.K. Denmead, G.W. Tweedale, and A.F. Wilson, eds), 319-362, Geol. Soc. Aust. Qld. Div., Brisbane, 1974.

Day, R.W., C.G. Murray and W.G. Whitaker, The eastern part of the Tasman orogenic zone, Tectonophysics. 48, 327-364, 1978.

Day, R.W., W.G. Whitaker, C.G. Murray, I.H. Wilson, and K.G. Grimes, Queensland Geology, A companion volume to the 1:2,500,000 scale geological map (1975), Geol. Surv. Qld., Publ., 383, 194 p., 1983.

de Keyser, F., The Palmerville Fault - "fundamental" structure in north Queensland, Geol. Soc. Aust. J., 10, 273-278, 1963.

Dickins, J.M., and E.J. Malone, Geology of the Bowen Basin, Queensland, Aust. Bur. Miner. Resour. Geol. Geophys., Bull., 130, 1973.

Dickinson, W.R., Widths of modern arc-trench gaps proportional to past duration of igneous activity in associated magmatic arcs, J. Geophys. Res., 78, 3376-3389, 1973.

Douglas, J.G., and J.A. Ferguson (eds), Geology of Victoria, Geol. Soc. Aust., Spec. Publ., 5, 1976.

Doutch, H.F., and E. Nicholas, The Phanerozoic sedimentary basins of Australia and their tectonic implication, Tectonophysics, 48, 365-388, 1978.

Edwards, A.C., Tectonic implications of the immobile trace-element geochemistry of mafic rocks bounding the Wonominta Block, Geol. Soc. Aust. J., 25, 459-465, 1979.

Evans, P.R., Petroleum geology of western New South Wales, APEA J., 17, 42-49, 1977.

Evans, P.R., and Y. Roberts, Evolution of central eastern Australia during the late Palaeozoic and early Mesozoic. Geol. Soc. Aust. J., 16, 325-340, 1980.

Exon, N.F., The geological evolution of the southern Taroom Trough and the overlying Surat Basin, APEA J., 14, 50-58, 1974.

Fagan, R.K., S-type granite genesis and emplacement in N.E. Victoria and its implications, Aust. Bur. Min. Resourc. Geol. Geophys., Records 1979/2, 29-30, 1979.

Fergusson, C.L., An ancient accretionary terrain in eastern New England-evidence from the Coffs Harbour Block, in: New England Geology (P.G. Flood, and B. Runnegar, ed), Univ. New England, Armidale, 63-70, 1982a.

Fergusson, C.L., Structure of the Late Palaeozoic Coffs Harbour Beds, northeastern New South Wales, Geol. Soc. Aust. J., 29, 25-40, 1982b.

Fergusson, C.L., R.A.F. Cas, W.J. Collins, G.Y. Craig, K.A.W. Crook, C.McA. Powell, P.A. Scott, and G.C. Young, The Upper Devonian Boyd Volcanic Complex, Eden, New South Wales, Geol. Soc. Aust. J., 26, 87-105, 1979.

Fleming, P.J.G., R.W. Day, C.G. Murray, and W.G. Whitaker, Late Palaeozoic invertebrate macrofossils in the Neranleigh-Fernvale Beds, Qld. Gov. Min. J., 75, 104-107, 1974.

Flint, R.B., and A.J. Parker (compilers), Tectonic Map of South Australia, 1:2,000,000/ scale, S. Aust. Dept. Mines & Energy, Adelaide, 1982.

Flood, P.G., and C.L. Fergusson, Tectono-stratigraphic units and structure of the Texas-Coffs Harbour region, in: New England Geology, (P.G. Flood, and B. Runnegar, ed), 71-78, Univ. New England, Armidale, 1982.

Galloway, M.C., Adavale, Qld, 1:250,000 Geological Series-Explanatory Notes, Sheet SG.55-5, Aust. Bur. Min. Res. Geol. Geophys., Canberra, 28 p., map, 1970.
Gee, R.D., Geological atlas 1 mile series, zone 7, Sheet 22 (8015N), Burnie, Explan. Rep. Geol. Surv., Dept. Mines Tasm., 1977.
Geological Society of Australia, Tectonic Map of Australia and New Guinea, 1:5,000,000, Sydney, 1971.
Gilligan, L.B., and E. Scheibner, Lachlan Fold Belt in New South Wales, Tectonophysics, 48, 217-265, 1978.
Gilligan, L.B., E.A. Felton, and F. Olgers, The regional setting of the Woodlawn deposit, Geol. Soc. Aust. J., 26, 135-140, 1979.
Glen, R.A., The Mulga Downs Group and its relation to the Amphitheatre Group southwest of Cobar, Geol. Surv. New South Wales, Q. Notes, 36, 1-10, 1979.
Glen, R.A., Nature of late-Early to Middle Devonian tectonism in the Buckambool area, Cobar, New South Wales, Geol. Soc. Aust. J., 127-138, 1982a.
Glen, R.A., The Amphitheatre Group, Cobar, New South Wales: preliminary results of new mapping and implications for ore search, N.S.W. Geol. Surv., Q. Notes, 49, 1-14, 1982b.
Glen, R.A., D.J. Pogson, and E. Scheibner, Geology of the Darling and Barka Basins, New South Wales, ESCAP Basin Atlas, in prep.
Green, D.C., Radiometric evidence of the Kanimblan Orogeny in southeastern Queensland and the age of the Neranleigh-Fernvale Group Geol. Soc. Aust. J., 20, 153-160, 1973.
Harrington, H.J., The Tasman Geosyncline in Australia, in: The Tasman Geosyncline (A.K. Denmead, G.W. Tweedale, and A.F. Wilson, eds), 383-407, Geol. Soc. Aust., Qld. Div., Brisbane, 1974.
Harrington, H.J., Correlation of the Permian and Triassic Gympie Terrane of Queensland with the Brook Street and Maitai Terranes of New Zealand, in: Permian Geology of Queensland, 431-436, Geol. Soc. Aust., Qld. Div., Brisbane, 1983.
Harrington, H.J., K.L. Burns, and B.R. Thompson, Gambier-Beaconsfield and Gambier-Sorell Fracture Zones and the movement of plates in the Australia Antarctica-New Zealand region, Nature 245, 109-112, 1973.
Harrison, T.M., and I. McDougall, Excess ^{40}Ar in metamorphic rocks from Broken Hill, New South Wales: implications for ^{40}Ar/^{39}Ar age spectra and thermal history in the region, Earth Planet Sci. Lett., 123-149, 1981.
Hawthorne, W.L., Regional coal geology of the Bowen Basin, in Economic Geology of Australia and Papua New Guinea, 2. Coal (D.M. Tranes and K. King, eds), Australas. Inst. Min. Metallurgy, Melbourne, 68-77, 1975.
Henderson, R.A., and P.J. Stephenson, eds, The Geology and Geophysics of Northeastern Australia, Geol. Soc. Aust., Qld. Div., 468 p., 1980.
Herbert, C., and R. Helby (eds), A guide to the Sydney Basin, N.S.W. Geol. Surv., Bull., 26, 1980.
Herbert, C., Palaeodrainage patterns in the southern Sydney Basin, N.S.W. Geol. Surv., Records, 14, 5-18, 1972.
Herbert, C., Depositional development of the Sydney Basin, N.S.W. Geol. Surv., Bull., 26, 11-52, 1980.
Hill, D., Geology, in: Handbook of Queensland, Australas. Assoc. Adv. Sci., Brisbane, 13-24, 1951.
Hills, E.S., Cauldron subsidences, granitic rocks and crustal fracturing in S.E. Australia, Geol. Rundsch., 47, 543-561, 1959.
Howell, D.G., E.R. Schermer, D.L. Jones, Zvi Ben-Avraham, and E. Scheibner, Preliminary terrane map of the Circum-Pacific Region, Am. Assoc. Petr. Geol., Tulsa, 1985.
Johnson, R.W., D.E. Mackenzie, and I.E.M. Smith, Delayed partial melting of subduction-modified mantle in Papua New Guinea, Tectonophysics, 46, 197-216, 1978.
Kilpatrick, D.J., and D.P. Fleming, Lower Ordovician sediments in the Wagga Trough: discovery of early Bendigonian graptolites near Eskdale, northeast Victoria, Jour. Geol. Soc. Aust., 27, 69-73, 1980.
Kirkegaard, A.G., Structural elements of the northern part of the Tasman Geosyncline in: The Tasman Geosyncline-a Symposium (A.K. Denmead, G.W. Tweedale, and A.F. Wilson, eds), 47-62, Geol. Soc. Aust., Qld. Div., Brisbane, 1974.
Kirkegaard, A.G., R.D. Shaw, and C.G. Murray, Geology of the Rockhampton and Port Clinton 1:250,000 Sheet areas, Qld. Geol. Surv., Rep., 38, 1970.
Korsch, R.J., A framework for the Paleozoic geology of the southern part of the New England Geosyncline, Geol. Soc. Austr., J., 25, 339-355, 1977.
Korsch, R., N.R. Archer, and G.W. McConachy, The Demon Fault, J. Proc. Roy. Soc. N.S.W., 111, 101-106, 1978.
Korsch, R.J., and H.J. Harrington, Stratigraphic and structural synthesis of the New England Orogen, Geol. Soc. Aust., J., 28, 205-220, 1981.
Kroenke, L.W., and J. Dupont, Subduction-obduction: a possible north-south transitions along the west flank of the Three Kings Rise, Geo-Marine Lett., 2, 011-016, 1982.
Lambeck, K., Structure and evolution of the Amadeus Basin, Central Australia, Geol. Soc. Aust. Abstr., 9, 109-111, 1983.
Lanphere, M.A., and J.J. Hockley, The age of nephrite occurrences in the Great Serpentinite Belt of New South Wales, Geol. Soc. Aust., J., 23, 15-17, 1976.
Leake, B.E., Granite emplacement: the granites of Ireland and their origin, in: "Crustal Evolu-

tion in Northwestern Britain and Adjacent Regions" (D.R. Bowes and B.E. Leake, eds), 221-248, Spec. Iss. Geol., J., 10, 1978.
Leitch, E.C., The geological development of the southern part of the New England Fold Belt, Geol. Soc. Austr., J., 21, 133-156, 1974.
Leitch, E.C., Plate tectonic interpretation of the Paleozoic history of the New England Fold Belt, Bull. Geol. Soc. Am., 86, 141-144, 1975.
Leitch, E.C., Structural succession in a Late Paleozoic slate belt and its tectonic significance, Tectonophysics, 47, 311-323, 1978.
Leitch, E.C., Crustal development in New England, in: New England Geology (P.G. Flood, and B. Runnegar, ed), 9-16, Univ. New England, Armidale, 1982.
Leitch, E.C., and P.A. Cawood, Olistoliths and debris flow deposits at ancient consuming plate margins: an eastern Australian example, Sediment. Geol., 25, 5-22, 1980.
Le Pichon, X., J. Francheteau, and J. Bonnin, Plate Tectonics, Elsevier, Amsterdam, 1973.
McIlveen, G.R., The Eden-Comerong-Yalwal rift zone and the contained gold mineralization, N.S.W. Geol. Surv., Rec., 16, 245-277, 1974.
McPhie. J., The Coombadja Volcanic Complex: A Late Permian Cauldron, northeastern New South Wales, in: New England Geology, Voisey Symposium 1982, Univ. New England, Armidale, 221-227, 1982.
Malone, E.J., Depositional evolution of the Bowen Basin, Geol. Soc. Aust., 11, 263-282, 1964.
Malone, E.J., F. Olgers, and A.G. Kirkegaard, The geology of the Duaringa and Saint Lawrence 1:250,000 Sheet areas, Queensland, Aust. Bur. Min. Resour. Geol. Geophys., Report, 121, 1969.
Marsden, M.A.H., The Devonian history of northeastern Australia, Geol. Soc. Aust., J., 19, 125-162, 1972.
Marsden, M.A.H., Upper Devonian-Carboniferous, Geol. Soc. Aust., Spec. Publ., 5, 77-124, 1976.
Mayne, S.J., E. Nicholas, A.L. Bigg-Wither, J. Rasidi, and M.J. Raine, Geology of the Sydney Basin-A Review, Aust. Bur. Min. Resour. Geol. Geophys., Bull., 149, 1974.
Milnes, A.R., W. Compston, and B. Daily, Pre- to syn-tectonic emplacement of early Palaeozoic granites in southeastern South Australia, Geol. Soc. Aust., J., 24, 87-106, 1977.
Murray, C.G., Alpine-type ultramafics in the northern part of the Tasman Geosyncline-possible remnants of Palaeozoic ocean floor: in The Tasman Geosyncline-a Symposium, Geol. Soc. Aust., Qld. Div., Brisbane, 161-181, 1974.
Murray, C.G., and A.G. Kirkegaard, The Thomson Orogen of the Tasman Orogenic Zone, Tectonophysics, 48, 299-325, 1978.
Murray, C.G., G.R. McClung, W.G. Whitaker, and P. R. Degeling, Geology of Late Palaeozoic sequences at Mount Barney, Queensland and Paddys Flat, New South Wales, Qld. Gov. Min., J., 82, 203-213, 1981.
Murray, C.G., and W.G. Whitaker, A review of the stratigraphy, structure and regional tectonic setting of the Brisbane Metamorphics, in: New England Geology (P.G. Flood and B. Runnegar, eds), 79-94, Univ. New England, Armidale, 1982.
Nur, A., and Z. Ben-Avraham, Lost Pacifica continent, Nature, 270, 41-43, 1977.
Olenin, V.B., The principles of classification of oil and gas basins, Aust. Oil Gas, J., 404-446, February 1967.
Olgers, F., Geology of the Drummond Basin, Queensland, Aust. Bur. Miner. Resour. Geol. Geophys. Bull., 132, 1972.
Olgers, F., and P.G. Flood, An angular Permian/ Carboniferous unconformity in southeastern Queensland and northeastern New South Wales, Geol. Soc. Aust., J., 17, 81-85, 1970.
Olgers, F., A.W. Webb, J.A. Smit, and B.A. Coxhead, The geology of the Goganto Range, Queensland, Aust. Bur. Min. Resour. Geol. Geophys., Records 1964/55 (unpubl.), 1964.
Oversby, B., Palaeozoic plate tectonics in the southern Tasman Geosynbcline, Nature Phys. Sci. 234, 45-48, 1971.
Oversby, B., L.P. Black, and J.W. Sheraton, Late Palaeozoic continental volcanism in northeastern Queensland, in: The geology and geophysics of Northeastern Queensland (R.A. Henderson and P.J. Stephenson, eds), 247-268, Geol. Soc. Aust., Qld. Div., Brisbane, 1980.
Owen, M., and D. Wyborn, Geology and geochemistry of the Tantangara and Brindabella 1:100,000 Sheet Areas, New South Wales and Australian Capital Territory, Aust. Bur. Miner. Resour. Geol. Geophys., Bull., 204, 1979.
Packham, G.H., Sedimentary history of the Tasman Geosyncline in southeastern Australia, Int. Geol. Cong., 21, 74-83, 1960.
Packham, G.H., The lower and middle Palaeozoic stratigraphy and sedimentary tectonics of the Sofala-Hill End-Euchareena region, N.S.W. Linn. Soc. N.S.W., Proc., 93, 111-163, 1968.
Packham, G.H., (ed), The geology of New South Wales, Geol. Soc. Aust., J., 16, 1969.
Packham, G.H., A speculative Phanerozoic history of the south-west Pacific, in: The Western Pacific; Island Arcs, Marginal Seas, Geochemistry, (P.J. Coleman, ed), 369-388, Univ. of W.A. Press, Perth, 1973.
Parker, A.J., Tectonic development of the Adelaide Fold Belt, Geol. Soc. Aust., Abstr., 10, 23-28, 1983.
Parkin, L.E., (ed), Handbook of South Australian Geology, Geology Survey of South Australia, Adelaide, 1969.
Pickett, J.W., The Silurian System in New South Wales, N.S.W. Geol. Surv., Bull., 29, 1982.
Pitcher, W.S., Granite type and tectonic environment, in: "Mountain Building Processes" (K.J. Hsu, ed), Academic Press, London, 19-40, 1982.
Plumb, K.A., The tectonic evolution of Australia, Earth-Sci. Rev., 14, 205-249, 1979.
Pogson, D.J., Stratigraphy, structure and tectonics: Nymagee-Melrose, Central Western New

South Sales, N.S.W. Inst. Techn. Sydney, Msc Thesis (unpubl.), 193 p., 1982.

Pogson, D.J., and E.A. Felton, Reappraisal of geology, Cobar-Canbelego-Mineral Hill region, central western New South Wales, N.S.W. Geol. Surv., Q. Notes, 33, 1-14, 1978.

Pogson, D.J. and E. Scheibner, Pre-Upper Cambrian sediments east of the Cooper Mine Range, New South Wales, N.S.W. Geol. Surv., Q. Notes, 4, 3-8, 1971.

Powell, C.McA., A critical appraisal of the tectonic history of the Hill End Trough and its margins, Aust. Soc. Explor. Geophys., Bull., 7, 14-18, 1976.

Powell, C.McA., Geology of N.S.W., South Coast, Geol. Soc. Aust., S.G.T.S.C., Field Guide No. 1, 118 p., 1983a.

Powell, C.McA., Tectonic relationship between the Late Ordovician and Late Silurian palaeogeographies of southeastern Australia, Geol. Soc. Aust., J., 30, 353-373, 1983b.

Powell, C.McA., Terminal fold-belt deformation: Relationship of mid-Carboniferous megakinks in the Tasman fold belt to coeval thrusts in cratonic Australia, Geology, 12(9), 546-549, 1984a.

Powell, C.McA., Uluru and Adelaidean Regimes (f) Ordovician to earliest Silurian marginal sea and island arc, in: Phanerozoic earth history of Australia (J.J. Veevers, ed), Oxford Univ. Press, Oxford, 290-303, 1984b.

Powell, C.McA., and D.R. Edgecombe, Mid-Devonian movements in the northeastern Lachlan Fold Belt, Geol. Soc. Austr., J., 25, 165-184, 1978.

Powell, C.McA., D.R. Edgecombe, N.M. Henry, and J.G. Jones, Timing of regional deformation of the Hill End Trough; a re-assessment, Geol. Soc. Austr., J., 23, 407-421, 1977.

Powell, C.McA. and C.L. Fergusson, The relationship of structures across the Lambian unconformity near Taralga, New South Wales, Geol. Soc. Austr., J., 26, 209-219, 1979.

Powell, C.McA., C.L. Fergusson and A.J. Williams, Structural relationships across the Lambian Unconformity in the Hervey Range-Parkes area, N.S.W., Linn. Soc. N.S.W., Proc., 104, 1980.

Powell, C.McA., G. Neef, D. Crane, P.A. Jell, and G. Percival, Significance of Late Cambrian (Idamean) fossils in the Cupola Creek Formation, northwestern New South Wales, Linn. Soc. N.W.S., Proc., 106, 127-150, 1982.

Preiss, W.V., Depositional and tectonic contrasts between Burra Group and Umberatana Group sedimentation, Geol. Soc. Aust., Abstr., 10, 13-16, 1983.

Queensland Geological Survey, Queensland Geology, Scale 1:2,500,000, Dept. Mines, Brisbane, 1975.

Raheim, A., and W. Compston, Correlations between metamorphic events and Rb-Sr ages in meta-sediments and eclogite from Western Tasmania, Lithos, 10, 271-289, 1977.

Richards, D.N.G., Palaeozoic granitoids of Northeastern Australia, in: The geology and geophysics of Northeastern Australia, (R.A. Henderson and P.J. Stephenson, eds), p. 229-246, Geol. Soc. Aust., Qld. Div., Brisbane, 1980.

Richards, Y.R., and O.P. Singleton, Palaeozoic Victoria, Australia: igneous rocks, ages and their interpretation, Geol. Soc. Aust., J., 27, 167-186, 1981.

Roberts, J., P.J. Jones, J.S. Jell, T.B.H. Jenkins, M.A.H. Marsden, R.G. McKellar, B.C. McKelvey, G. Seddon, Correlation of the Upper Devonian rocks of Australia, Geol. Soc. Aust., J., 18, 467-490, 1972.

Roberts, J., and B.A. Engel, Carboniferous paleogeography of the Yarrol and New England orogens, eastern Australia, J. Geol. Soc. Aust., 27, 167-186, 1980.

Runnegar, B.N., The Permian faunas of northern New South Wales and the connection between the Sydney and Bowen Basin, Geol. Soc. Aust., J., 16, 697-710, 1970.

Runnegar, B.N., and J.A. Ferguson, Stratigraphy of the Permian and Lower Triassic marine sediments of the Gympie district, Queensland, Qld. Univ. Press, Dept. Geol., Brisbane, 6(9), 1969.

Rutland, R.W.R., Orogenic evolution of Australia, Earth-Sci. Rev., 12, 161-196, 1976.

Rutland, R.W.R., A.J. Parker, G.M. Pitt, W.V. Preiss and B. Murrell, The Precambrian of South Australia, in: Precambrian of the Southern Hemisphere (D.R. Hunter, ed), 309-360, Elsevier, Amsterdam, 1981.

Scheibner, E., The Kanmantoo Pre-Cratonic Province in New South Wales, N.S.W. Geol. Surv., Q. Notes, 7, 1-10, 1972a.

Scheibner, E., Actualistic models in tectonic mapping, Int. Geol. Congr., 24th Montreal, Rep. 3, 405-422, 1972b.

Scheibner, E., A plate tectonic model of the Palaeozoic tectonic history of New South Wales, Geol. Soc. Aust., J., 20, 405-426, 1974a.

Scheibner, E., Fossil fracture zones, segmentation and correlation problems in the Tasman Fold Belt System, in: The Tasman Geosyncline (A.K. Denmead, G.W. Tweedale, and A.F. Wilson, eds), 65-98, Geol. Soc. Aust., Qld. Div., Brisbane, 1974b.

Scheibner, E., Explanatory notes on the Tectonic Map of New South Wales, N.S.W. Geol. Surv., Sydney, 283 p., 1976.

Scheibner, E., ed, The Phanerozoic structure of Australia and variations in tectonic style, Tectonophysics, 48, 153-430, 1978.

Scheibner, E., Some aspects of the geotectonic development of the Lachlan Fold Belt, N.S.W. Geol. Surv., Report GS 1982/069 (unpubl.) (text of public lecture), 1982.

Scheibner, E., Tectonic development of the southern part of the Cobar Trough in the Mount Hope area, Geol. Soc. Aust., Abstr., 9, 302-304, 1983a.

Scheibner, E., Suspect terranes in the Tasman Fold Belt System (Eastern Australia), Proc. Circum-Pacif. Terrane Conf., Stanford Univ. Publ., Geol. Ser., XVIII, 170-174, 1983b.

Scheibner, E., Suspect terranes in the Tasman

Fold Belt System, Eastern Australia, in: Proc. Circum-Pacif. Terrane Conf., Stanford Univ., Aug. 1983, Am. Assoc. Petr. Geol., 1985.

Scheibner, E., and R.A. Glen, The Peel Thrust and its tectonic history, N.S.W. Geol. Surv., Q. Notes, 8, 2-14, 1972.

Scheibner, E., and J.A. Pearce, Eruptive environments and inferred exploration potential of metabasalts from New South Wales, J. Geochem. Explor., 10, 63-74, 1978.

Scheibner, E., and B.P.J. Stevens, The Lachlan River Lineament and its relationship to metallic deposits, N.S.W. Geol. Surv., Q. Notes, 14, 8-18, 1974.

Shaw, S.E., and R.H. Flood, The New England Batholith, eastern Australia: geochemical variations in time and space, J. Geophys. Res., 86, 10530-10544, 1981.

Sherwin, L., Age of the Nelungoloo Volcanics near Parkes, N.S.W. Geol. Surv., Q. Notes, 35, 15-18, 1979.

Solomon, M., and J.R. Griffiths, Tectonic evolution of the Tasman Orogenic Zone, eastern Australia, Nature Phys. Sci., 237, 3-6, 1972.

Sprigg, R.C., Sedimentation in the Adelaide Geosyncline and the formation of the Continental terrace, Sir Douglas Mawson Anniversary Volume (M.F. Glaessner, and R.C. Sprigg, eds), Univ. Adelaide, S. Aust., 153-159, 1952.

Steiner, J., The Merrimbula Group of the Eden-Merrimbula area, N.S.W. R. Soc., N.S.W., J. Proc., 108, 37-51, 1975.

Suppel, D.W., and P.R. Degeling, Comments on magmatism and metallogenesis of the Lachlan Fold Belt, Geol. Soc. Aust., Abst., 6, 4-5, 1982.

Thomas, B.M., D.G. Osborne, and A.J. Wright, Hydrocarbon habitat of the Surat/Bowen Basin, APEA J., 22, 213-226, 1982.

Thomas, W.A., Continental margins, orogenic belts, and intracratonic structures, Geology, 11, 270-272, 1983.

Thomson, J., Drake 1:1000,000 Geological Sheet, N.S.W. Geol. Surv., Sydney, 1973.

Tucker, D.H., and I.G. Hone, Preliminary Magnetic Map of Onshore Australia, Scale 1:2,500,000, Aust. Bur. Min. Resour. Geol. Geophys., Canberra, 1984.

Vallance, T.G., Plutonic and metamorphic rocks, Geol. Soc. Aust., J., 16, 180-200, 1969.

VandenBerg, A.H.M., The Tasman Fold Belt system in Victoria, Tectonophysics, 48, 267-297, 1978.

Veevers, J.J., ed, Phanerozoic Earth History of Australia, Oxford Science Publ., Oxford, 418 p., 1984.

Veevers, J.J., and D. Cotterill, Western margin of Australia-a Mesozoic analog of the East African rift system, Geology, 4, 713-717, 1976.

Veevers, J.J., J.G. Jones, and C.McA. Powell, Tectonic framework of Australia's sedimentary basins, APEA J., 22, 283-300, 1982.

Voisey, A.H., Australian Geosynclines, Aust. J. Sci., 22, 188-198, 1959.

von der Borch, C.C., Evolution of Late Proterozoic to Early Palaeozoic Adelaide Foldbelt, Australia: comparisons with post-Permian rifts and passive margins, Tectonophysics, 70, 115-134, 1980.

Webb, A.W., R.P. Coats, C.M. Fanning, and R.B. Flint, Geochronological framework of the Adelaide Geosyncline, Geol. Soc. Aust., Abstr., 10, 7-9, 1983.

Webby, B.D., History of the Ordovician continental platform shelf margin of Australia, Geol. Soc. Aust., J., 25, 41-63, 1978.

Weber, C.R., and E. Scheibner, The origin of some Permo-Triassic metal zones in the New England region, New South Wales, N.S.W. Geol. Surv., Q. Notes, 26, 1-14, 1977.

White, A.J.R., and B.W. Chappell, Ultrametamorphism and granitoid genesis, Tectonophysics, 43, 7-22, 1977.

White, A.J.R., and B.W. Chappell, Granitoid types and their distribution in the Lachlan Fold Belt southeastern Australia, Geol. Soc. Am., Mem., 159, 21-34, 1983.

Williams, H., Tasman Fold Belt system in Tasmania, Tectonophysics, 48, 159-205, 1978.

Williams, H., and R.D. Hatcher Jr., Suspect terranes and accretionary history of the Appalachia Orogen, Geology, 10, 530-536, 1982.

Wopfner, H., Depositional history and tectonics of South Australian sedimentary basins, Min. Res. Rev., S. Aust., 133, 32-50, 1972.

Wyborn, D., B.W. Chappell, and R.M. Johnston, Three S-type volcanic suites from the Lachlan Fold Belt, Southeastern Australia, J. Geophys. Res., 86, B11, 10335-10348, 1981.

Wyborn, L.A.I., Aspects of the geology of the Snowy Mountains region and their implications for the tectonic evolution of the Lachlan Fold Belt, Ph.D Thesis, Aust. at. Uiv., (upubl.), 1977.